Essential
AS Biology
for OCR

Glenn and Susan Toole

Published in 2004 by:
Nelson Thornes Ltd
Delta Place
27 Bath Road
CHELTENHAM
GL53 7TH
United Kingdom

04 05 06 07 08 / 10 9 8 7 6 5 4 3 2 1

A catalogue record for this book is available from the British Library

ISBN 0 7487 8511 6

Illustrations by Ian Foulis & Associates, Plymouth, Devon

Page make-up by Tech-Set Ltd

Printed in Croatia by Zrinski

Contents

Part 3   Human health and disease

Skills

Introduction

This book has been written to meet the Oxford, Cambridge and RSA (OCR) specification for Advanced Supplementary (AS) Biology. It matches the OCR specification and has therefore been endorsed by OCR. The subject content is divided into three parts, each corresponding to the three theory modules of the OCR specification. Each part is sub-divided into chapters that exactly correspond to the sub-divisions of each of the modules. The layout of the book is designed to present information in a clear and easy-to-access way. Its features include:

- **Double-page units** to cover one specific topic and so divide the material into manageable portions.

- **Numbered headed sections** to further sub-divide the topic into bite-sized pieces. This makes accessing material simpler and allows for easy cross-referencing of information.

- **Extensive use of bullet points** to produce easy-to-follow lists of information. They permit quick referencing and often have key introductory words that make learning, and hence revision, easier.

- **Full-colour diagrams** to illustrate points made in the text, to aid understanding and to improve clarity.

- **Colour photographs** to further improve understanding and to add realism to the information and ideas.

- **Accessible language** to make comprehension easier.

- **Bold type** to emphasise key words in the text and make revision more effective.

- **Purple type** to highlight biological words that are not defined within a specific topic, but which can be found in the glossary. This enables the reader easily to access a full explanation of biological terms used in the text.

- **Comprehensive glossary** to provide a clear and concise definition of over 250 biological terms used throughout the book.

- **Extensive cross-referencing** to link different topics and provide a fully integrated understanding of biology as a whole.

- **Summary tests** to provide a quick check on how well the factual content of each unit has been learned. Answers are provided at the back of the book so that an accurately completed test forms a concise summary of the information in each unit.

- **OCR examination questions** to provide practice in the type of questions to be expected in the final examination. Arranged at the end of each chapter, these questions test the full range of skills expected at AS, including application of knowledge, understanding, analysis, synthesis and evaluation.

- **Extension material** to expand knowledge beyond the strict boundaries of the specification. The intention is both to help you obtain higher grades and to widen your horizons in order to stimulate an interest in broader aspects of biology.

We trust that you will enjoy using this book and find it interesting and informative. We hope that it will stimulate an interest in biology that encourages you to pursue your study further. Above all, we hope it will contribute to your success in the AS examinations.

Glenn and Susan Toole

Foundation biology

1

Cell structure

1.1

Cells and microscopy

Table 1.1 *A brief history of microscopy and cell theory*

Date	Discovery/Invention
Before 1600	Lenses were developed that could magnify objects up to five times
Early 1600s	Microscopes using two lenses (compound microscopes) were developed to increase magnification
1632–1723	Dutchman Anton van Leeuwenhoek developed better-quality single lenses with which he observed single-celled organisms, blood cells and spermatozoa
1665	Englishman Robert Hooke used a compound microscope to observe that cork was made up of many tiny compartments. He called these 'cells' because they reminded him of the small, almost identical, rooms called cells, in which monks lived
1838	German botanist Matthias Schleiden proposed the theory that all plants are composed of cells
1839	German zoologist Theodor Schwann extended the cell theory to include all animals

Cork cells as drawn by the English physicist Robert Hooke in 1665

Have you ever wished that you could see just that little bit better? How annoying it is to be not quite able to see the numbers on the doors of houses when you are looking for a particular address, or to pick out a friend in a large crowd. Early scientists must have felt much the same as they struggled to make out the details of the objects they were studying. Imagine their joy at the development of the first glass lenses and then the compound light microscope – a whole new world was revealed to them.

In time, however, frustration returned as they tried to discover the sub-cellular detail that was beyond the limit of the light microscope. Their curiosity was satisfied by the development of the electron microscope in 1933 – an instrument that is now capable of magnifying objects up to 500 000 times, whereas the best light microscope achieves only 1500 times.

Thanks to microscopes, we now take it for granted that living organisms are composed of cells. This fact came, not as the result of a single piece of research, but rather from a series of discoveries made over more than two centuries (table 1.1).

1.1.1 The cell theory

The cell is the basic unit of life and current cell theory states that:

- All living organisms are made up of one cell (**unicellular**) or more cells (**multicellular**).
- Metabolic processes take place within cells.
- New cells are derived from existing ones.
- Cells possess the genetic material of an organism which is passed from parent cells to daughter cells.
- A cell is the smallest unit of an organism capable of surviving independently.

Cells are necessary because the chemical reactions of vital processes, such as respiration and photosynthesis, require molecules to come into contact with one another. Keeping the required molecules within the boundary of a cell membrane, rather than letting them disperse freely throughout the organism, ensures that they react more effectively. In addition, each cell membrane can control, to some extent, which molecules to allow in and which to exclude. In this way different cells can carry out different functions. We shall see later, in units 1.5 and 1.6, that, even within a cell, there are different sub-units, called organelles, which are specialised to perform a particular function.

1.1.2 Microscopy

Microscopes are instruments that magnify the image of an object in some way. A simple convex glass lens can act as a magnifying glass but such lenses work more effectively if they are in a compound light microscope. The relatively long wavelength of light rays means that a light microscope can only distinguish between two objects if they are 0.2µm, or further, apart. This restriction can be overcome by using beams of electrons rather than beams of light. With their shorter wavelengths, the beam of electrons in the electron microscope can distinguish two objects as close together as 0.1nm.

However good a microscope is, it will only be effective if the material to be viewed under it is properly prepared. This often involves the material being stained in some way to make the parts more easily visible.

1.1.3 Magnification

The magnification of an object is how many times bigger the image is when compared to the original object.

$$\text{magnification} = \frac{\text{size of image}}{\text{size of object}}$$

In practice, it is more likely that you will be asked to calculate the size of an object when you know the size of the image and the magnification. In this case:

$$\text{size of object} = \frac{\text{size of image}}{\text{magnification}}$$

The important thing to remember when calculating the magnification is to ensure that the units of length (table 1.2) are the same for both the object and the image.

Imagine, for example, that you know an object is actually 100nm in length and you are asked how much it is magnified in a photograph. You should first measure the object in the photograph. Suppose it is 10mm long. The magnification is:

$$\frac{\text{size of image}}{\text{size of object}} = \frac{10\text{mm}}{100\text{nm}}$$

Now convert the measurements to the same units – normally the smallest – which in this case is nanometres. There are 10 000 000 nanometres in 10 millimetres and therefore the magnification is:

$$\frac{\text{size of image}}{\text{size of object}} = \frac{10\,000\,000\text{nm}}{100\text{nm}} = \frac{100\,000}{1}$$

$$= 100\,000 \text{ times}$$

1.1.4 Resolution

The **resolution**, or **resolving power**, of a microscope is the minimum distance apart that two objects can be in order for them to appear as separate items. Whatever the type of microscope, the resolving power depends on the wavelength or form of radiation used. In a light microscope it is about 0.2µm – any two objects which are 0.2µm or more apart will be seen separately, but any objects closer than 0.2µm will appear as a single item. In other words, greater resolution means greater clarity. The greater the resolution, the more clear and precise is the image produced.

Increasing the magnification will increase the size of an image, but does not always increase the resolution. Every microscope has a limit of resolution. Up to this point increasing the magnification **will** reveal more detail but beyond this point increasing the magnification **will not** – the object, while appearing larger, will just be more blurred.

Table 1.2 Units of length

Unit	Symbol	Equivalent in metres
kilometre	km	10^3
metre	m	1
millimetre	mm	10^{-3}
micrometre	µm	10^{-6}
nanometre	nm	10^{-9}

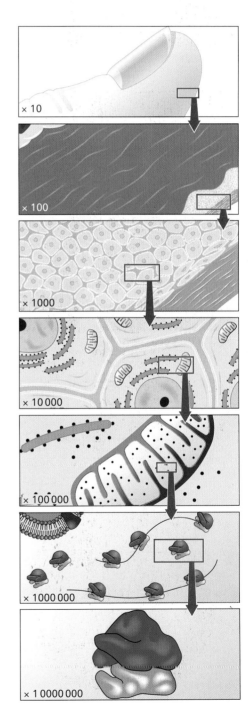

Fig 1.1 The effect of progressive magnification of a portion of human skin

The light microscope

In its simplest form, the light microscope consists of a single lens which operates as a magnifying glass. More effective is the **compound light microscope** (Fig 1.2), which has three systems of lenses.

- **The condenser lenses** are located beneath the microscope stage and can be adjusted in height to ensure that light is focused on the specimen being examined. This allows the resolving power of the microscope to be used to its full effect.
- **The objective lenses** produce an initial magnified image of the specimen.
- **The eyepiece lenses** further magnify the image produced by the objective lenses.

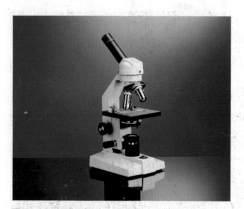

A compound light microscope

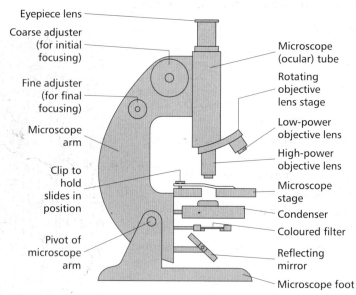

Eyepiece lens
Coarse adjuster (for initial focusing)
Fine adjuster (for final focusing)
Microscope arm
Clip to hold slides in position
Pivot of microscope arm

Microscope (ocular) tube
Rotating objective lens stage
Low-power objective lens
High-power objective lens
Microscope stage
Condenser
Coloured filter
Reflecting mirror
Microscope foot

Fig 1.2 *A compound light microscope*

1.2.1 An animal cell as seen under a light microscope

When viewed under a light microscope, a typical animal cell is made up of:

- **Cell surface (plasma) membrane** – a thin protein and **phospholipid** bilayer that controls the movement of materials in and out of the cell (section 1.7.1).
- **Cytoplasm** – watery material with a jelly-like consistency.
- **Nucleus** – made up of jelly-like **nucleoplasm**, the **nucleolus** and **chromatin** (section 1.5.1).
- **Mitochondria** – tiny rod-shaped structures within which the reactions of aerobic respiration occur (section 1.5.3).
- **Centriole** – a hollow cylinder involved in cell division (section 1.7.3).
- **Cytoplasmic granules** – small particles that act as a store of carbohydrate.

The structure of a typical animal cell as seen under a light microscope is illustrated in figure 1.3.

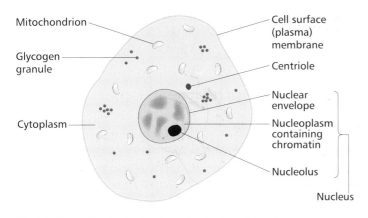

Fig 1.3 *Generalised animal cell as seen under a light microscope*

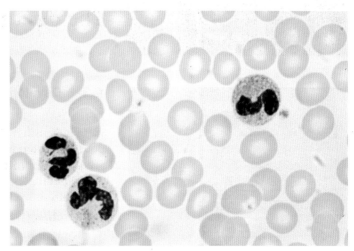

Animal cells (blood cells) as seen under a light microscope

1.2.2 A plant cell as seen under a light microscope

A typical plant cell seen under a light microscope comprises:

- **Cell wall** – a tough, yet elastic wall of cellulose that surrounds the cell.
- **Cell surface (plasma) membrane** – a thin protein and phospholipid bilayer that controls movement of materials in and out of the cell (section 1.7.1).
- **Cytoplasm** – a thin jelly-like layer just inside the cell surface membrane.
- **Nucleus** – made up of jelly-like nucleoplasm, the nucleolus and thread-like chromatin (section 1.5.1).
- **Vacuole** – contains a solution of mineral salts and sugars known as **cell sap**.
- **Tonoplast** – a thin protein and phospholipid bilayer that encloses the vacuole.
- **Mitochondria** – tiny rod-shaped structures involved in aerobic respiration (section 1.5.3).
- **Starch grains** – rounded bodies that store carbohydrate.

The structure of a typical plant cell as seen under a light microscope is illustrated in figure 1.4.

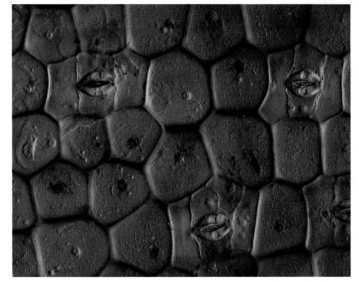

Plant cells (epidermal cells) as seen under a light microscope

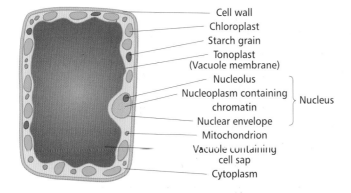

Fig 1.4 *Generalised plant cell as seen under a light microscope*

SUMMARY TEST 1.2

Viewed under a light microscope, an animal cell is seen to comprise a cell surface membrane made up of protein and (**1**), cytoplasm, a nucleus with thread-like (**2**), rod-shaped structures involved in respiration called (**3**), centrioles and (**4**) granules which store carbohydrate; in plant cells these stores are seen as (**5**) grains. Plant cells are also surrounded by a (**6**), which is made of the tough, yet elastic, material called (**7**). In the centre of a plant cell is a vacuole which contains a solution of salts and sugars called (**8**). This vacuole is surrounded by a membrane known as the (**9**).

1.3.1 Development of the electron microscope

In the 19th century scientists fully exploited the use of the light microscope but were increasingly aware of its limitations due to its poor resolving power. This poor resolution is the result of the relatively long wavelength of light and so the search for a form of electromagnetic radiation with a shorter wavelength began. X-rays, in theory, should increase magnification but cannot easily be focused. In the 1930s, however, a microscope was developed that used a beam of **electrons**. This new microscope had two main advantages:

- The electron beam had a very short wavelength and the microscope could therefore resolve objects well – it had a high resolving power.
- As electrons are negatively charged the beam could be focused using electromagnets (Fig 1.5).

The first electron microscopes could only resolve objects that were 100nm apart – just twice the resolving power of a light microscope. The best modern electron microscopes, by contrast, can resolve objects that are just 0.1nm apart – 2000 times better than a light microscope. A comparison of the advantages and disadvantages of the light and electron microscopes is given in table 1.3.

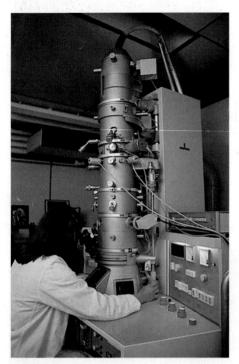

An electron microscope

Table 1.3 *Comparison of advantages and disadvantages of the light and electron microscopes*

Light microscope	Electron microscope
Advantages	**Disadvantages**
Small and portable – can be used almost anywhere	Very large and must be operated in special rooms
Unaffected by magnetic fields	Affected by magnetic fields
Preparation of material is relatively quick and simple, requiring only a little expertise	Preparation of material is lengthy and requires considerable expertise and sometimes complex equipment
Material rarely distorted by preparation	Preparation of material may distort it
Natural colour of the material can be observed	All images are in black and white
Cheap to purchase and operate	Expensive to purchase and operate
Disadvantages	**Advantages**
Magnifies objects up to 1500× only	Magnifies objects more than 500 000×
The depth of field is restricted	It is possible to investigate a greater depth of field

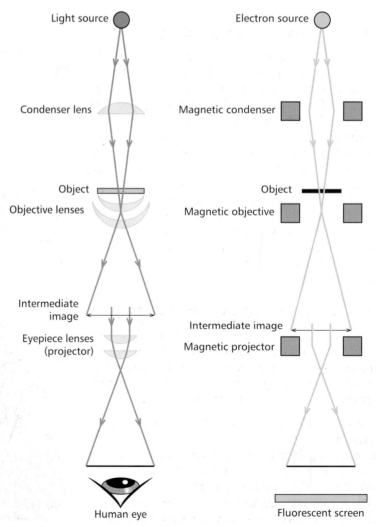

Fig 1.5 *Comparison of radiation pathways in light and electron microscopes*

Because electrons are absorbed by the molecules in air, a near-vacuum has to be created within the chamber of an electron microscope in order for it to work effectively. There are two types of electron microscope:

- the transmission electron microscope (TEM)
- the scanning electron microscope (SEM).

1.3.2 The transmission electron microscope (TEM)

The **transmission electron microscope (TEM)** consists of an electron gun that produces a beam of electrons by heating a tungsten filament. The beam is then 'focused' onto the specimen by means of the electromagnets which make up the condenser. The greater the electrical current applied to these electromagnets, the more the beam is deflected. After passing through the condenser, the image is enlarged by passing it through objective and projector lenses. The human eye cannot detect electrons and so the image is made visible by directing the electron beam onto a fluorescent screen. In a TEM, the beam passes through a thin section of the specimen. Parts of this specimen absorb electrons and therefore appear dark. Other parts of the specimen allow the electrons to pass through and hit the screen, which fluoresces, and so appears bright. The image produced on the screen can be photographed to give a **photomicrograph**.

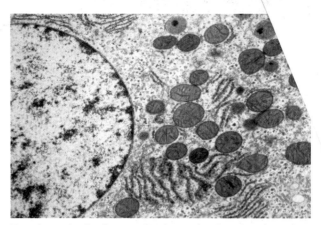

Part of an animal cell as seen under an electron microscope

1.3.3 The scanning electron microscope (SEM)

One disadvantage of the TEM is that specimens must be extremely thin to allow electrons to penetrate. The result is therefore a flat, two-dimensional image. Basically similar to a TEM, the **scanning electron microscope (SEM)** overcomes this problem by directing a beam of electrons onto the surface of the specimen from above, rather than penetrating it from below. The specimen must first be dried and coated with a metal (to produce secondary electrons). The beam is then passed back and forth across a portion of the specimen in a regular pattern. The electrons are scattered by the specimen and the pattern of this scattering depends on the contours of the specimen surface. By computer analysis of the pattern of scattered electrons and secondary electrons produced, a three-dimensional image can be built up. The basic SEM has a lower resolving power than a TEM, around 20nm, but still 10 times better than a light microscope.

1.3.4 The environmental scanning electron microscope

One major disadvantage of electron microscopes is that the removal of water from the specimen during its preparation and the vacuum in which it is observed prevent the specimen being observed in its natural 'wet' state. This disadvantage can be overcome by using an **environmental scanning electron microscope** in which:

- the specimen is kept at a much lower vacuum than the rest of the instrument chamber
- a secondary electron detector is used which operates in a gas vapour such as water. Less preparation of the specimen is necessary as little of the primary electron beam reaches the specimen. The few secondary electrons that are emitted ionise the gas, thus producing **environmental electrons** that boost the image that is produced.

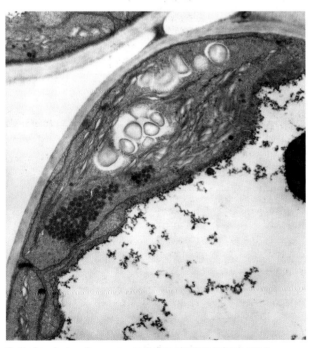

Part of a plant cell as seen under an electron microscope

Each cell of an organism can be regarded as a metabolic compartment designed to perform a particular function. Depending on that function, each cell type has an internal structure that suits it for the job it does. This is known as the **ultrastructure** of the cell. To illustrate the range of cell ultrastructure, the plant cell and animal cell shown in this unit are 'generalised'. They represent a combination of many different types of cell rather than any cell in particular – indeed, no cell actually displays **all** the features shown here.

1.4.1 The animal cell

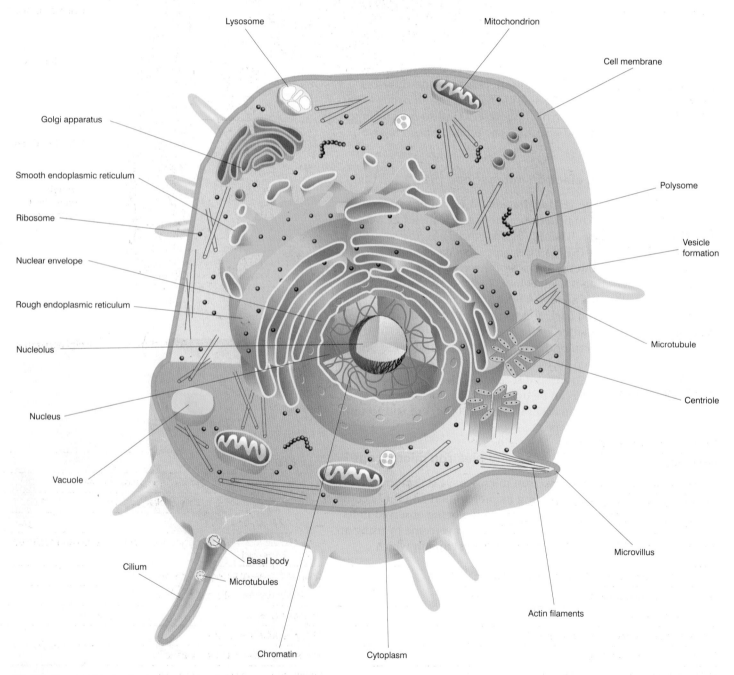

Fig 1.6 *A generalised animal cell*

1.4.2 The plant cell

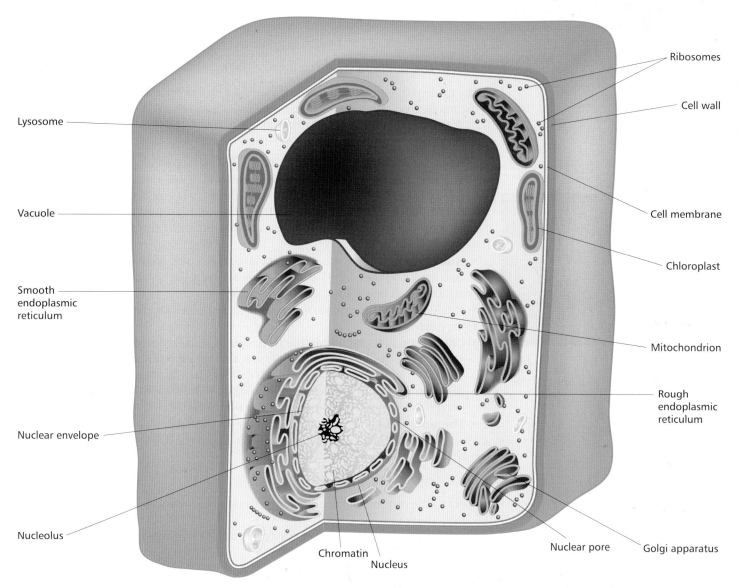

Lysosome

Vacuole

Smooth
endoplasmic
reticulum

Nuclear envelope

Nucleolus

Ribosomes

Cell wall

Cell membrane

Chloroplast

Mitochondrion

Rough
endoplasmic
reticulum

Chromatin

Nucleus

Nuclear pore

Golgi apparatus

Fig 1.7 *A generalised plant cell*

Table 1.4 *Differences between plant and animal cells*

Plant cells	Animal cells
Tough, slightly elastic cellulose cell wall present (in addition to the cell membrane)	Cell wall absent – only a membrane surrounds the cell
Pits and **plasmodesmata** present in the cell wall	No cell wall and therefore no pits or plasmodesmata
Chloroplasts present in large numbers in most cells	Chloroplasts absent
Mature cells normally have a large single, central vacuole filled with cell sap	Vacuoles, e.g. contractile vacuoles, if present, are small and scattered throughout the cell
Cytoplasm normally confined to a thin layer at the edge of the cell	Cytoplasm present throughout the cell
Nucleus at the edge of the cell	Nucleus anywhere in the cell, but often central
Centrioles absent in higher plants	Contrioloc procont
Cilia and flagella absent in higher plants	Cilia or flagella often present
Starch grains used for storage	Glycogen granules used for storage
Only some cells are capable of division	Almost all cells are capable of division

1.5.1 The nucleus

The nucleus, the most prominent feature of a **eukaryotic** cell when viewed under the microscope, may vary in shape, size and position from cell to cell. In contrast, its function remains essentially the same – to retain the organism's hereditary material and to control the cell's activities. Usually spherical and between 10 and 20μm in diameter, the nucleus has a number of parts:

- **The nuclear envelope** is a double membrane which surrounds the nucleus. Its outer membrane is continuous with the endoplasmic reticulum of the cell and often has ribosomes on its surface. It serves to control the entry and exit of materials in and out of the nucleus and to contain the reactions taking place within it.
- **Nuclear pores** allow the passage of large molecules such as messenger RNA (section 5.1.5) out of the nucleus. There are typically around 3000 in each nucleus, each being 40–100nm in diameter.
- **Nucleoplasm** is the granular, jelly-like material which makes up the bulk of the nucleus.
- **Chromatin** is found within the nucleoplasm and is composed of DNA and associated proteins. This is the diffuse form that chromosomes take up when the cell is not dividing. When the cell divides, the chromatin condenses into chromosomes (unit 6.1). Chromatin comprises **heterochromatin** – densely packed DNA which stains darkly, and **euchromatin**, which stains less heavily.
- **The nucleolus** is a small spherical body (occasionally there is more than one) within the nucleoplasm. It manufactures ribosomal RNA (section 5.1.3) and assembles the ribosomes.

The functions of the nucleus are to:

- act as the control centre of the cell through the production of mRNA and protein synthesis
- retain the genetic material of the cell in the form of DNA / chromosomes
- manufacture rRNA and ribosomes
- start the process of cell division.

1.5.2 Chloroplasts

Chloroplasts are found in eukaryotic cells which photosynthesise. They are flat discs, usually 2–10μm in diameter and 1μm thick (Fig 1.8), and are made up of a number of parts:

- **The chloroplast envelope** is a double membrane, the inner one of which is folded into a series of lamellae. It controls the entry and exit of substances in and out of the chloroplast.
- **The stroma** is a colourless, gelatinous matrix which contains the enzymes necessary for the light independent stage of photosynthesis. Small amounts of DNA and oil droplets are also found in the stroma.
- **The grana** are structures that look like a stack of coins. There are typically 50 grana in a chloroplast, and each is made up of up to 100 stacked, flattened sacs called **thylakoids**, or **lamellae** (Fig 1.8). It is to the thylakoids that the chlorophyll molecules are attached. The grana therefore carry out the light dependent stage of photosynthesis.
- **Starch grains** act as temporary stores of the carbohydrate that is produced during photosynthesis.

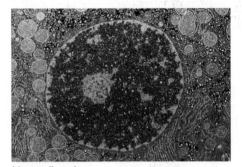

Liver cell nucleus

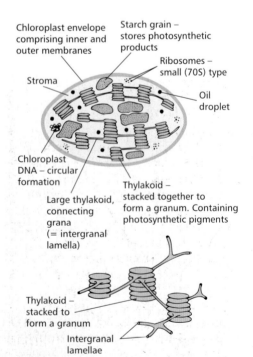

Chloroplast envelope comprising inner and outer membranes

Starch grain – stores photosynthetic products

Ribosomes – small (70S) type

Stroma

Oil droplet

Chloroplast DNA – circular formation

Thylakoid – stacked together to form a granum. Containing photosynthetic pigments

Large thylakoid, connecting grana (= intergranal lamella)

Thylakoid – stacked to form a granum

Intergranal lamellae

Fig 1.8 *Structure of chloroplasts*

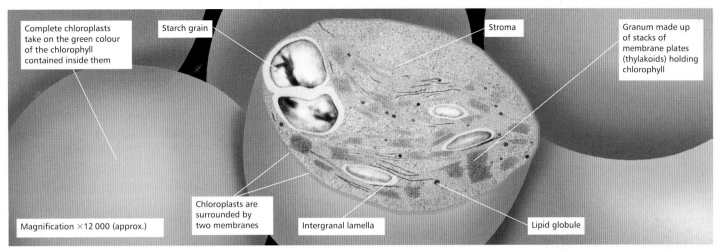

Fig 1.9 *Chloroplast structure*

Labels on figure:
- Complete chloroplasts take on the green colour of the chlorophyll contained inside them
- Starch grain
- Stroma
- Granum made up of stacks of membrane plates (thylakoids) holding chlorophyll
- Magnification ×12 000 (approx.)
- Chloroplasts are surrounded by two membranes
- Intergranal lamella
- Lipid globule

1.5.3 The mitochondrion

Present in all but a few eukaryotic cells, mitochondria (Fig 1.10) are rod-shaped and 1–10μm in diameter. They are made up of a number of parts:

- **A double membrane** surrounds the organelle, the outer one controlling the entry and exit of material. The inner membrane is folded to form extensions known as cristae.
- **Cristae** are shelf-like extensions of the inner membrane, some of which extend across the whole width of the mitochondrion. These provide a large surface area for the attachment of structures called **stalked (elementary) particles**, that are about 4nm high (Fig 1.11). The stalked particles contain enzymes involved in the synthesis of ATP.
- **The matrix** makes up the remainder of the mitochondrion. It is a semi-rigid material containing protein, lipids and traces of DNA that allow them to control the production of their own proteins. The enzymes involved in **Krebs cycle** are found in the matrix, as are mitochondrial ribosomes.

Functions of mitochondria

Mitochondria act as the sites for the Krebs cycle and **oxidative phosphorylation** stages of respiration. They are therefore responsible for the production of energy-rich ATP molecules from carbohydrates. Because of this, the number of mitochondria, their size and the number of cristae all increase in cells that have a high level of metabolic activity and therefore need a good supply of ATP. Such cells include those of the muscles and the liver.

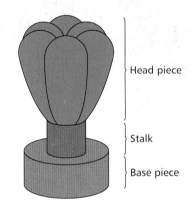

Fig 1.11 *Stalked particle*

Labels: Head piece, Stalk, Base piece

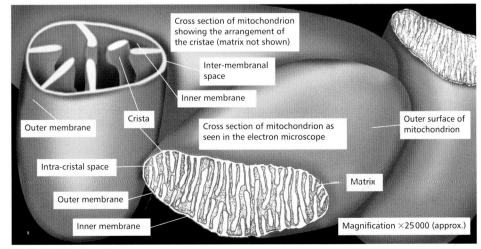

Fig 1.10 *Mitochondria*

Labels on figure:
- Cross section of mitochondrion showing the arrangement of the cristae (matrix not shown)
- Inter-membranal space
- Inner membrane
- Outer membrane
- Crista
- Cross section of mitochondrion as seen in the electron microscope
- Outer surface of mitochondrion
- Intra-cristal space
- Outer membrane
- Inner membrane
- Matrix
- Magnification ×25 000 (approx.)

Endoplasmic reticulum and ribosomes, Golgi apparatus and lysosomes

1.6.1 Endoplasmic reticulum

The endoplasmic reticulum (ER) is an elaborate, three-dimensional system of sheet-like membranes spreading through the cytoplasm of cells. It is continuous with the nuclear membrane. The membranes enclose flattened sacs called cisternae (Fig 1.12). There are two types of ER:

- **Rough endoplasmic reticulum (RER)** has ribosomes present on the outer surfaces of the membranes.
- **Smooth endoplasmic reticulum (SER)** lacks ribosomes on its surface and is often more tubular in appearance.

Functions of endoplasmic reticulum

- Provides a large surface area for the synthesis of proteins (RER).
- Provides a pathway for the transport of materials, especially proteins, throughout the cell (RER).
- Synthesises, stores and transports lipids (SER).
- Synthesises, stores and transports carbohydrates (SER).
- Contains lytic enzymes (SER of liver cells).

It follows that cells that need to manufacture and store large quantities of carbohydrates, proteins and lipids have a very extensive ER. Such cells include liver and secretory cells.

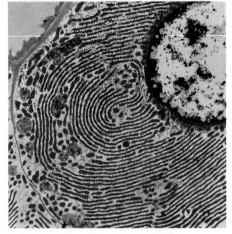
Endoplasmic reticulum

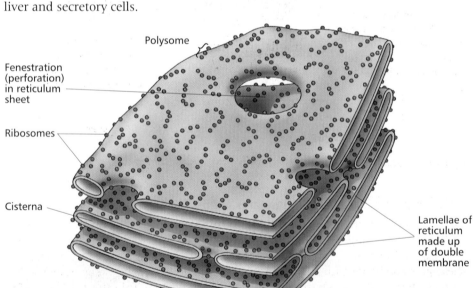

Fig 1.12 *Structure of rough endoplasmic reticulum*

1.6.2 Ribosomes

Ribosomes are small cytoplasmic granules found in all cells. They may occur in the cytoplasm or be associated with the RER. There are two types, depending on the cells in which they are found:

- **80S type**, found in **eukaryotic cells**, is around 25nm in diameter.
- **70S type**, found in **prokaryotic cells** (unit 1.8), is slightly smaller.

Each ribosome has two sub-units – one large and one small (Fig 1.13) – each of which contains ribosomal RNA and protein. Despite their small size, they occur in such vast numbers that they can account for up to 25% of the dry mass of a cell. Ribosomes are important in protein synthesis (units 5.6 and 5.7).

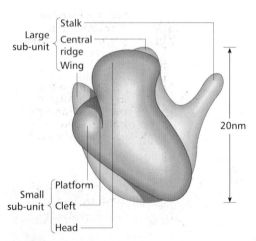

Fig 1.13 *Structure of a ribosome*

1.6.3 Golgi apparatus

The **Golgi apparatus (Golgi body)** occurs in almost all eukaryotic cells and is similar to SER in structure (section 1.6.1), except that it is more compact. It consists of a stack of membranes which make up flattened sacs, or **cisternae**, and associated hollow **vesicles** (Fig 1.14). The proteins and lipids produced by the ER are passed through the Golgi apparatus in strict sequence. The Golgi modifies these proteins, often adding non-protein components such as carbohydrate to them. It also 'labels' them, allowing them to be accurately sorted and sent to their correct destinations. Once sorted, the modified proteins and lipids are transported in vesicles which are regularly pinched off from the ends of the Golgi cisternae (Fig 1.14). These vesicles move to the cell surface, where they fuse with the membrane and release their contents to the outside.

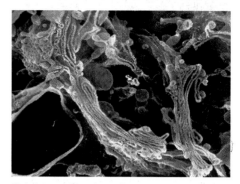

Golgi apparatus of an olfactory bulb cell

Functions of the Golgi apparatus

In general, the Golgi apparatus acts as the cell's post office, receiving, sorting and delivering proteins and lipids. More specifically, it:

- adds carbohydrates to proteins to form glycoproteins such as mucin
- produces secretory enzymes such as those secreted by the pancreas
- secretes carbohydrates such as those used in making cell walls in plants
- transports, modifies and stores lipids
- forms lysosomes (section 1.6.4).

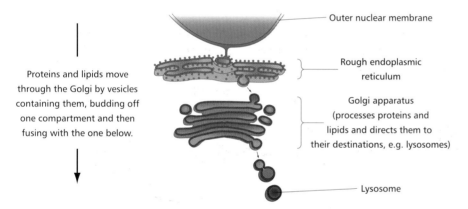

Proteins and lipids move through the Golgi by vesicles containing them, budding off one compartment and then fusing with the one below.

Outer nuclear membrane

Rough endoplasmic reticulum

Golgi apparatus (processes proteins and lipids and directs them to their destinations, e.g. lysosomes)

Lysosome

Fig 1.14 The Golgi apparatus and its relationship to the nucleus, ER and lysosomes

1.6.4 Lysosomes

Lysosomes are formed when the vesicles produced by the Golgi apparatus include within them enzymes such as proteases and lipases. Up to 50 such enzymes may be contained in a single lysosome. Up to 1.0μm in diameter, lysosomes isolate these potentially harmful enzymes from the rest of the cell, before releasing them, either to the outside or into a **phagocytic** vesicle within the cell (Fig 1.15).

Functions of lysosomes

Lysosomes are used to destroy foreign material inside or outside the cell. More particularly, they:

- break down material ingested by phagocytic cells such as white blood cells or *Amoeba spp*.
- release enzymes to the outside of the cell **(exocytosis)** in order to destroy material around the cell
- digest worn out organelles **(autophagy)** so that the useful chemicals of which they are made can be re-used
- completely break down cells after they have died **(autolysis)**.

Given the roles that lysosomes perform, it is not surprising that they are especially abundant in secretory and phagocytic cells.

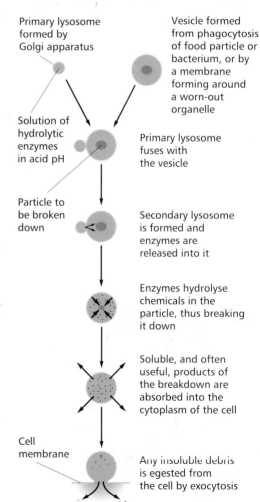

Primary lysosome formed by Golgi apparatus

Vesicle formed from phagocytosis of food particle or bacterium, or by a membrane forming around a worn-out organelle

Solution of hydrolytic enzymes in acid pH

Primary lysosome fuses with the vesicle

Particle to be broken down

Secondary lysosome is formed and enzymes are released into it

Enzymes hydrolyse chemicals in the particle, thus breaking it down

Soluble, and often useful, products of the breakdown are absorbed into the cytoplasm of the cell

Cell membrane

Any insoluble debris is egested from the cell by exocytosis

Fig 1.15 The functioning of a lysosome

Cell surface (plasma) membrane, cilia and centrioles

1.7.1 Cell surface (plasma) membrane

The cell surface membrane is the boundary between the cell cytoplasm and the environment. It controls the movement of substances into and out of the cell, permanently excluding some and permanently containing others; yet others may cross the membrane on one occasion and be prevented from doing so at another time. The cell surface membrane is therefore said to be **partially permeable**. It is made up of proteins (45%) and phospholipids (45%), the remainder (10%) being cholesterol, glycolipids and glycoproteins. The arrangement of these molecules is described and illustrated in unit 4.1.

Under the electron microscope, the cell membrane appears as a three-layered structure consisting of two dark outer layers enclosing a pale central one (see photo).

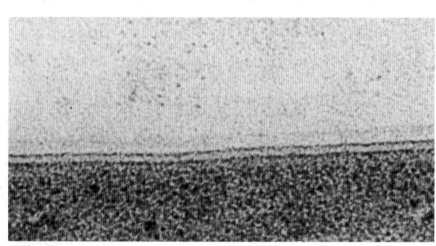

Cell surface membrane

Functions of the cell surface membrane

The cell surface membrane:

- controls movement of substances in and out of the cell
- forms a recognition site so that the body's immune system can recognise its own cells
- acts as a receptor site for the attachment of specific hormones and neurotransmitters
- in animal cells, may be folded to form microvilli to provide a larger surface area for the absorption of substances
- helps cells attach to one another and so form tissues.

1.7.2 Cilia

Cilia (Fig 1.16) are the threads that extend from the cell surface. They are typically 3–4μm long, but may be up to 10μm, and are around 0.2μm in diameter. The base of the cilium, the basal body, is embedded in the cytoplasm of the cell. The basal body contains nine sets of three microtubules. The rest of the cilium extends from the basal body and is made up of two central, single microtubules, surrounded by a ring of nine pairs of microtubules with a protein

LS Basal region of a cilium

Peripheral microtubules

A   B   Plasma membrane

Central microtubules

Basal plate (derived from, or represents, the centrosome)

Basal body

Strand (connects adjacent basal bodies)

Rootlet (only present in certain cells)

TS Cilium (section A/B)

One of nine paired peripheral microtubules

Extension of paired peripheral microtubules – increasingly well developed towards base of the cilium

One of two central microtubules (present only in motile cilia)

Plasma membrane

Fig 1.16 *Structure of a cilium*

extension on each pair. Only a few types of cells possess cilia, but where they are found they occur in large numbers – around 200 on a ciliated epithelial cell of the human trachea for example. These groups of cilia work together in a coordinated manner to produce a wave-like motion across the surface of the cell.

Functions of cilia

The functions of the cilia are:

- to move an entire organism, e.g. cilia on the surface of the **protoctist** *Paramecium* propel it through the water
- to move material within an organism, e.g. cilia lining the respiratory tract of mammals move mucus towards the throat, and cilia lining the oviduct move the secondary oocyte (ovum) towards the uterus.

1.7.3 Centrioles

Centrioles are found in almost all animal cells, as well as in the cells of certain algae and fungi, but not in plant cells. They are hollow cylinders $0.5\mu m$ in length and $0.2\mu m$ in diameter. Their internal structure is the same as the basal body of a cilium, comprising nine sets of three microtubules (Fig 1.17). There are two centrioles in a cell and they lie at right angles to one another near to the nucleus.

Functions of centrioles

The centrioles have two main functions.

- The microtubules within the centrioles form the spindle fibres during nuclear division and so position and move the chromosomes during the process.
- The centrioles may be involved in the formation of microtubules that make up the cytoskeleton of the cell.

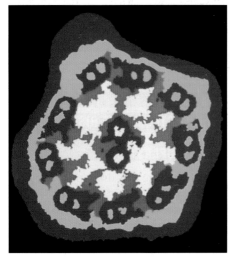

Cilium in TS as seen under an electron microscope

Fig 1.17 Centrioles

Centrioles in transverse (left) and longitudinal (right) section as seen under an electron microscope

SUMMARY TEST 1.7

The cell surface membrane is made up of 45% phospholipid and 45% **(1)**; the remaining 10% comprises glycolipids, **(2)** and **(3)**. Cilia are threads around $3–4\mu m$ in length that extend from the cell surface. They are made up of a **(4)** which is found in the cytoplasm of the cell, from which extends the rest of the cilium, comprising a series of **(5)** arranged in an outer ring of **(6)** pairs and a total of **(7)** single microtubules in the centre. In animal cells there is a total of **(8)** centrioles. These are important in **(9)**, where they form **(10)** fibres that help position and move **(11)** during the process.

Prokaryotic cells

EXTENSION

DID BACTERIA GIVE RISE TO MITOCHONDRIA AND CHLOROPLASTS?

From reading this unit you will be aware that the fundamental difference between prokaryotic cells and eukaryotic cells is the presence of membrane-bounded organelles within the latter. How then did these organelles arise within eukaryotic cells? One possibility is that they arose from invaginations of the cell surface membrane, which then became pinched off to give a separate membrane structure within the main cell. This, however, does not explain why organelles such as mitochondria and chloroplasts possess ribosomes and circular pieces of DNA, and are able to replicate themselves.

Another theory, called the **endosymbiosis** theory, has therefore been put forward. This proposes that certain bacterial cells invaded others and lived within them in a relationship that benefited both. We now call this type of relationship mutualism. For example, a photosynthetic bacterium living inside a non-photosynthetic one, could provide carbohydrate and oxygen to the host cell, in return for carbon dioxide, minerals and protection from the larger host cell. In time, the relationship became so mutually dependent that neither could survive without the other, and so the cell became a single entity, rather than two separate ones living together.

Although cells come in a bewildering variety of size, shape and function, they nevertheless fall into two basic groups:

- **Prokaryotic cells** ('pro' = before, 'karyote' = nucleus) have no nucleus or nuclear membranes.
- **Eukaryotic cells** ('eu' = true, 'karyote' = nucleus) have a nucleus bounded by nuclear membranes.

Other differences between prokaryotic and eukaryotic cells are listed in table 1.5.

Table 1.5 *Comparison of prokaryotic and eukaryotic cells*

Prokaryotic cells	Eukaryotic cells
No true nucleus, only diffuse area(s) of nucleoplasm with no nuclear envelope	Distinct nucleus, with a nuclear envelope
No nucleolus	Nucleolus is present
Circular strands of DNA but no chromosomes	Chromosomes present in which DNA is located
No membrane-bounded organelles	Membrane-bounded organelles such as mitochondria are present
No chloroplasts, only photosynthetic lamellae in some bacteria	Chloroplasts present in plants and algae
Ribosomes are smaller (70S type)	Ribosomes are larger (80S type)
Flagella (if present) lack internal 9+2 microtubule arrangement	Flagella, where present, have a 9+2 internal microtubule arrangement
No endoplasmic reticulum or associated Golgi apparatus and lysosomes	Endoplasmic reticulum present along with Golgi apparatus and lysosomes
Cell wall made of peptidoglycan	Where present, cell wall is made mostly of cellulose or chitin

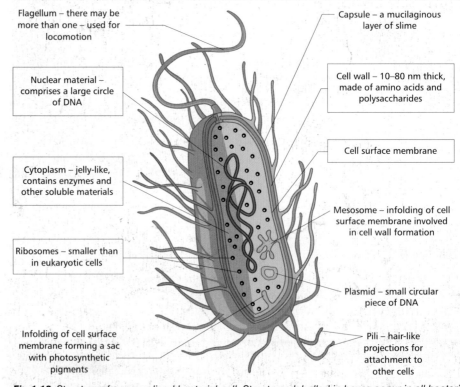

Fig 1.18 *Structure of a generalised bacterial cell. Structures labelled in boxes occur in all bacteria whereas the others occur only in certain species*

1.8.1 Structure of a bacterial cell

Bacteria occur in every habitat in the world; they are versatile, adaptable and successful. Much of their success is a result of their small size, normally in the range 0.1–10µm in length. Their cellular structure is relatively simple (Fig 1.18). All bacteria possess a cell wall which is made up of **peptidoglycan** (murein) – a polysaccharide cross-linked by peptide molecules. Around this wall, many bacteria further protect themselves by secreting a **capsule** of mucilaginous slime. Hair-like structures, made of protein and called **pili**, extend through the cell wall in some species. These enable the bacteria to stick to one another or to other surfaces. **Flagella** occur in certain types of bacteria. These lack microtubules and so do not beat, which they do in eukaryotic cells. Their rigid corkscrew shape and rotating base, however, cause bacteria to spin through fluids.

Within the cytoplasm of bacterial cells are scattered **ribosomes** (70S type). These are smaller than those of eukaryotic cells (80S type), but nevertheless serve the same function in protein synthesis. **Glycogen granules** and **oil droplets** are used for storage. Infoldings of the cell, known as **mesosomes**, often occur. These provide a large surface area for the attachment of respiratory enzymes. In photosynthetic bacteria there are lamellae called **thylakoids**, which contain the enzymes and bacterial chlorophyll essential to photosynthesis. The genetic material in bacteria is in the form of a circular strand of DNA. Separate from this, and not necessary for growth and metabolism, are smaller circular pieces of DNA called **plasmids**. These can reproduce themselves independently and may give the bacterium resistance to harmful chemicals such as **antibiotics**. Plasmids are used extensively as vectors (carriers of genetic information) in **genetic engineering** (section 5.8.3).

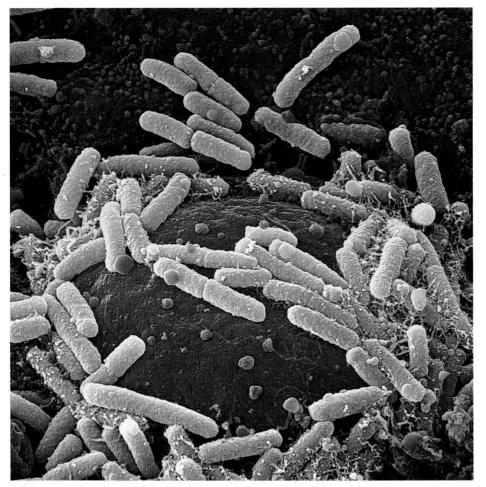

SEM of a colony of the rod-shaped bacterium – Escherichia coli

Table 1.6 *Roles of structures found in a bacterial cell*

Cell structure	Role
Cell wall	Physical barrier which protects against mechanical damage and excludes certain substances
Capsule	Protects bacterium from other cells, e.g. white blood cells, and also helps groups of bacteria to stick together for further protection
Cell surface (plasma) membrane	Acts as a differentially permeable layer which controls the entry and exit of chemicals
Mesosome	Provides a large surface area for the attachment of respiratory enzymes
Flagellum	Aids movement of bacterium because its rigid, corkscrew shape and rotating base help the cell spin through fluids
Pili	Help cells stick to one another or to other surfaces
Circular DNA	Possesses the genetic information for the replication of bacterial cells
Plasmids	Possess genes which aid the survival of bacteria in adverse conditions, e.g. produce enzymes which break down antibodies
Ribosomes	Site of protein synthesis (70S type)
Glycogen granules	Store carbohydrates for breakdown during respiration to provide energy
Lipid droplets	Store lipids as a more concentrated, longer-term, store for conversion to carbohydrate and use in respiration
Photosynthetic lamellae (thylakoids)	Contain enzymes and bacterial chlorophyll and therefore carry out photosynthesis

SUMMARY TEST 1.8

Prokaryotic cells lack a distinct **(1)** and include the group of organisms called **(2)**. Their DNA is **(3)** in shape and known as a **(4)** and their ribosomes are smaller than in **(5)** cells and are known as the **(6)** type. If photosynthesis takes place it does so in lamellae called **(7)** rather than in a true chloroplast.

(a) Squamous epithelium as seen under a light microscope

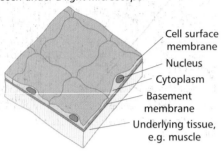

Cell surface membrane
Nucleus
Cytoplasm
Basement membrane
Underlying tissue, e.g. muscle

(b) Single squamous epithelium cell as seen under an electron microscope

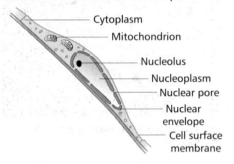

Cytoplasm
Mitochondrion
Nucleolus
Nucleoplasm
Nuclear pore
Nuclear envelope
Cell surface membrane

Fig 1.19 Squamous epithelium

(a) Ciliated columnar epithelium as seen under a light microscope

Cilia
Cell surface membrane
Cytoplasm
Nucleus
Basement membrane
Underlying tissue, e.g. muscle

(b) A ciliated columnar epithelial cell as seen under an electron microscope

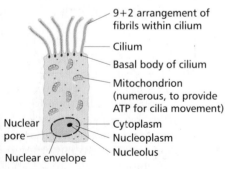

9+2 arrangement of fibrils within cilium
Cilium
Basal body of cilium
Mitochondrion (numerous, to provide ATP for cilia movement)
Nuclear pore
Cytoplasm
Nucleoplasm
Nucleolus
Nuclear envelope

Fig 1.20 Ciliated columnar epithelium

1.9.1 Tissues

Single-celled organisms such as the **protoctist** *Amoeba* carry out all essential life functions within the one cell. Although they perform all functions well enough, they cannot be totally efficient at all of them because each function requires different conditions and these may conflict. One activity may be best carried out by a long thin cell and another by a spherical one. Compromises have to be made because no single cell can provide optimum conditions for all life functions at the same time. For this reason, the cells of multicellular organisms are each differently adapted to perform a particular role. This **division of labour** leads to cells becoming specialised in structure to suit the role that they carry out. This is known as **cell differentiation**. In becoming specialised, the cells lose the ability to perform other functions and so become dependent on other cells to carry out these activities for them.

In order to work efficiently, cells are usually grouped together. Such a collection of similar cells that perform a specific function is known as a **tissue**. Examples of tissues in animals include **epithelial tissues**, such as squamous epithelium and ciliated epithelium. In plants, examples include the two **vascular tissues**, xylem and phloem.

1.9.2 Epithelial tissues

Epithelial tissues are animal tissues that:

- lie on a basement membrane of **collagen** and **glycoproteins** that is secreted by the cells beneath the epithelium. The basement membrane holds the cells together
- are further held together by an adhesive substance between the cells (intercellular substance)
- line the surfaces of organs
- often have a protective or secretory function.

An epithelial tissue may be one cell thick (**simple epithelium**) or more than one cell thick (**compound epithelium**). Examples of simple epithelia include:

- **Squamous epithelium** – which forms a single layer of thin, irregular, closely packed cells each attached to the basement membrane (Fig 1.19). They may be likened to irregular floor tiles (the cells) attached to floorboards (basement membrane). The cells are so thin that the central nucleus often forms a bump in the surface. They therefore form an ideal diffusion surface and are found in the alveoli of the lungs, the walls of blood capillaries and the renal (Bowman's) capsule in the kidneys. Squamous epithelium also lines ducts, arteries and veins, where it provides a relatively friction-free surface to allow the smooth flow of fluids within them.
- **Ciliated epithelium** – which forms a single layer of cube-shaped (cuboid) cells or longer narrow (columnar) cells, each attached to the basement membrane. At the end opposite to the basement membrane, the cells possess cilia (section 1.7.2) which are used to move materials along the surfaces that they line, e.g. the secondary oocyte (ovum) along the oviduct, mucus along the trachea and bronchi. The structure of ciliated epithelium is shown in figure 1.20.

1.9.3 Vascular tissues

The tissue in plants responsible for transporting liquids is called vascular tissue and comprises two main types:

- **Xylem** is made up of a number of cell types and is used to transport water and mineral ions throughout the plant and to give structural support. Its basic structure is shown in the photo opposite and full details of the different cell types are given in unit 10.4.1.
- **Phloem** also has a number of cell types and its function is to transport organic material throughout the plant. The photo shows its basic structure; full details are given in unit 10.2.1.

1.9.4 Organs

Just as cells are grouped into tissues, so tissues are grouped together into organs. An **organ** is a combination of tissues that are coordinated to perform a variety of functions, although they often have one predominant major physiological function. Organs are themselves grouped into **organ systems**.

In animals, for example, the stomach is an organ that carries out the digestion of certain types of food. It is made up of tissues such as:

- **muscle** – to churn and mix the stomach contents
- **epithelium** – to protect the stomach wall and produce secretions
- **connective tissue** – to hold together the other tissues.

The stomach, along with the other organs like the duodenum, ileum, pancreas and liver, forms the organ system known as the digestive system.

In plants, a leaf is an organ made up of the following tissues:

- **palisade mesophyll** – which carries out photosynthesis
- **spongy mesophyll** – adapted for gaseous diffusion
- **epidermis** – to protect the leaf and allow gaseous diffusion
- **phloem** – to transport organic materials away from the leaf
- **xylem** – to transport water and ions into the leaf.

A plan of the tissues of a dicotyledonous leaf is given in figure 1.21.

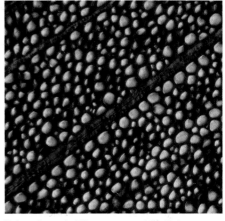

Xylem tissue in TS as seen under a light microscope

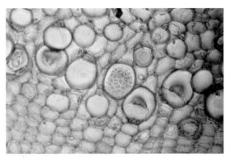

Phloem tissue in TS as seen under a light microscope

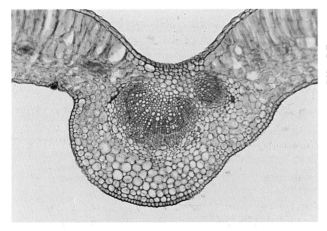

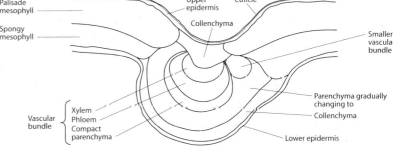

Fig 1.21 *Plan diagram of the tissues shown in the above photograph of a TS through the midrib of a dicotyledonous leaf*

1 a The figure is a drawing of an animal cell **nucleus** as seen using an electron microscope.

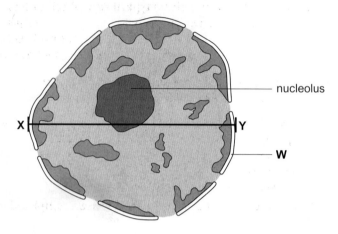

(i) Name the structure labelled **W**. *(1 mark)*

(ii) The actual diameter of the nucleus, measured along the line **XY**, is 7μm.

Calculate the magnification of the nucleus. Show your working.

(2 marks)

b Each part of a cell is specialised to carry out a particular function.

Below is a list of parts of a cell, labelled **A** to **F**. Each of the lists of statements, numbered **1** to **6**, refers to one of these parts of the cell.

A nucleus	**1**	where some lipids, including steroids, are made
B mitochondrion	**2**	controls entry of substances into the cell
C plasma (cell surface) membrane	**3**	controls the activities of the cell
D chloroplast	**4**	where polypeptides are made
E smooth endoplasmic reticulum	**5**	where photosynthesis takes place
F ribosomes	**6**	where aerobic respiration takes place

Match a statement to each part of the cell, e.g. **A = 3**

(5 marks)
(Total 8 marks)
OCR 2801 Jan 2003, B (BF), No.1

2 a Explain what is meant by the term *tissue*. *(2 marks)*

The diagram shows cells from two types of epithelial tissue, **A** and **B**, as seen under the electron microscope. The cells are not drawn to the same scale.

A

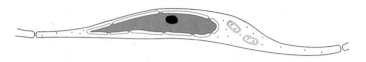

B

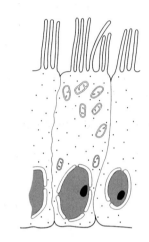

b (i) Name the types of epithelial tissue **A** and **B**.
(2 marks)

(ii) Explain why the cells of tissue **B** contain many more mitochondria than those in tissue **A**.
(2 marks)

c State **two** ways in which cells of tissues **A** and **B** differ from prokaryotic cells. *(2 marks)*
(Total 8 marks)
OCR 2801 May 2002, B (BF), No.1

3 The figure is a diagram of a bacterium as seen using an electron microscope.

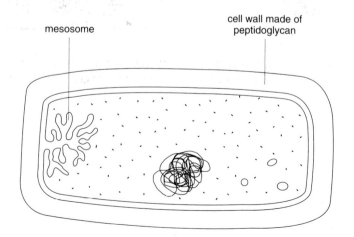

a The bacterium contains DNA, as do eukaryotic cells.
  (i) State **two other** ways in which the structure of the cell in the figure is similar to a typical animal cell. *(2 marks)*
  (ii) Describe how bacterial DNA differs from that found in eukaryotic cells. *(2 marks)*

b Some bacteria similar to that shown in the figure can cause disease. Antibiotics are often given to patients who are suffering from diseases caused by bacteria. Examples of the mode of action of two antibiotics are given below.

Antibiotic 1   binds to the enzyme RNA polymerase in bacteria, preventing transcription.

Antibiotic 2   prevents the formation of peptide cross links between peptidoglycan chains in the cell wall.

  (i) Explain why the prevention of transcription leads to the death of bacteria. *(3 marks)*
  (ii) Suggest how the action of antibiotic 2 on cell walls leads to the death of bacteria. *(2 marks)*
(Total: 10 marks)
OCR 2801 Jun 2003, B (BF), No.3

4 The figure is an electron micrograph of part of a leaf mesophyll cell.

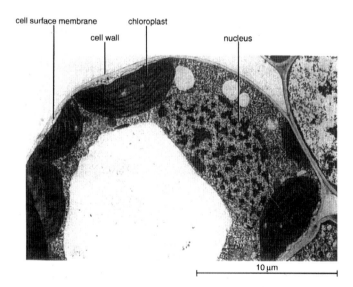

a With reference to the figure, state **two** reasons why the cell is considered to be a eukaryotic cell. *(2 marks)*

b State the functions of the following structures in this cell
  (i) chloroplast; *(1 mark)*
  (ii) cell surface membrane. *(1 mark)*

c Calculate the magnification of the figure. (Show your working.) *(2 marks)*

d State **one** way in which a typical animal cell differs from the cell shown in the figure. *(1 mark)*
(Total 7 marks)
OCR 2801 Jun 2001, B (BF), No.1

5 The table below compares the features of typical eukaryotic and prokaryotic cells.

a Complete the table by placing one of the following, as appropriate, in each empty box in the table:
a tick (✔)
a cross (✗)
the words 'sometimes present'

One of the two boxes in each row has been completed for you.

	eukaryotic cell	prokaryotic cell
cell wall	sometimes present	✔
nuclear envelope	✔	
Golgi apparatus		✗
flagellum	sometimes present	
ribosomes		✔
carries out respiration	✔	
chloroplast	sometimes present	

(6 marks)

b (i) Explain the meaning of the term *tissue*. *(2 marks)*
  (ii) State **one** example of a plant tissue. *(1 mark)*
  (iii) State **one** example of an animal tissue. *(1 mark)*
(Total 10 marks)
OCR 2801 Jan 2001, B (BF), No.5

6 a State **three structural** features of prokaryotic cells. *(3 marks)*

b Describe the functions in eukaryotic cells of *lysosomes*, *ribosomes*, and *centrioles*.

(In this question, 1 mark is available for the quality of written communication.)

(7 marks)
(Total 10 marks)
OCR 2801, 2000 Specimen paper, B (BF), No.2

2.1   Introduction to biological molecules

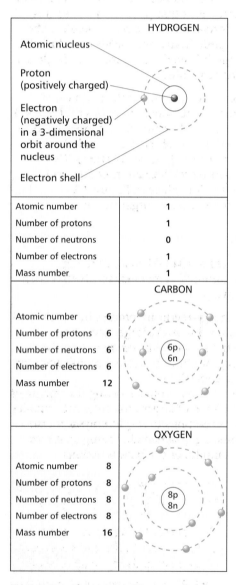

Atomic number	1
Number of protons	1
Number of neutrons	0
Number of electrons	1
Mass number	1

Atomic number	6
Number of protons	6
Number of neutrons	6
Number of electrons	6
Mass number	12

Atomic number	8
Number of protons	8
Number of neutrons	8
Number of electrons	8
Mass number	16

Fig 2.1 *Atomic structure of three commonly occurring biological elements*

Biological molecules are particular groups of chemicals that are found in living organisms. Their study is known as **molecular biology**. All molecules, whether biological or not, are made up of units called atoms.

2.1.1   Atoms

Atoms are the smallest units of a chemical element that can exist independently. An atom comprises a nucleus that contains particles called protons and neutrons (the hydrogen atom is the only exception as it has no neutrons). Tiny particles called electrons orbit the nucleus of the atom. The main features of these sub-atomic particles are:

- **Neutrons** – occur in the nucleus of an atom and have the same mass as protons but no electrical charge.
- **Protons** – occur in the nucleus of an atom and have the same mass as neutrons but do have a positive charge.
- **Electrons** – orbit in fixed shells around the nucleus but a long way from it. They have such a small mass that their contribution to the overall mass of the atom is negligible. They are, however, negatively charged and their number determines the chemical properties of an atom.

In an atom the number of protons and electrons is the same and therefore there is no overall charge. The **mass number** of an atom is the total number of protons and neutrons in a given atom. The atomic structure of three biological elements is given in figure 2.1.

2.1.2   Oxidation, reduction and the formation of ions

If an atom loses or receives an electron it becomes an ion.

- The loss of an electron is called **oxidation** and leads to the formation of a positive ion, e.g. the loss of an electron from a hydrogen atom produces a positively charged hydrogen ion, written as H^+.
- The receiving of an electron is called **reduction** and leads to the formation of a negative ion, e.g. if a chlorine atom receives an additional electron it becomes a negatively charged chloride ion, written as Cl^-.

More than one electron may be lost or received, e.g. the loss of two electrons from a calcium atom forms the calcium ion, Ca^{2+}. Ions may be made up of more than one type of atom, e.g. a sulphate ion is formed when one sulphur atom and four oxygen atoms receive two electrons and form the sulphate ion, SO_4^{2-}.

2.1.3   Isotopes

While the number of protons in an element always remains the same, the number of neutrons can vary. The different types of the atom so produced are called **isotopes**. Isotopes of any one element have the same chemical properties but differ in mass. Each type is therefore recognised by its different mass number. Isotopes, especially radioactive ones, are very useful in biology for tracing the route of certain elements in biological processes and for dating fossils.

2.1.4 Bonding and the formation of molecules

Atoms may combine with each other in a number of ways:

- **Covalent bonding** – atoms share a pair of electrons in their outer shells. As a result the outer shell of both atoms is filled and a more stable compound, called a **molecule**, is formed.
- **Ionic bonding** – ions with opposite charges attract one another. This electrostatic attraction is known as an ionic bond. For example the positively charged sodium Na^+ and negatively charged chloride Cl^- form an ionic bond to make sodium chloride. Ionic bonds are weaker than covalent bonds.
- **Hydrogen bonding** – the electrons within a molecule are not evenly distributed but tend to group at one position. This region is more negatively charged than the rest of the molecule. A molecule with an uneven distribution of charge is said to be **polarised**, i.e. it is a **polar molecule**. The negative region of one polarised molecule and the positively charged region of another attract each other. A weak electrostatic bond is formed between the two. Although each bond is individually weak, they can collectively form important forces that alter the physical properties of molecules. This is especially true in the case of water.

2.1.5 Polymerisation and the formation of macromolecules

Certain molecules can be linked together to form long chains of similar sub-units. These chains form large molecules, or **macromolecules**, known as **polymers** and the process by which they are formed is therefore called **polymerisation**. The sub-units of a polymer are usually based on carbon. Many, such as polythene and polyesters, are industrially produced. Others, like polysaccharides, polypeptides and polynucleotides, are made naturally by living organisms. The basic sub-unit of a polysaccharide is a monosaccharide or single sugar (unit 2.2), e.g. glucose.

Polynucleotides are formed from mononucleotide sub-units (unit 5.1). Polypeptides are formed by linking together peptides which have amino acids as their basic sub-unit (unit 2.6).

2.1.6 Condensation and hydrolysis reactions

In the formation of polymers by polymerisation in organisms, each time a new sub-unit is attached a molecule of water is formed. Reactions that produce water in this way are termed **condensation reactions**. Therefore the formation of a polypeptide from amino acids and that of the polysaccharide starch from the monosaccharide glucose are both condensation reactions.

Polymers can be broken down through the addition of water. Water molecules break down the bonds that link the sub-units of a polymer, thereby splitting the molecule into its constituent parts. This type of reaction is called **hydrolysis** ('hydro' = water; 'lysis' = splitting). Thus polypeptides can be hydrolysed into amino acids, and starch can be hydrolysed into glucose. Figure 2.2 summarises atomic and molecular organisation.

2.1.7 Metabolism

All the chemical processes that take place in living organisms are collectively called metabolism. Metabolism can be divided into two parts:

- **Anabolism** – an energy-requiring process in which small molecules are combined to make larger ones. The condensation reactions that build polymers from basic sub-units, e.g. polypeptides formed from amino acids, are examples of anabolic reactions.
- **Catabolism** – chemical reactions involving the release of energy in the breakdown of larger molecules into smaller ones. The hydrolysis reactions that split polymers into their basic sub-units, e.g. polypeptides being split into amino acids, are examples of catabolic reactions.

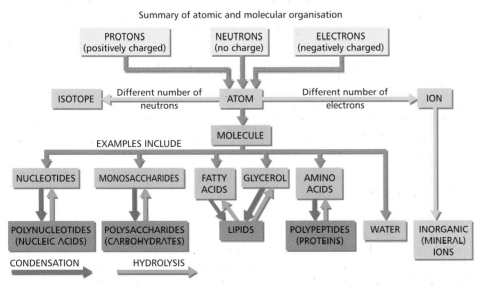

Summary of atomic and molecular organisation

Fig 2.2 Summary of atomic and molecular organisation

REMEMBERING OXIDATION AND REDUCTION

To avoid confusing the two terms – simply look at the first vowel in each of the key words in the table below:

Table 2.1 Remembering oxidation and reduction

Process	Oxidation	Reduction
Change of electrons in atoms	**L**o**s**es	**R**eceives
Atom becomes	**O**xidised	**R**educed
Ion produced	**P**ositive	**N**egative
First vowel	**O**	**e**

A useful acronym is OIL RIG – Oxidation Is Loss, Reduction Is Gain

Carbohydrates – monosaccharides and disaccharides

Table 2.2 *Types of monosaccharide*

Formula	Name	Examples
$C_3H_6O_3$ (n = 3)	Triose	Glyceraldehyde
$C_5H_{10}O_5$ (n = 5)	Pentose	Ribose Deoxyribose
$C_6H_{12}O_6$ (n = 6)	Hexose	Glucose Fructose Galactose

Table 2.3 *Types of disaccharide*

glucose	+	glucose	=	maltose
glucose	+	fructose	=	sucrose
glucose	+	galactose	=	lactose

SEMI-QUANTITATIVE NATURE OF THE BENEDICTS TEST

Table 2.4 shows the relationship between the concentration of reducing sugar and the colour of the solution and precipitate formed during the Benedict's test. The differences in colour mean that the Benedict's test is **semi-quantitative**, i.e. it can be used to estimate the approximate amount of reducing sugar in a sample. First a range of colour standards is produced by preparing a series of glucose solutions of known concentration. To an equal volume of each is added the same volume of Benedict's reagent and they are then heated for the same length of time before being cooled to room temperature. An equal volume of an unknown sample is then treated in the same way and the colour compared with that of the colour standards. As shown in the table, samples that turn red contain more reducing sugar than those that turn yellow.

Table 2.4 *The Benedict's test*

Concentration of reducing sugar	Colour of solution and precipitate
None	Blue
Very low	Green
Low	Yellow
Medium	Brown
High	Red

A further extension of this experiment would be to carry out the reducing sugar test and then to filter the suspensions. The precipitate can then be dried and weighed. The greater the mass of precipitate, the more reducing sugar is present.

As the word suggests, carbohydrates are carbon molecules (carbo) combined with water (hydrate); their general formula is $C_x(H_2O)_y$. Like many organic molecules, they are made up of individual units called **monomers**, which can be combined to form larger units called **polymers**. In carbohydrates the basic unit is a sugar, or **saccharide**. A single unit is called a **monosaccharide**, pairs of which can be combined to form a **disaccharide**. Monosaccharides are usually combined in much larger numbers to form **polysaccharides** (see units 2.3 and 2.4).

2.2.1 Monosaccharides

Monosaccharides are sweet-tasting, soluble substances that have the general formula $(CH_2O)_n$. While 'n' can be any number from 3 to 7, the three most common groups of monosaccharides are shown in table 2.2.

2.2.2 Structure of monosaccharides

Perhaps the best-known monosaccharide is **glucose**. This molecule is a hexose (6-carbon) sugar and has the formula $C_6H_{12}O_6$. However, the atoms of carbon, hydrogen and oxygen can be arranged in many different ways. Although the molecular arrangement is often shown as a straight chain for convenience, the atoms actually form a ring which can take a number of forms, as shown in figure 2.3. Different molecular structures are given different names, e.g. glucose, fructose and galactose, and further differences are shown by a letter before the molecule's name, e.g. α-glucose, β-glucose. You will see from figure 2.3 that the hydroxyl group -OH on carbon atom 1 is at the bottom of the ring in α-glucose and at the top of the ring in β-glucose. Although some of these differences are small, they often give the resulting molecules very different properties (see extension box in unit 2.3).

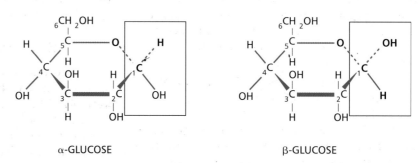

Fig 2.3 *Molecular arrangement of α-glucose and β-glucose*

2.2.3 Disaccharides

When combined in pairs, monosaccharides form a **disaccharide**. As table 2.3 shows, the two monosaccharides that combine can be the same or different. When they join, a molecule of water is removed and the reaction is therefore called a **condensation reaction**. The bond that is formed is called a **glycosidic bond**. As this bond is between carbon atom 1 of one monosaccharide and carbon atom 4 of the other, it is known as a 1,4 glycosidic bond. Figure 2.4 illustrates the formation and breaking of a 1,4 glycosidic bond.

When water is added to a disaccharide under suitable conditions, it breaks the glycosidic bond into its constituent monosaccharides. This is called **hydrolysis** (breakdown by water). The breakdown is very slow, however, unless it is catalysed by the appropriate enzyme.

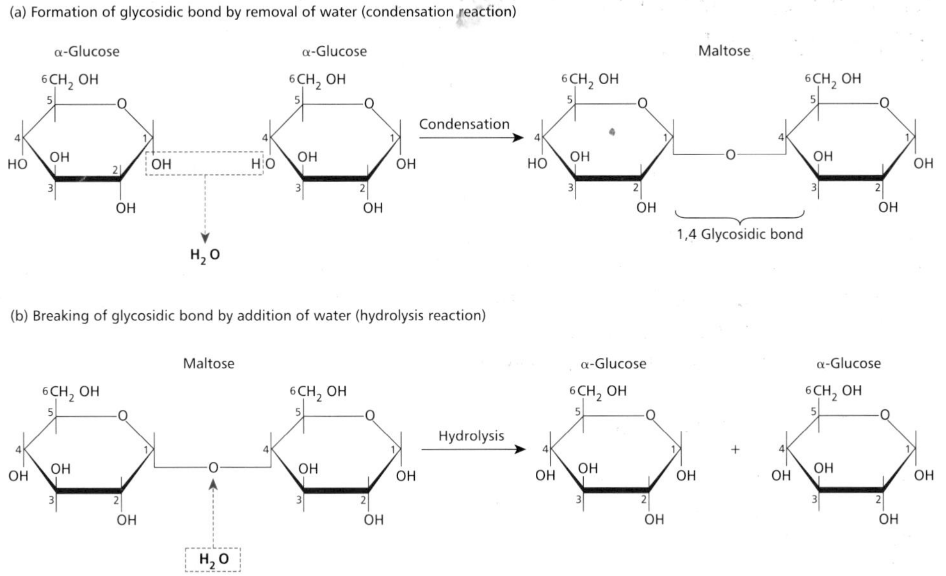

(a) Formation of glycosidic bond by removal of water (condensation reaction)

(b) Breaking of glycosidic bond by addition of water (hydrolysis reaction)

Fig 2.4 Formation and breaking of glycosidic bond in the interconversion of α-glucose and maltose (some carbon atoms have been omitted for simplicity)

WHAT IS A REDUCING SUGAR?

You may recall from section 2.1.2 that the receiving of an **electron** is called reduction. Any substance that provides electrons to carry out reduction is called a reducing agent, i.e. an agent that reduces other substances. Benedict's reagent contains copper II sulphate. The Cu^{2+} ions from the copper sulphate are reduced by electrons from the $-C=O$ group found in certain sugars to form Cu^+ ions. This results in the change to copper I oxide that forms a red precipitate. The sugars that contribute electrons are called reducing sugars.

2.2.4 Tests for reducing and non-reducing sugars

All monosaccharides and some disaccharides (e.g. maltose) are reducing sugars. The test for a reducing sugar is known as the **Benedict's test**. When a reducing sugar is heated with an alkaline solution of copper II sulphate (Benedict's reagent) it forms an insoluble precipitate of copper I oxide. The colour of the precipitate changes from green through yellow, orange and brown to deep red, depending on the quantity of reducing sugar present (see table 2.4). Some disaccharides, such as sucrose, are non-reducing sugars. There is no direct test for a non-reducing sugar, but they can be identified by first hydrolysing them with a dilute acid and then detecting the resulting reducing sugars by the Benedict's test. The process is as follows.

- Heat a sample with Benedict's reagent in a water bath. If there is no change (solution remains blue), then no reducing sugar is present.
- Heat the sample in a water bath for 5 minutes with dilute hydrochloric acid to hydrolyse the non-reducing sugar, then neutralise with sodium hydrogencarbonate and allow to cool.
- Re-test the resulting solution by heating in a water bath with Benedict's reagent, which will now turn yellow/brown/red due to the reducing sugars made from hydrolysis of the non-reducing sugar.

Instead of using acid hydrolysis to determine the presence of sucrose, an alternative is to hydrolyse the sucrose into reducing sugars using the enzyme, sucrase. Take 2cm³ of the solution to be tested, add one drop of 5–10% sucrase (invertase) concentrate and leave it for 30 minutes. After this the solution can be tested for the presence of reducing sugar using the Benedict's test.

2.2.5 Roles of monosaccharides and disaccharides

Monosaccharides and disaccharides function as respiratory substrates that are broken down to provide energy in the form of **ATP** for carrying out living processes. They are particularly useful because they have a large number of C-H groups and these can be easily oxidised, yielding a lot of energy. Table 2.5 lists the roles of the most important sugars

Table 2.5 Roles of sugars

Name of carbohydrate	Function
Ribose / deoxyribose	Makes up part of nucleotides and as such gives structural support to the nucleic acids RNA and DNA. Constituent of hydrogen carriers such as NAD, NADP and FAD. Constituent of ATP
Glucose	Major respiratory substrate in plants and animals. Used in the synthesis of disaccharides and polysaccharides. Constituent of nectar. Major transporter of carbohydrate in mammals
Galactose	Respiratory substrate. Used in the synthesis of lactose
Fructose	Respiratory substrate. Constituent of nectar. Sweetens fruits, which attracts animals and aids seed dispersal
Sucrose	Respiratory substrate. Form in which most carbohydrate is transported in plants. Storage material in some plants
Lactose	Respiratory substrate. Mammalian milk contains 5% lactose and is therefore a major carbohydrate source for sucklings
Maltose	Respiratory substrate

Carbohydrates – starch and glycogen

Starch and glycogen are examples of **polysaccharides**. Polysaccharides are polymers, formed from the combining together of many monosaccharide units. The monosaccharides are joined by glycosidic bonds that are formed by **condensation reactions**. The resulting chain may vary in length and be branched and folded in various ways. All these features affect the properties of the polysaccharide that is formed. As polysaccharides are very large molecules (**macromolecules**), they are insoluble – a feature which suits them for storage. When they are **hydrolysed**, polysaccharides break down into monosaccharides or disaccharides. Some polysaccharides, such as cellulose (unit 2.4), are not used for storage, but give structural support to plant cells.

2.3.1 Starch

Starch is a polysaccharide that is found in many parts of a plant in the form of small granules, or grains, e.g. starch grains in chloroplasts. Especially large amounts occur in seeds and storage organs such as potato tubers. It forms an important component of food and is the major energy source in most diets. Apart from the starch produced for eating, about 30 million tonnes are extracted from plants across the world for other purposes. These include wallpaper pastes, paper coatings, textiles, paints, cosmetics and medicines. Starch is a mixture of two substances – amylose and amylopectin.

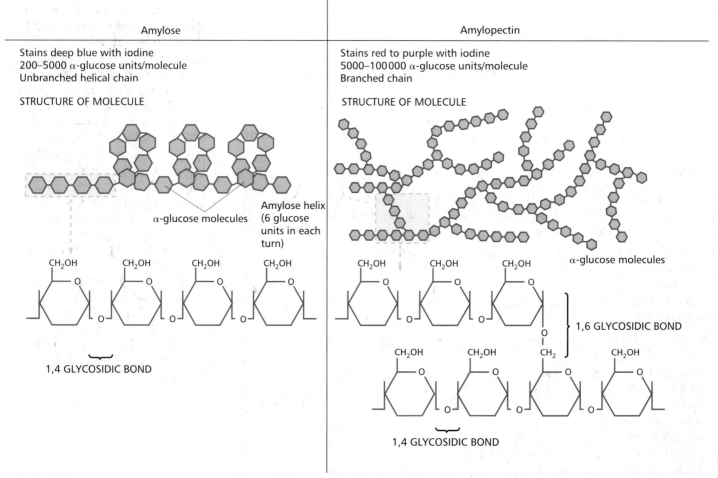

Amylose	Amylopectin
Stains deep blue with iodine 200–5000 α-glucose units/molecule Unbranched helical chain	Stains red to purple with iodine 5000–100 000 α-glucose units/molecule Branched chain

STRUCTURE OF MOLECULE

α-glucose molecules

Amylose helix (6 glucose units in each turn)

CH_2OH CH_2OH CH_2OH CH_2OH

1,4 GLYCOSIDIC BOND

STRUCTURE OF MOLECULE

α-glucose molecules

CH_2OH CH_2OH CH_2OH

1,6 GLYCOSIDIC BOND

CH_2OH CH_2OH CH_2 CH_2OH

1,4 GLYCOSIDIC BOND

Fig 2.5 *Comparison of amylose and amylopectin*

- **Amylose** is composed of between 200 and 5000 α-glucose units, which are joined in a straight chain by 1,4 glycosidic bonds. This chain is then wound into a tight coil which makes the molecule more compact and therefore it can be stored more efficiently as it takes up less space.
- **Amylopectin** is made up of between 5000 and 100 000 α-glucose units joined to each other by 1, 4 and 1,6 glycosidic bonds.

A comparison of amylose and amylopectin is given in figure 2.5. About 80% of starch is amylopectin and the remaining 20% is amylose. The main role of starch is for energy storage, something it is especially suited for because:

- It is insoluble and therefore does not have any **osmotic** influence within cells, i.e. it does not tend to draw water into the cells.
- Being insoluble, it does not easily diffuse out of cells.
- It is compact, so a lot of it can be stored in a small space.
- When hydrolysed it forms glucose, which is both easily transported and readily used in respiration, to provide energy in the form of ATP.

Starch is never found in animal cells. Instead a similar polysaccharide, called **glycogen**, serves the same role.

2.3.2 Glycogen

Glycogen is very similar in structure to amylopectin but has shorter chains and is more highly branched. It is sometimes called 'animal starch' because it is the major carbohydrate storage product of animals, in which it is stored as small granules mainly in the muscles and the liver. Its structure suits it for storage for the same reasons as those given for starch (section 2.3.1) except that, because it is made up of smaller chains, it is even more readily hydrolysed to α-glucose. Glycogen is found in animal cells and fungi but never in plant cells.

2.3.3 Test for starch

Starch is easily detected by its ability to turn the iodine in potassium iodide solution from a yellow colour to blue-black. The colouration is due to the iodine molecules becoming fixed in the centre of the helix of each starch molecule (Fig 2.6). It is important that this test is carried out at room temperature (or below), as high temperatures cause the starch helix to unwind, releasing the iodine, which then assumes its usual yellow colouration.

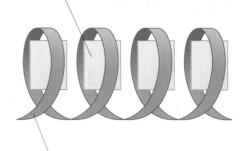

Iodine molecule in the centre of the starch helix

Starch helix formed by α-glucose molecules (6 per turn of helix). The dimensions of the centre are just sufficient to fit iodine molecules within it

Fig 2.6 *Amylose – iodine staining reaction*

SUMMARY TEST 2.3

From the following list of carbohydrates choose one or more which most closely fit each of the statements 1–9 below. Each carbohydrate may be used once, more than once or not at all..

- α-glucose
- β-glucose
- amylose
- amylopectin
- glycogen

1 Stains deep blue with iodine
2 Is known as 'animal starch'
3 Found in plants and possesses 1,6 glycosidic bonds
4 Monosaccharide found in starch
5 Possesses 1,4 glycosidic bonds
6 Makes up 80% of starch
7 Can be hydrolysed
8 Branched molecule formed by condensation
9 Easily diffuses in and out of cells

Carbohydrates – cellulose

Cellulose is a polysaccharide that makes up around 50% of all organic carbon and is therefore the most abundant molecule on earth.

2.4.1 Structure of cellulose

Cellulose differs from starch and glycogen in one major respect – it is made of monomers of β-glucose rather than α-glucose. This seemingly small variation produces fundamental differences in the structure and function of this polysaccharide. The main reason for this is that, in the β-glucose units, the positions of the -H group and the -OH group on carbon atom 1 are reversed (Fig 2.7). In β-glucose the -OH group is above, rather than below, the ring. This means that, in order for 1,4 glycosidic links to form, each β-glucose molecule must be rotated by 180° compared to its neighbour. The result is that carbon atom 6 (the one forming part of the -CH₂OH group) on each β-glucose molecule alternates between being above and below the chain (Fig 2.7). Rather than forming a coiled chain like starch, cellulose has straight, unbranched chains. These run parallel to one another, allowing hydrogen bonds (section 2.1.4) to form cross-linkages between adjacent chains. While each individual hydrogen bond adds very little to the strength of the molecule, the sheer overall number of them makes a considerable contribution to strengthening cellulose and making it the valuable structural material it is. The arrangement of β-glucose chains in a cellulose molecule is shown in figure 2.8.

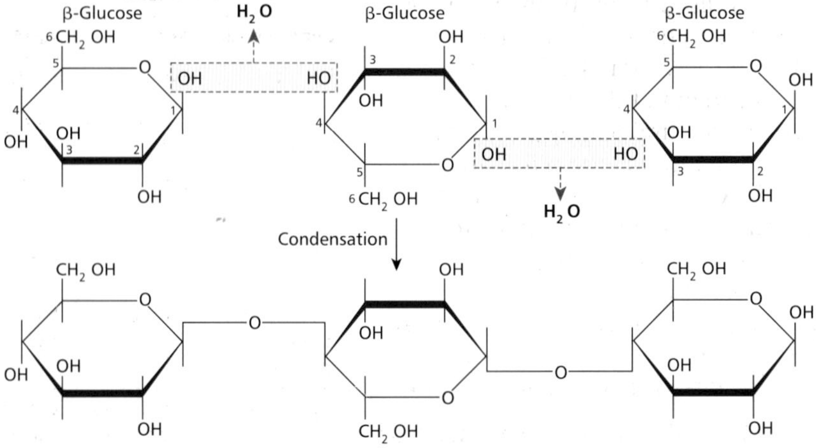

Fig 2.7 *Formation of 1,4 glycosidic bonds between three β-glucose molecules (some carbon molecules have been omitted for simplicity)*

Being composed of fewer β-glucose units, the chain, unlike that of starch, has adjacent glucose molecules rotated by 180°. This allows hydrogen bonds to be formed between the hydroxyl (–OH) groups on adjacent parallel chains which help to give cellulose its structural stability.

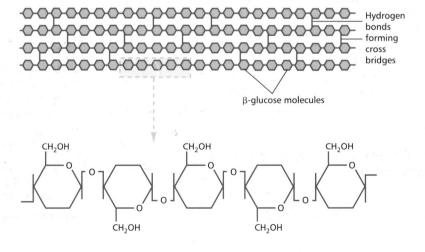

Fig 2.8 *Structure of the cellulose molecule*

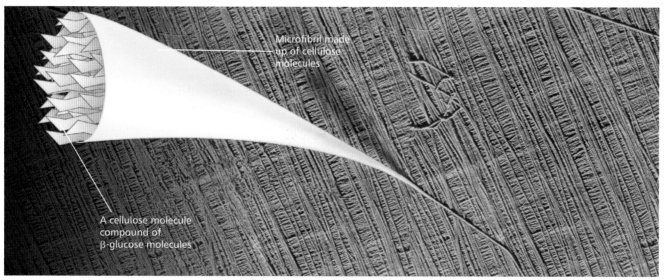

Fig 2.9 *Structure of a cellulose microfibril*

The cellulose molecules are grouped together to form microfibrils which, in turn, are arranged in parallel groups called fibres (Fig 2.9).

2.4.2 Functions of cellulose

Cellulose frequently makes up between 20% and 50% of plant cell walls. It is, however, not part of the living cell but rather a non-living covering that encases the **protoplast** within. The cellulose cell wall is therefore freely permeable, allowing materials to get in and out of the living cell. Cellulose performs a mainly structural role by providing rigidity to the plant cell. It also prevents the cell from bursting as water enters it by **osmosis**. It does this by the cellulose cell wall exerting an inward pressure that stops any further influx of water. As a result, living plant cells are turgid and push against one another, making herbaceous parts of the plant semi-rigid. This is especially important in maintaining stems and leaves in a turgid state so that they can provide the maximum surface area for photosynthesis.

The molecular stability of the cellulose molecule means it is well suited to giving structural support. However, it is also difficult to digest. It is therefore not a useful food for animals, which rarely produce cellulose-digesting enzymes. Some animals get round this by forming **mutualistic** relationships with cellulose-digesting microorganisms in their intestines. The structural strength of cellulose has been made use of by humans. Cotton and rayon used in fabrics are largely cellulose. Cellophane used in packaging and celluloid used in photographic films are also derived from cellulose. Paper is perhaps the best-known cellulose product.

2.4.3 Comparison of cellulose and other carbohydrates

Table 2.6 compares cellulose with the other polysaccharides, amylose, amylopectin and glycogen.

Table 2.6 *Comparison of the polysaccharides amylose, amylopectin, glycogen and cellulose*

Characteristic	Amylose	Amylopectin	Glycogen	Cellulose
Found in	Plants	Plants	Animals and fungi	Plants
Found as	Grains	Grains	Tiny granules	Fibres
Function	Energy store	Energy store	Energy store	Structural support
Basic monomer unit	α-glucose	α-glucose	α-glucose	β-glucose
Type of bond between monomer units	1,4 glycosidic	1,4 and 1,6 glycosidic	1,4 and 1,6 glycosidic	1,4 glycosidic
Type of chain	Unbranched and helical (coiled)	Long, relatively few branches with some coiling	Short, relatively many branches with some coiling	Long, unbranched straight chains with no coiling

SUMMARY TEST 2.4

Cellulose is made up of (**1**) monomers joined together by (**2**) links. It forms straight, unbranched chains that run parallel to each other and are cross-linked by (**3**). These cellulose molecules are then grouped together to form (**4**), which in turn are grouped into fibres. Cellulose performs a (**5**) function in plants by giving a plant cell rigidity. It also prevents cells from (**6**) when water enters by (**7**). In this way it keeps herbaceous parts of plants (**8**) so that they provide the maximum surface area for (**9**).

Lipids make up a varied and diverse group of substances that share the following characteristics:

- They contain carbon, hydrogen and oxygen.
- The proportion of oxygen to carbon and hydrogen is smaller than in carbohydrates.
- They are insoluble in water.
- They are soluble in organic solvents such as alcohols and acetone.

The main groups of lipids are **triglycerides (fats and oils)**, **phospholipids** and **waxes**. Other forms include steroids and cholesterol.

2.5.1 Triglycerides (fats and oils)

There is no fundamental chemical difference between a fat and an oil. Fats are solid at room temperature (10–20°C), whereas oils are liquid. Triglycerides are so called because they have three (tri) fatty acids combined with glycerol (glyceride). Each fatty acid forms an ester bond with glycerol in a **condensation reaction** (Fig 2.10). Hydrolysis of a triglyceride therefore produces glycerol and three fatty acids.

The three fatty acids may all be the same, thereby forming a simple triglyceride, or they may be different, in which case a mixed triglyceride is produced. In either case it is a condensation reaction.

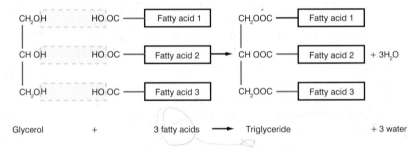

Fig 2.10 *Formation of a triglyceride*

2.5.2 Fatty acids

As the glycerol molecule in all fatty acids is the same, the differences in the properties of different fats and oils come from variations in the fatty acids. There are over 70 fatty acids and all have a carboxyl (-COOH) group with a hydrocarbon chain attached. This chain may possess no double bonds and is then described as **saturated**, because all the carbon atoms are linked to the maximum possible number of hydrogen atoms, i.e. they are saturated with hydrogen atoms. If there is a single double bond, it is **mono-unsaturated**; if more than one double bond is present, it is **polyunsaturated**.

2.5.3 Phospholipids

Phospholipids are similar to lipids except that one of the fatty acid molecules is replaced by a phosphate molecule (Fig 2.11). Whereas fatty acid molecules repel water (are **hydrophobic**), phosphate molecules attract water (are **hydrophilic**).

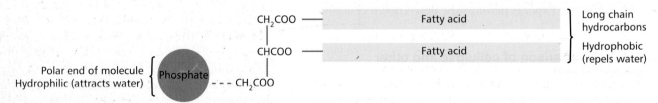

Fig 2.11 *Structure of a phospholipid*

Phospholipids are important components of cell membranes. Both the inside of a cell and the environment outside are watery, and the phospholipids in cell membranes form a double layer, with the hydrophilic heads of the molecules pointing into either the watery environment outside the membrane or the watery medium inside the cell. The hydrophobic tails point into the middle of the membrane (Fig 2.12). This **bilayer** arrangement makes cell membranes fluid and easily traversed by lipid-soluble substances.

2.5.4 Waxes

Waxes are similar to triglycerides, except that glycerol is replaced by a longer chain alcohol, making them more complex structures. They are more solid than fats at room temperature and their major role is to produce a protective waterproofing cuticle on the surfaces of leaves and insects.

2.5.5 Roles of lipids

Due to their variety of types and forms, lipids perform many different functions in living organisms. These include:

- **An energy source** – lipids provide more than twice as much energy as carbohydrate when they are oxidised. This makes them excellent stores of energy, especially in animals and in plant seeds, both of which have to move, or be moved, from place to place and therefore need to keep their mass to a minimum. Being insoluble in water, lipids are not easily leached from cells.
- **Insulation** – fats are slow conductors of heat and so are often stored beneath the body surface in endothermic (maintaining a constant body temperature) animals, which helps retain body heat. This subcutaneous fat is especially important in aquatic organisms, such as whales and seals, where hair is ineffective.
- **Protection** – fat is often stored around delicate organs such as the kidney, where it acts as packing material to protect the organ from physical damage.
- **Waterproofing** – water conservation is important to all terrestrial organisms, but especially those in hot and dry regions of the world. Both plants and insects have waxy cuticles that conserve water, while mammals produce an oily secretion from the sebaceous glands in the skin. The hydrophobic fatty acids repel water and so lipids are especially efficient at preventing water loss. In addition, the **oxidation** of lipids in respiration provides more water than the oxidation of an equivalent mass of carbohydrate – a considerable bonus where water is in short supply, e.g. desert-living organisms.
- **Buoyancy** – lipids are less dense than water and so aquatic animals that have fat for insulation enjoy the added advantage of being more buoyant. This is important because aquatic mammals breathe air and need to come to the surface to obtain it. Aquatic birds and insects use oil droplets to aid buoyancy.
- **Cell membranes** – phospholipids are important in cell membranes, contributing to their flexibility and the transfer of lipid-soluble substances across them.

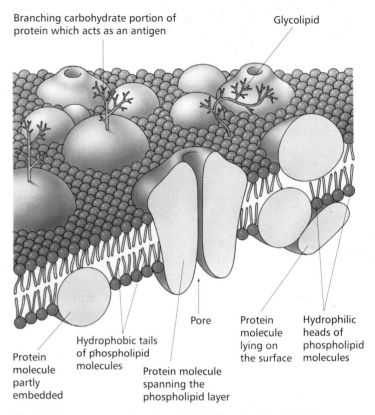

Branching carbohydrate portion of protein which acts as an antigen

Glycolipid

Pore

Protein molecule lying on the surface

Hydrophilic heads of phospholipid molecules

Hydrophobic tails of phospholipid molecules

Protein molecule partly embedded

Protein molecule spanning the phospholipid layer

Fig 2.12 *Arrangement of the phospholipid bilayer in the cell surface membrane*

TEST FOR LIPIDS

The test for lipids is known as the **emulsion test** and is carried out as follows.

- Take a completely dry and grease-free test tube.
- Add about $2cm^3$ of the sample being tested and add $5cm^3$ of ethanol.
- Shake the tube thoroughly to dissolve any lipid in the sample.
- Add $5cm^3$ of water and shake gently.
- A cloudy-white colour indicates the presence of a lipid.
- As a control, repeat the procedures using water instead of the sample; the final solution should remain clear.

The cloudy colour is due to any lipid in the sample being finely dispersed in the water to form an emulsion. Light passing through this emulsion is refracted as it passes from oil droplets to water droplets, making it appear cloudy.

SUMMARY TEST 2.5

Fats and oils make up a group of lipids called **(1)** which when hydrolysed form **(2)** and fatty acids. A fatty acid with more than one double bond is called **(3)**. In a phospholipid the number of fatty acids is **(4)**; these are called **(5)** because they repel water. Lipids are important as an **(6)** in seeds and as **(7)** in endothermic animals. Around the kidneys they perform the role of **(8)** whereas in plant and insect cuticles they are used for **(9)**.

Amino acids and polypeptides

Amino acids are the basic monomer units which combine to make up proteins. Around 100 amino acids have been identified, of which 20 occur naturally in proteins.

2.6.1 Structure of an amino acid

Every amino acid has a central carbon atom to which are attached four different chemical groups:

- **amino group** ($-NH_2$) – a basic group from which part of the name amino acid is derived
- **carboxyl group** (-COOH) – an acid group which gives the amino acid the rest of its name
- **hydrogen atom** (-H)
- **R group** – a variety of different chemical groups ranging from a single hydrogen atom, as in glycine, to a double ring structure, as in tyrosine. Each amino acid has a different R group.

The general structure of an amino acid is shown in figure 2.13.

As the carboxyl group is acidic and the amino group is basic, an amino acid is both an acid and a base – it is said to be **amphoteric**. Amphoteric compounds act as **buffer** solutions in that they resist the tendency to alter their pH, despite the addition of acids or bases. This property is important in cells because it helps them maintain the stable pH that is necessary for the efficient functioning of enzymes.

Green plants can manufacture all the amino acids they need by combining nitrates, which they absorb from the soil, with various carbohydrates that they make during photosynthesis. Animals, however, need to obtain eight amino acids from their food. These are called **essential amino acids**. The remaining twelve amino acids, called **non-essential amino acids**, can be manufactured from the eight essential ones if they are not provided in adequate amounts in the diet.

2.6.2 Formation and breakage of a peptide bond

In the same way that monosaccharide monomers combine to form disaccharides (section 2.2.3), so amino acid monomers can combine to form a **dipeptide**. The process is essentially the same – namely the removal of a water molecule in a condensation reaction (section 2.1.6). The water is made by combining an -OH from the carboxyl group of one amino acid with an -H from the amino group of another amino acid. The two amino acids then become linked by a new **covalent bond** between the carbon atom of one amino acid and the nitrogen atom of the other. The formation of a peptide bond is illustrated in figure 2.14. In the same way as a glycosidic bond of a disaccharide can be broken by the addition of water (hydrolysis), so the peptide bond of a dipeptide can also be broken by hydrolysis (section 2.1.6), to give its two constituent amino acids.

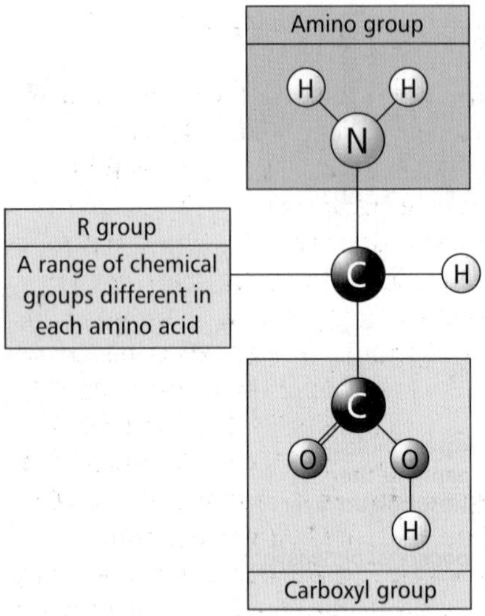

Fig 2.13 *General structure of an amino acid*

Amino group

R group
A range of chemical groups different in each amino acid

Carboxyl group

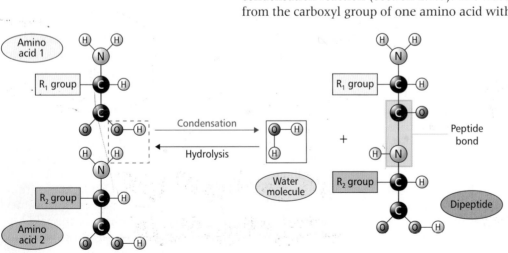

Fig 2.14 *Formation of a peptide bond*

Amino acid 1

R_1 group

Hydrolysis

Condensation

Water molecule

R_2 group

Amino acid 2

R_1 group

Peptide bond

R_2 group

Dipeptide

2.6.3 The primary structure of proteins – polypeptides

Through a series of condensation reactions, many amino acid monomers can be joined together in a process called **polymerisation**. The resulting chain of many hundreds of amino acids is called a **polypeptide**. The sequence of amino acids in a polypeptide chain forms the **primary structure** of any protein. As polypeptides have many (usually hundreds) of the 20 naturally occurring amino acids joined in any sequence, it follows that there is an almost limitless number of possible combinations, and therefore types, of primary protein structure. A simple protein may consist of a single polypeptide chain. More commonly, however, a protein is made up of a number of polypeptide chains.

2.6.4 The secondary structure of proteins

The linked amino acids that make up a polypeptide possess both -NH and -C=O groups on either side of every peptide bond. Both these groups are polar, i.e. their **electrons** are unevenly distributed. As a result, the hydrogen of the -NH group has an overall positive charge while the O of the -C=O group has an overall negative charge. These two groups therefore readily form hydrogen bonds (section 2.1.4). This causes the long polypeptide chain to be twisted into one of two basic three-dimensional shapes:

- **α-helix** – the polypeptide chain is coiled into a cylindrical shape
- **β-pleated sheet** – different polypeptide chains become linked in parallel flat sheets. Sometimes the chains run in opposite directions (antiparallel) and if a chain is very long it may fold back on itself to give two parallel portions (β-turn).

Fig 2.15 illustrates these two basic types of protein structure. Which secondary shape is adopted depends upon the various R groups present on the amino acids. Indeed, some proteins show no regular arrangement at all.

SUMMARY TEST 2.6

Amino acids always contain an acid carboxyl group, which has the chemical formula (**1**), as well as a basic group called an (**2**) group. Amino acids therefore have both acidic and basic properties and are said to be (**3**). Any two amino acids can combine in a (**4**) reaction to form a (**5**) bond between them. Many amino acids can combine to form a polypeptide chain which can become altered in shape due to (**6**) bonds formed between certain groups. These secondary shapes may be cylindrical, in which case they are called (**7**), or flat, in which case they are called (**8**).

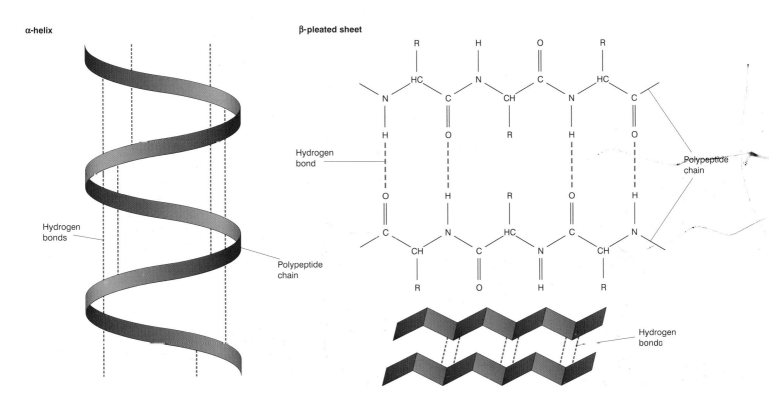

Fig 2.15 *Structure of the α-helix and the β-pleated sheet*

Protein structure

Proteins are large molecules with relative molecular masses ranging from many thousands up to 40 million. While the types of carbohydrates and lipids of all organisms are relatively few and very similar, their proteins are numerous (around 10 000 types in humans) and differ from species to species. The shape of any one type of protein molecule differs from that of other proteins. As the amino acids of which they are made differ only in their R groups, it is these that determine the shape, and therefore the functions, of a protein. We have already seen in unit 2.6 how the primary and secondary shapes of a protein are determined. We shall now see how these structures are moulded into the tertiary and quaternary structures that make up a protein's final shape, or **configuration**.

2.7.1 Tertiary structure of proteins

The α-helices or β-pleated sheets of the secondary protein structure can be twisted and folded even more to give the complex, and often unique, three-dimensional structure of each protein (Fig 2.17). This is known as the tertiary structure and is the result of four possible types of bonds that can arise between each amino acid. In order of relative strengths, these are:

- **Disulphide bridges** are found between sulphur atoms in the molecules of the amino acid cysteine. They are **covalent bonds** and, as such, form very strong links which make the tertiary protein structure very stable.
- **Ionic bonds** occur between any carboxyl and amino groups that have not been involved in forming peptide bonds. These groups ionise to give $-NH_3^+$ and $-COO^-$ groups, which then form bonds due to their mutual attraction. These bonds are weaker than disulphide bridges and can be broken by changes in pH.
- **Hydrogen bonds** result from the attraction between the electronegative oxygen atoms on the -CO groups and the electropositive H atoms on either the -OH or -NH groups. Although they are individually weaker than ionic bonds, their large number makes them an important factor in maintaining the tertiary structure of a protein.
- **Hydrophobic interactions** are due to certain non-polar R groups in amino acids that have side groups which repel water. As a result they may fold or twist the polypeptide chain as they take up a position towards the centre of the protein, further away from the watery medium outside.

Fig 2.16 illustrates how each of these bonds is formed.

2.7.2 Quaternary structure of proteins

Large proteins often form complex molecules containing a number of individual polypeptide chains that are linked in various ways. There may also be non-protein (prosthetic groups) groups associated with the molecules (Fig 2.17). Examples of quaternary structure are illustrated by the blood protein haemoglobin (section 8.6.1) and the hormone insulin.

(a) **Disulphide bridges** – *covalent bond between R groups of cysteine amino acids*

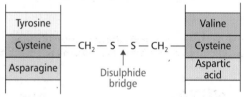

(b) **Ionic bonds** – *between NH_3^+ and COO^- ions on basic amino acids such as asparagine and acid ones such as aspartic acid*

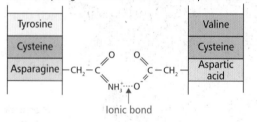

(c) **Hydrophobic interactions** – *between non-polar R groups such as those on the amino acids tyrosine and valine*

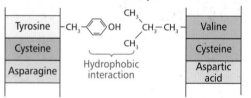

(d) **Hydrogen bonds** – *between electronegative oxygen atoms on CO groups and electropositive H atoms on NH groups*

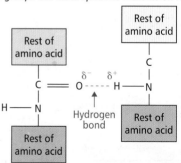

Fig 2.16 *Types of bond that determine the shape of a protein*

2.7.3 Test for proteins

The most reliable protein test is the **Biuret test**, which detects peptide links. It is performed as follows.

- To a sample of the solution thought to contain protein, add an equal volume of sodium hydroxide solution at room temperature.

- Add a few drops of very dilute (0.05%) copper II sulphate solution and mix gently.
- A purple colouration indicates the presence of peptide bonds and hence a protein.
- A control should be carried out by performing the above stages, but with water replacing the sample under test.

(a) *The primary structure of a protein is the sequence of amino acids found in its polypeptide chains. This sequence determines its properties and shape. Following the elucidation of the amino acid sequence of the hormone insulin by Frederick Sanger in 1954, the primary structure of many other proteins is now known.*

Lysine	Aspartic acid	Cysteine	Alanine	Tyrosine	Lysine	Glutamic acid	Valine	Glycine

(b) *The secondary structure is the shape which the polypeptide chain forms as a result of hydrogen bonding. This is most often a spiral known as the α-helix, although other configurations occur.*

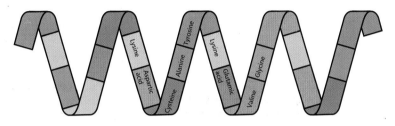

(c) *The tertiary structure is due to the bending and twisting of the polypeptide helix into a compact structure. All three types of bond, disulphide, ionic and hydrogen, contribute to the maintenance of the tertiary structure.*

(d) *The quaternary structure arises from the combination of a number of different polypeptide chains and associated non-protein (prosthetic) groups into a large, complex protein molecule.*

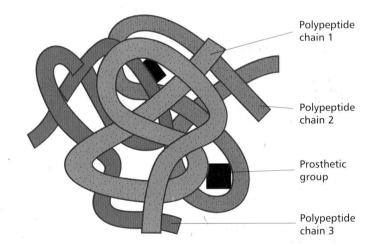

Polypeptide chain 1

Polypeptide chain 2

Prosthetic group

Polypeptide chain 3

Fig 2.17 *Structure of proteins*

SUMMARY TEST 2.7

The primary structure of proteins is determined by the sequence of (1) which make up the (2) chain. The secondary structure results from coiling or folding of the chain due to (3) formed between the -NH and the (4) group of the (5) bonds. Three additional bonds cause further twisting and folding of the chain. The first of these bonds arises between (6) atoms in cysteine molecules and is called (7). The second type is called (8) and results from electrostatic forces between carboxyl and amino groups of amino acids. Thirdly there are forces due to amino acid side groups which repel water and these are called (9). The strongest of these bonds are the (10), followed by (11), while (12) are the weakest of the four. The quaternary structure of proteins results from a number of chains combining, sometimes also incorporating non-protein groups known as (13) groups.

Fibrous and globular proteins

Proteins perform many different roles in living organisms. In one form or another they are essential for the efficient functioning of every characteristic of life. Their roles depend on their molecular configurations, which are of two basic types.

- **Fibrous proteins**, such as collagen and keratin, have structural functions.
- **Globular proteins**, such as enzymes, haemoglobin and insulin, carry out metabolic functions.

Table 2.7 lists the differences between fibrous and globular proteins.

2.8.1 The fibrous protein collagen

Fibrous proteins form long chains which run parallel to one another. These chains are linked by cross-bridges and so form very stable molecules. One example is **collagen**, a protein found in tissues requiring physical strength, e.g. **tendons**, walls of blood vessels, bone and the fibres that hold teeth in place. Collagen is extremely strong and stable, being able to withstand immense pulling forces without stretching. At the same time it is flexible, so that, while the collagen in a tendon transmits the pull of a muscle to the bone without stretching, it can still bend around a joint as it flexes during movement. The ability of collagen to do this is the result of the following features of its structure.

- Its primary structure is largely a repeat of the amino acid sequence glycine–proline–alanine, which forms an unbranched polypeptide chain.
- The collagen molecule is made up of three such polypeptide chains wound together in the same way as individual fibres are wound together in a rope.
- As the three main amino acids present are relatively small and compact, the helix produced is very tightly wound. Larger amino acids would have led to a more loosely wound, and therefore less strong, helix,
- The triple-stranded molecules run parallel to others, forming even stronger units called fibres.
- The collagen molecules in the fibres are held together by cross-linkages formed between lysine amino acids of adjacent fibres. This adds greater strength and stability to the structure.
- The points where one collagen molecule ends and the next begins are spread throughout the structure. If they were all joined together in the same region this would be a weak point and therefore prone to breaking under tension.

The structure of collagen is illustrated in figure 2.18.

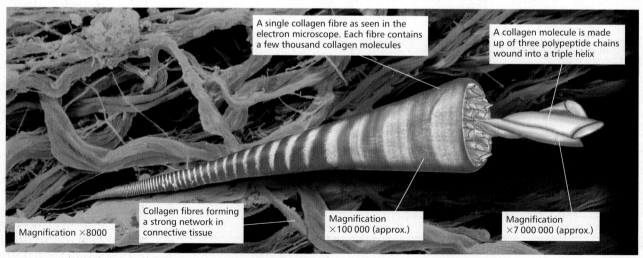

A single collagen fibre as seen in the electron microscope. Each fibre contains a few thousand collagen molecules

A collagen molecule is made up of three polypeptide chains wound into a triple helix

Collagen fibres forming a strong network in connective tissue

Magnification ×8000

Magnification ×100 000 (approx.)

Magnification ×7 000 000 (approx.)

Fig 2.18 *Fine structure of the fibrous protein collagen*

2.8.2 The globular protein haemoglobin

The sequence of amino acids in globular proteins is far more varied, and they form a more compact structure than a fibrous protein. If the polypeptide chains of a fibrous protein are thought of as string twisted into a rope, then a globular protein is like the same string rolled into a ball. One example of a globular protein is haemoglobin (Fig 2.19 and section 8.6.1). This is an oxygen-carrying respiratory pigment found in most animal groups. Its ability to transport oxygen is related to its structural features, which in adult humans, include the following:

- It has a primary structure of four polypeptide chains.
- Two of these chains, called α-polypeptides, are identical and each consists of 141 amino acids.
- The other two chains, called β-polypeptides, also form an identical pair and each consists of 146 amino acids.
- Each polypeptide chain is folded into a compact shape and all four are linked together to form an almost spherical haemoglobin molecule.
- Hydrophobic interactions (section 2.7.1) of groups within the haemoglobin molecule help to maintain its precise shape – an important factor in its ability to carry oxygen.
- Hydrophilic groups in the molecule tend to orient themselves to point outwards. This enables haemoglobin to mix more readily with a watery medium.
- Associated with each polypeptide is a haem group – which contains a ferrous (Fe^{2+}) ion. Non-protein groups such as this are called **prosthetic groups** and they from an important and integral part of the protein molecule.
- Each Fe^{2+} ion can combine with a single oxygen molecule (O_2), making a total of four O_2 molecules that can be carried by a single haemoglobin molecule in humans. (Other types of haemoglobin have different numbers of haem groups and so carry different numbers of O_2 molecules.)

When haemoglobin combines with oxygen it forms a molecule called **oxyhaemoglobin** and changes colour from purplish to bright red. The structure of a haemoglobin molecule is shown in figure 2.19.

Globular proteins such as haemoglobin and enzymes have a very specific shape. Even slight changes to their structure can make such molecules far less efficient at carrying out their functions. In the case of haemoglobin a slight alteration in shape as a result of the condition sickle cell anaemia, makes it far less able to transport oxygen.

Table 2.8 lists some other important proteins and their functions.

Table 2.8 Protein functions

Protein	Function of protein
Trypsin and pepsin	Digestion of proteins/ polypeptides
Myoglobin	Stores oxygen in muscle
Actin and myosin	Needed for contraction of muscle
Antibodies	Defend against bacterial invasion
Gluten	Storage protein in seeds
Histone	Gives structural support to chromosomes

Four polypeptide chains make up the haemoglobin molecule. Each molecule contains 574 amino acids

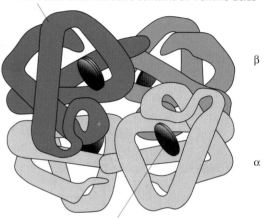

β

α

Each chain is attached to a haem group that can combine with oxygen

Fig 2.19 Quaternary structure of a haemoglobin molecule

Table 2.7 Comparison of fibrous and globular proteins

Fibrous proteins	Globular proteins
Repetitive regular sequences of amino acids	Irregular amino acid sequences
Actual sequences may vary slightly between two examples of the same protein	Sequence highly specific and never varies between two examples of the same protein
Polypeptide chains form long parallel strands	Polypeptide chains folded into a spherical shape
Length of chain may vary in two examples of the same protein	Length always identical in two examples of the same protein
Stable structure	Relatively unstable structure
Insoluble	Soluble – forms colloidal suspensions
Support and structural functions	Metabolic functions
Examples include collagen and keratin	Examples include all enzymes, some hormones (e.g. insulin) and haemoglobin

SUMMARY TEST 2.8

Proteins are of two basic types: fibrous proteins such as collagen and **(1)** proteins such as the blood pigment haemoglobin. Proteins like actin and myosin have a structural function and are therefore examples of a **(2)** protein. Collagen is a fibrous protein with the repeating amino acid sequence of **(3)**. It is found in structures such as **(4)**, which attach muscle to bone, where its properties of **(5)**, and **(6)** suit it to its role. A single haemoglobin molecule is made up of polypeptides, which total **(7)** in number. Each contains a **(8)** group that contains a single **(9)** ion, to which can be attached a single oxygen molecule.

2.9 Water and inorganic ions

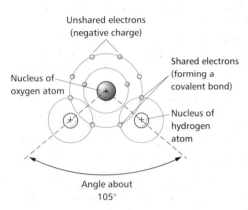

Fig 2.20 A water molecule

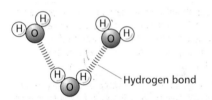

Fig 2.21 Water molecules showing hydrogen bonding

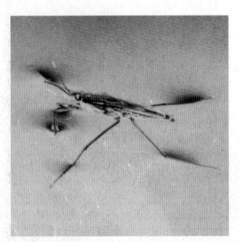

Due to surface tension, pond skaters walk on water

Although water is the most abundant liquid on Earth, it is certainly no ordinary molecule. Its unusual properties are due to its dipolar nature and the subsequent hydrogen bonding that this allows.

2.9.1 The dipolar water molecule

A water molecule is made up of two atoms of hydrogen and one of oxygen as shown in figure 2.20. The atoms form a triangular shape. Although the molecule has no overall charge, the distribution of negatively charged electrons is uneven because the oxygen atom draws them away from the hydrogen atoms. The oxygen atom therefore has a slight negative charge, while the hydrogen atoms have a slight positive one. In other words, the water molecule has both positive and negative poles and is therefore described as **dipolar**.

2.9.2 Water and hydrogen bonding

Different poles attract, and therefore the positive pole of one water molecule will be attracted to the negative pole of another water molecule. The attractive force between these opposite charges is called a **hydrogen bond** (Fig 2.21). Although each bond is fairly weak (about one-tenth as strong as a **covalent bond**), together they form important forces that cause the water molecules to stick together, giving water its unusual properties.

2.9.3 Specific heat capacity of water

Because water molecules stick together, it takes more energy (heat) to separate them than would be needed if they did not bond to one another. For this reason the boiling point of water is higher than expected. Without its hydrogen bonding, water would be a gas (water vapour) at the temperatures commonly found on Earth and life as we know it would not exist. For the same reason, it takes more energy to heat a given mass of water, i.e. water has a high **specific heat capacity**. Water therefore acts as a buffer against sudden temperature variations, making the aquatic environment a temperature-stable one.

2.9.4 Latent heat of vaporisation of water

Hydrogen bonding between water molecules means that it requires a lot of energy to evaporate one gram of water. This energy is called the **latent heat of vaporisation**. Evaporation of water such as sweat in mammals is therefore a very effective means of cooling because body heat is used to evaporate the water.

2.9.5 Cohesion and surface tension in water

The tendency of molecules to stick together is known as **cohesion**. With its hydrogen bonding, water has large cohesive forces and these allow it to be pulled up through a tube, such as a **xylem vessel** in plants. In the same way, water molecules at the surface of a body of water tend to be pulled back into the body of water rather than escaping from it. This force is called **surface tension** and means that the water surface acts like a skin and is strong enough to support small organisms such as pond skaters (see photo).

2.9.6 The density of water

Most substances are at their least dense when a gas and at their most dense when a solid, with the liquid phase having an intermediate density. Water is different. Water is actually less dense in the form of ice than when it is a liquid. This property is crucial to the survival of aquatic organisms as it means that ponds, lakes etc. freeze from the top down rather than from the bottom up. The ice formed at the top then acts as an insulating layer that delays the freezing of the water beneath it. Large bodies of water almost never freeze completely, allowing their inhabitants to survive.

2.9.7 The importance of water to living organisms

Water is the main constituent of all organisms – up to 98% of a jellyfish is water and mammals are typically 65% water. Water is also where life on Earth arose and it is the environment in which many species still live. It is important for other reasons too.

Water in metabolism
- Water is used to break down many complex molecules by **hydrolysis**, e.g. proteins to amino acids.
- Chemical reactions take place in an aqueous medium.
- Water is a major raw material in photosynthesis.

Water as a solvent
Water readily dissolves other substances and is used for:
- transport, e.g. blood
- removal of wastes, e.g. ammonia, urea
- secretions, e.g. digestive juices, tears.

Water as a lubricant
Water in combination with certain proteins makes a good lubricant. Examples include:
- mucus to aid external movement, e.g. in an earthworm moving through the soil, or internal movement, e.g. in the gut and vagina
- fluids to reduce friction, e.g. the synovial fluid in joints, the pleural fluid around the lungs and the perivisceral fluid around internal organs.

Water giving support
Water is not easily compressed and therefore is used in:
- the hydrostatic skeleton of animals such as earthworms
- the amniotic fluid to support the fetus
- creating the turgor pressure that supports the leaves of plants.

Other important features of water
- Its evaporation cools organisms and allows them to control their temperature.
- It is transparent and therefore aquatic plants can photosynthesise.

2.9.8 The roles of inorganic ions

Although needed only in tiny amounts, inorganic ions are essential to the normal functioning of organisms. The roles of eight of the most important ions are shown in table 2.9.

Table 2.9 *Roles of some important ions*

Ion	Functions
Nitrate NO_3^-	A component of amino acids, proteins, vitamins, coenzymes, **nucleotides** and chlorophyll. Some hormones contain nitrogen, e.g. auxins in plants and insulin in animals
Phosphate PO_4^{3-}	A component of nucleotides, ATP and some proteins. Used in the phosphorylation of sugars in respiration. A major constituent of bone and teeth. A component of cell membranes, in the form of **phospholipids**
Potassium K^+	Helps to maintain the electrical, osmotic and anion/cation balance across cell membranes. Assists active transport of certain materials across the cell membrane. Necessary for protein synthesis and is a co-factor in photosynthesis and respiration. A constituent of sap vacuoles in plants and so helps to maintain turgidity
Calcium Ca^{2+}	In plants, calcium pectate is a major component of the middle lamella of cell walls. It also aids the **translocation** of carbohydrates and amino acids. In animals, it is the main constituent of bones, teeth and shells. Needed for the clotting of blood and the contraction of muscle
Sodium Na^+	Helps to maintain the electrical, osmotic and anion/cation balance across cell membranes. Assists **active transport** of certain materials across the cell membrane. A constituent of the sap vacuole in plants and so helps maintain turgidity
Chloride Cl^-	Helps to maintain the electrical, osmotic and anion/cation balance across cell membranes. Needed for the formation of hydrochloric acid in gastric juice. Assists in the transport of carbon dioxide by blood (**chloride shift**)
Magnesium Mg^{2+}	A constituent of chlorophyll. An activator for some enzymes, e.g. ATPase. A component of bones and teeth
Iron Fe^{2+} or Fe^{3+}	Found in electron carriers used in respiration and photosynthesis. Required to make chlorophyll. Forms part of the haemoglobin molecule

SUMMARY TEST 2.9

A water molecule is said to be **(1)** because it has a positive and a negative pole as a result of the uneven distribution of **(2)** within it. This creates attractive forces called **(3)** between water molecules, causing them to stick together. This stickiness, otherwise called **(4)**, of water means that its molecules are pulled inwards at its surface. This force is called **(5)**. Water is able to split large molecules into smaller ones by a process known as **(6)**. Water is the raw material for the process of **(7)** in green plants and acts as a **(8)** in the synovial joints of animals. Three ions found in bone are **(9)**, **(10)** and **(11)**, whereas the three needed for maintaining an osmotic balance across cell membranes are **(12)**, **(13)** and **(14)**. The element found in both vitamins and insulin is **(15)**, whereas **(16)** is found in teeth but not in ATP or chlorophyll.

1 a Complete the following passage.

Glycogen is a type of carbohydrate known as a polysaccharide. In mammals, it is stored in particular parts of the body, namely the cells of the ...... and .... Glycogen is a polymer of ......-glucose molecules. It forms coiled chains, in which the glucose molecules are joined by 1, ... links. Many side chains are attached to the coiled chains, by 1, ... links. Starch is also a polysaccharide. It has two components, ... and ....

(7 marks)

b The table below refers to chemical tests for biological molecules. Complete the table.

method	biological molecule tested for	observation if biological molecule is	
		present	absent
add a few drops of iodine solution			
add alcohol and shake; pour into water			

(4 marks)

c A student was carrying out some tests to identify substances present in a solution. It was known that there were **two** different food substances in the solution. **Three** different tests were carried out, each one on a **separate** 10cm³ sample. The tests that were carried out and the results are shown in the table below.

	test	result
1	the biuret test was carried out	a purple colour was observed
2	the Benedict's test was carried out	the blue colour of the reagent did not change on heating
3	dilute hydrochloric acid was added and the mixture was boiled; it was allowed to cool, neutralised and then test 2 was carried out	the blue colour of the reagent changed to a brick-red precipitate on heating

State the **two** food substances present in the student's solution. (2 marks)

(Total 13 marks)

OCR 2801 Jan 2003, B (BF), No.2

2 Food packaging often includes information on the nutritional value of the food product. The information on the side of a cereal packet indicates that the cereal contains iron, which 'Helps the body's use of oxygen, carrying it to all the cells of the body.'

a (i) Name the iron-containing molecule in the human body that transports oxygen. (1 mark)
(ii) Outline briefly how this molecule carries out the function stated for iron on the side of the cereal packet. (2 marks)

The information on the cereal packet indicates that the cereal contains fat (triglyceride).

b Describe the molecular structure of a triglyceride. (3 marks)

Cereals may also contain sodium, potassium, magnesium, chloride and phosphate ions.

c Complete the table below by choosing **two** of these ions and stating a function for each in living organisms.

ion	function

(2 marks)
(Total 8 marks)

OCR 2801 May 2002, B (BF), No.2

3 There are many different polysaccharides found in plants and animals. Gum arabic is classed as a complex polysaccharide and is produced by the tree, *Acacia senegal*. It seeps out from the cut surface when the tree is damaged.

The molecules of gum arabic have a branched structure and are soluble in water. It is classed as a heteropolysaccharide, which means that it is made up of a number of different sugars. Hydrolysis of gum arabic produces four different monosaccharides.

a Describe what happens during the hydrolysis of a polysaccharide molecule. (2 marks)

b Gum arabic is similar to other polysaccharides in a number of ways but also differs from them.

Complete the table opposite, comparing gum arabic with other polysaccharides.

	gum arabic	amylopectin (a component of starch)	cellulose	glycogen
branched structure	yes	yes		
heteropolysaccharide	yes		no	
found in animals/ plants	plants		plants	
function in organism	healing			energy store

(8 marks)

(Total 10 marks)

OCR 2801 Jun 2003, B (BF), No.1

4 a State **one** role for each of the following **in** living organisms.

   (i)   calcium   *(1 mark)*
   (ii) magnesium   *(1 mark)*
   (iii) phosphate   *(1 mark)*
   (iv) sodium   *(1 mark)*
   (v)   water   *(1 mark)*

A solution is thought to contain both sucrose and glucose. A student carried out a test and confirmed that a small amount of glucose was present in the solution.

b Describe how the student could determine that the solution also contains sucrose.   *(4 marks)*

c Describe the molecular structure of starch (amylose and amylopectin).

(In this question, 1 mark is available for the quality of written communication.)

(8 marks)

(Total 17 marks)

OCR 2801 Jun 2001, B (BF), No.3

5 a (i)   State the components needed to synthesise a triglyceride.   *(2 marks)*

   (ii) Name the chemical reaction by which these components are joined.   *(1 mark)*

b State **one** function of triglycerides in living organisms.   *(1 mark)*

Lipase is an enzyme that catalyses the hydrolysis of triglycerides. It is a soluble globular protein. The function of an enzyme depends upon the precise nature of its tertiary structure. The figure represents the structure of an enzyme. The black strips represent the disulphide bonds which help to stabilise its tertiary structure.

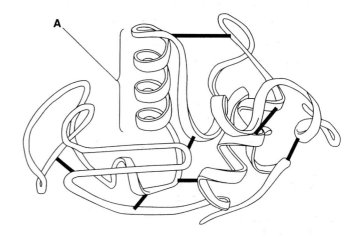

c (i)   Describe the nature of the disulphide bonds that help to stabilise the tertiary structure of a protein such as lipase.   *(2 marks)*

   (ii) Name **two other** types of bonding that help to stabilise tertiary structure.   *(2 marks)*

Region **A** on the figure is a secondary structure.

d Describe the nature of region **A**.   *(2 marks)*

(Total 10 marks)

OCR 2801 Jan 2001, B (BF), No.2

3

3.1

Enzymes

Enzyme structure and mode of action

Carbonic anhydrase is an important enzyme in the transport of carbon dioxide in the blood, where it converts dissolved carbon dioxide into carbonic acid, which then dissociates to hydrogencarbonate and hydrogen ions:

$$CO_2 + H_2O \rightarrow H_2CO_3$$
$$H_2CO_3 \rightarrow HCO_3^- + H^+$$

In the presence of carbonic anhydrase the reaction takes place more than 10 million times faster than it would without it. The active site of carbonic anhydrase is within a deep groove in the enzyme molecule. Three histidine amino acid molecules hold a zinc ion firmly in this groove. Adjacent amino acids recognise and bind a carbon dioxide molecule, while water is fixed onto the zinc ion and converted to a hydroxyl (-OH) group. This hydroxyl group is then positioned in exactly the right position for it to combine with CO_2 to form HCO_3^- – a hydrogencarbonate ion. Carbonic anhydrase therefore works by precisely orientating the substrates so that they easily react. Other enzymes use other methods. Some use electronegative amino acids to help 'stress' and break bonds, while others use charged amino acids to polarise substrates and cause them to react. Whatever the mechanism, the precise positioning of the substrates by the enzyme is always important.

Enzymes are globular proteins which catalyse metabolic reactions. A catalyst alters the rate of a chemical reaction without itself undergoing permanent change. It can therefore be used repeatedly and so is effective in tiny amounts. Enzymes do not make a reaction happen; they simply alter the speed of ones which already occur.

3.1.1 Enzyme structure

As globular proteins, enzymes have a specific three-dimensional shape which is determined by their sequence of amino acids (section 2.7.2). Despite their large overall size, enzyme molecules only have a small region that is functional. This is known as the **active site**. Only a few amino acids of the enzyme molecule make up this active site; the remainder are used to maintain its overall three-dimensional shape. The substrate molecule is held within the active site by bonds that temporarily form between the R groups of the amino acids of the active site and groups on the substrate molecule. This structure is known as the **enzyme–substrate complex** (Fig 3.1).

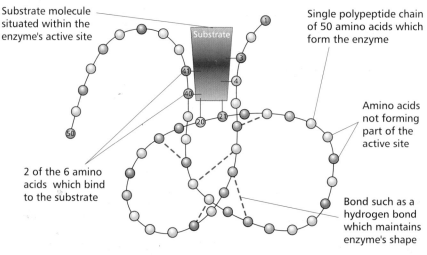

Fig 3.1 *Enzyme–substrate complex showing the six out of 50 enzyme amino acids that form the active site*

3.1.2 Enzymes and activation energy

Consider a typical chemical reaction:

$$\text{sucrose + water} \quad \rightarrow \quad \text{glucose + fructose}$$
$$\text{(substrates)} \qquad\qquad \text{(products)}$$

For such a reaction to occur naturally, the energy of the products (glucose + fructose) must be less than that of the substrates (sucrose + water). Such reactions, however, need an initial boost of energy to get them kick-started. This is known as the **activation energy**. It can be likened to a small stone lying on a hillside. If initial energy is applied to the stone, e.g. by giving it a push, then it will begin to move. It will move downhill, rather than uphill, because this lowers its potential energy (it is a fundamental law of thermodynamics that materials will naturally tend towards a state of low energy). Once set in motion, however, the stone gathers its own momentum and reaches the bottom with no further input of energy. This comparison shows how an initial input of energy (activation energy) can cause a reaction to continue on its own. In other words, there is an energy hill, or barrier, which must be overcome before the reaction can proceed. What enzymes do is to lower this activation energy level so that the reaction can happen more easily (Fig 3.2). For example, they allow many reactions to take place at a lower temperature than normal. As a result, some metabolic processes occur rapidly at the human body temperature of 37°C, which is relatively cool in terms of chemical reactions. These would proceed too slowly to sustain life as we know it were they to take place without enzymes.

3.1.3 How enzymes work

In one sense, enzymes operate in the same way as a key operates a lock: each key has a very specific shape which, on the whole, fits and operates only one lock. In the same way, a substrate will only fit the active site of one particular enzyme. Enzymes are therefore **specific** in the reactions that they catalyse. The shape of the substrate (key) exactly fits the active site of the enzyme (lock). This is known as the **lock and key theory** and explains, in a simple way, what exactly happens (Fig 3.3). In practice, the process is more refined: it is suggested that, unlike a rigid lock, the enzyme actually changes its form slightly to fit the shape of the substrate. In other words, it is flexible and moulds itself around the substrate, just as a glove moulds itself to the shape of someone's hand. The enzyme has a certain basic shape, just as a glove has, but this becomes slightly different as it alters in the presence of the substrate. As it alters its shape, the enzyme puts a strain on the substrate molecule and thereby lowers its activation energy. This whole process is called the **induced fit theory** of enzyme action.

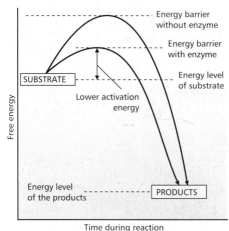

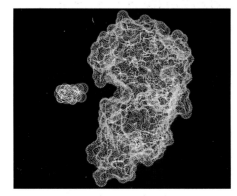

Fig 3.2 How enzymes lower activation energy

The ribonuclease A enzyme and its substrate

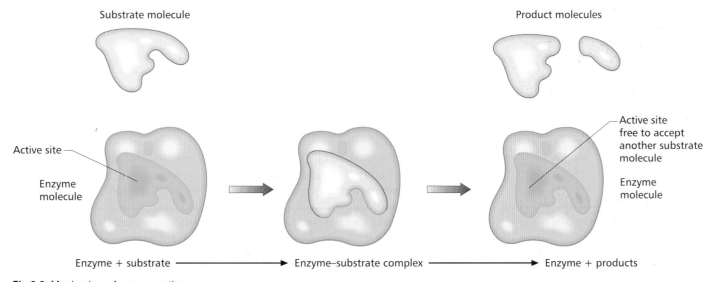

Fig 3.3 Mechanism of enzyme action

SUMMARY TEST 3.1

Enzymes act as biological **(1)**. They are **(2)** proteins that have a specific shape within which there is a functional portion known as the **(3)**. Enzymes lower the **(4)** of a reaction, allowing it to proceed at a lower temperature than it would normally. In an enzyme-controlled reaction, the general term for the substance on which the enzyme acts is **(5)** and the substances formed at the end of the reaction are known as the **(6)**. The enzyme molecule and the substance it acts on fit together very precisely, giving rise to the name **(7)** theory of enzyme action. In practice, the enzyme is thought to change shape slightly and so mould itself to the shape of the substance it acts on. This is called the **(8)** theory of enzyme action.

Effect of temperature and pH on enzyme action

Enzymes in the coloured bacteria and algae in this hot spring are not denatured even at temperatures in excess of 80°C whereas those in other organisms are denatured at temperatures of 40°C

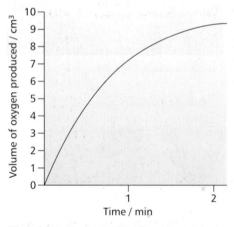

Fig 3.4 *Measurement of the formation of oxygen due to the action of catalase on hydrogen peroxide*

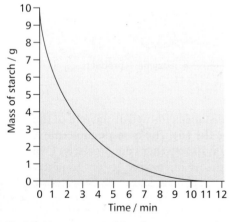

Fig 3.5 *Measurement of the disappearance of starch due to the action of amylase*

Before considering how pH and temperature affect enzymes, it is worth bearing in mind that, for an enzyme to work, it must:

* come into physical contact with its substrate
* have an **active site** which fits the substrate.

Almost all factors that influence the rate at which an enzyme works do so by affecting one or both of the above two circumstances. In order to investigate how enzymes are affected by various factors we need to be able to measure the reactions they catalyse.

3.2.1 Measuring enzyme-catalysed reactions

To measure the progress of an enzyme-catalysed reaction we usually measure its time-course, i.e. how long it takes for a particular event to run its course. The two 'events' most frequently measured are:

* **the formation of products** of the reaction, e.g. the volume of oxygen produced when catalase acts on hydrogen peroxide (Fig 3.4)
* **the disappearance of the substrate**, e.g. the reduction in concentration of starch when it is acted upon by amylase (Fig 3.5).

Although the graphs in figures 3.4 and 3.5 differ, the explanation for their shapes is the same:

* At first there is a lot of substrate (hydrogen peroxide / starch) but no product (water and oxygen / maltose).
* It is very easy for substrate molecules to come into contact with the empty active sites on the enzyme molecules.
* All enzyme active sites are filled and the substrate is rapidly broken down into its products.
* The amount of substrate decreases as it is broken down, resulting in an increase in the amount of product.
* As the reaction proceeds, there is less and less substrate and more and more product.
* It becomes more difficult for the substrate molecules to come into contact with the enzyme molecules because:
  – there are fewer substrate molecules around and so some active sites may not be filled at any one moment
  – the product molecules may 'get in the way' of substrate molecules reaching an active site.
* It therefore takes longer for substrate molecules to be broken down by the enzyme and so its rate of disappearance slows, and consequently the rate of formation of product also slows. Both graphs 'tail off.'
* The rate of reaction continues to slow until there is so little substrate that further decreases in its concentration cannot be measured.
* The graphs flatten out because all the substrate has been used up and so no new product can be produced.

3.2.2 Effect of temperature on enzymes

A rise in temperature increases the kinetic energy of molecules, which therefore move around more rapidly and collide with one another more often. In an enzyme-catalysed reaction, this means that the enzyme and substrate molecules come together more often in a given time, so that the rate of reaction is increased. Shown on a graph, this gives a rising curve. However, the temperature rise also increases the energy of the atoms that make up the enzyme molecule. Its atoms begin to vibrate and cause the hydrogen and other bonds that hold it in shape to break. Gradually, the shape of the active sites is disrupted. At first, the substrate fits less easily into the active site slowing the rate of reaction. For many human enzymes this may begin at temperatures of around 45°C. At some point, usually around 60°C, the enzyme is so disrupted that it stops working altogether. It is said to be **denatured**. Shown on a graph, the rate of this reaction follows a falling curve. The actual effect of temperature on the rate of an enzyme reaction is a combination of these two factors (Fig 3.6). The optimum working temperature differs from enzyme to enzyme. Some work best at around 10°C, while others continue to work well at 80°C. Many enzymes in the human body have an optimum temperature of 40°C, although others have a higher optimum, e.g. pepsin works fastest at 60°C. Our body temperatures have, however, evolved to be 37°C because:

- Although higher body temperatures would increase the metabolic rate slightly, the advantages are offset by the additional energy (food) that would be needed to maintain the higher temperature.
- Proteins, other than enzymes, may be denatured at higher temperatures.
- At higher temperatures, any further rise in temperature, e.g. during illness, might denature the enzymes.

Denaturing enzymes at high temperatures is used to prevent the spoilage (breakdown) by various enzymes found in food materials. This is the basis for heating food before canning or bottling it and for blanching vegetables before freezing.

3.2.3 Effect of pH on enzymes

The pH of a solution is a measure of its hydrogen ion concentration. Each enzyme has an optimum pH, i.e. a pH at which it works best (Fig 3.7). This is because the exact arrangement of the active site of an enzyme is partly fixed by hydrogen and ionic bonds between $-NH_2$ and $-COOH$ groups of the polypeptides that make up the enzyme. Even small changes in pH affect this bonding, causing changes of shape in the active site. In a similar way to a rise in temperature, this reduces the effectiveness of an enzyme and eventually causes it to stop working altogether, i.e. it becomes denatured. This is why foods can be preserved in vinegar: the low pH denatures the enzymes that would otherwise cause the food to break down. Solutions, known as **buffer solutions**, can be used to prevent fluctuations in pH. A buffer solution is a mixture of at least two chemicals that counteract the effect of acids and alkalis. A buffer solution does not therefore change its pH when a small amount of acid or alkali is added.

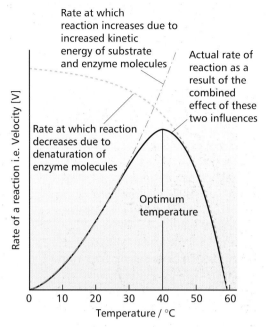

Fig 3.6 Effect of temperature on the rate of an enzyme-controlled reaction

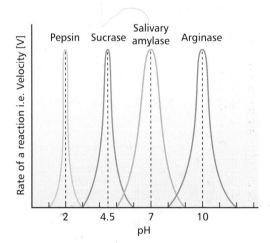

Fig 3.7 Effect of pH on the rate of an enzyme-controlled reaction

SUMMARY TEST 3.2

We can measure the progress of an enzyme-catalysed reaction by measuring its **(1)**. This is usually done by measuring either the **(2)** of the substrate or the formation of the **(3)**. For example, in the case of the enzyme amylase, we could either measure the rate at which **(4)** is produced or the rate at which **(5)** is used up. If the temperature is increased, the rate of enzyme action will **(6)** up to a point at which its molecular structure is disrupted and the shape of its **(7)** is altered so that the substrate no longer fits it. At this point the enzyme is said to be **(8)**. Many human enzymes have an optimum working temperature of **(9)**. Enzymes also have an optimum pH at which they operate best. Some, like pepsin, work best at a pH of **(10)** while others, such as **(11)**, function best in neutral conditions.

In addition to external factors such as temperature and pH, the substrate and enzyme concentrations affect the rate of enzyme-catalysed reactions. An enzyme reaction is always most rapid at first because the enzyme and substrate molecules can freely collide with one another. As the reaction proceeds, product molecules are formed and these reduce the number of collisions between enzyme and substrate molecules. You might say the product molecules 'get in the way'. The rate of reaction therefore slows. For this reason, we often measure the **initial rate of reaction** when studying enzyme-controlled reactions.

3.3.1 Effect of enzyme concentration on the rate of reaction

Once an **active site** on an enzyme has acted on its substrate, it is free to repeat the procedure on another substrate molecule. This means that enzymes are not used up in the reaction and therefore work efficiently at very low concentrations. In some cases, a single enzyme molecule can act on millions of substrate molecules in one minute.

As long as there is an excess of substrate, an increase in the amount of enzyme leads to a proportionate increase in the rate of reaction. A graph of the rate of reaction against enzyme concentration will initially show a proportionate increase. This is because there is more substrate than the enzyme's active sites can cope with. If we therefore increase the enzyme concentration, some of the excess substrate can now also be acted upon and the rate of reaction will increase. If, however, the substrate is limiting, i.e. there is not sufficient to supply all the enzyme's active sites at one time, then any increase in enzyme concentration will have no effect on the rate of reaction. The rate of reaction will therefore stabilise at a constant level, i.e. the graph will level off. This is because the available substrate is already being used as rapidly as it can be by the existing enzyme molecules. These events are summarised in figure 3.8.

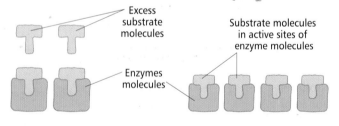

Low enzyme concentration

There are too few enzyme molecules to allow all substrate molecules to find an active site at one time. The rate of reaction is therefore only half the maximum possible for the number of substrate molecules available.

Intermediate enzyme concentration

With twice as many enzyme molecules available, all the substrate molecules can occupy an active site at the same time. The rate of reaction has doubled to its maximum because all active site are filled.

High enzyme concentration

The addition of further enzyme molecules has no effect as there are already enough active sites to accomodate all the available substrate molecules. There is no increase in the rate of reaction.

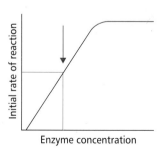

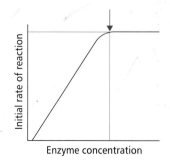

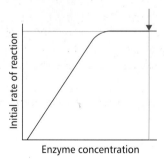

Fig 3.8 *Effect of enzyme concentration on the rate of enzyme action*

3.3.2 Effects of substrate concentration on the rate of enzyme action

If the amount of enzyme is fixed at a constant level and substrate is slowly added, the rate of reaction increases in proportion to the amount of substrate that is added. This is because the enzyme molecules have only a limited number of substrate molecules to collide with, and therefore the active sites of the enzymes are not working to full capacity. As more substrate is added, the active sites gradually become fully utilised, until the point where all of them are working as fast as they can. The rate of reaction is at its maximum (V_{max}). After that, the addition of more substrate will have no effect on the rate of reaction. In other words, when the substrate is in excess the rate of reaction levels off. A summary of the effect of substrate concentration on the rate of enzyme action is given in figure 3.9.

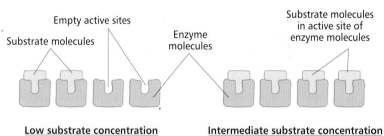

Low substrate concentration

There are too few substrate molecules to occupy all the available active sites. The rate of reaction is therefore only half the maximum possible for the number of enzyme molecules available.

Intermediate substrate concentration

With twice as many substrate molecules available, all the active sites are occupied at one time. The rate of reaction has doubled to its maximum because all the active sites are filled.

High substrate concentration

The addition of further substrate molecules has no effect as all active sites are already occupied at one time. There is no increase in the rate of reaction

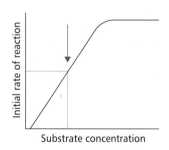
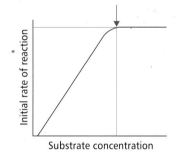
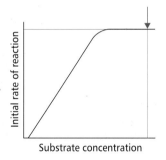

Fig 3.9 *Effect of substrate concentration on the rate of enzyme action*

SUMMARY TEST 3.3

Enzymes work fastest at the start of a process and this is called **(1)**. As the enzyme concentration increases, the rate of reaction **(2)**, provided there is excess substrate. The graph may later 'tail off' if the amount of substrate is limited because not all the **(3)** of the enzyme molecules are filled. If the substrate concentration of an enzyme-controlled reaction is halved then the rate of reaction will be **(4)**. When the substrate is in excess the rate of reaction will **(5)**.

3.4 Enzyme inhibition

Enzyme inhibitors are substances that directly or indirectly interfere with the functioning of the **active site** of an enzyme and so reduce its activity. Sometimes the inhibitor binds itself so strongly to the active site that it cannot be removed and so permanently prevents the enzyme functioning. These are so called **non-reversible** or **permanent inhibitors**; they include heavy metal ions such as mercury and silver. Most inhibitors only make temporary attachments to the active site. These are called **reversible inhibitors** and are of two types:

- **competitive** (active site directed) – inhibitor binds to the active site of the enzyme
- **non-competitive** (non-active site directed) – inhibitor binds to the enzyme at a position other than the active site.

3.4.1 Competitive (active site directed) inhibitors

Competitive inhibitors have a molecular shape similar to that of the substrate, which allows them to occupy the active site of an enzyme. They therefore compete with the substrate for the available active sites (Fig 3.10). It is the difference between the concentration of the inhibitor and the concentration of the substrate that determines the effect this has on enzyme activity: if the substrate concentration is increased, the effect of the inhibitor is reduced. The inhibitor is not permanently bound to the active site and so, when it leaves, another molecule can take its place. This could be a substrate or inhibitor molecule, depending on how much of each type is present. Sooner or later, all the substrate molecules will find an active site, but the greater the concentration of inhibitor, the longer this will take. Examples of competitive inhibitors include malonic acid, which inhibits succinic dehydrogenase in the Krebs cycle, and hirudin, which is used by leeches to inhibit thrombin and so prevent clotting when they take a blood meal.

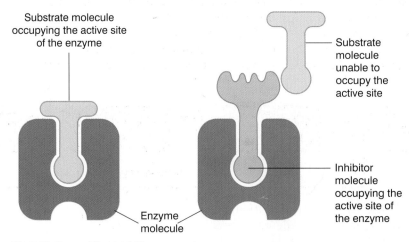

Substrate molecule occupying the active site of the enzyme

Substrate molecule unable to occupy the active site

Inhibitor molecule occupying the active site of the enzyme

Enzyme molecule

Fig 3.10 *Competitive inhibition*

3.4.2 Non-competitive (non-active site directed) inhibitors

Non-competitive inhibitors attach themselves to the enzyme at a site which is not the active site. This is known as the **allosteric site** (allosteric = 'at another place'). Upon attaching to the enzyme, the inhibitor alters the shape of the enzyme's active site in such a way that substrate molecules can no longer occupy it, and so the enzyme cannot function (Fig 3.11). As the substrate and the inhibitor are not competing for the same site, an increase in substrate concentration does not decrease the effect of the inhibitor (Fig 3.12). An example of a non-competitive inhibitor is cyanide, which inhibits the respiratory enzyme, cytochrome oxidase.

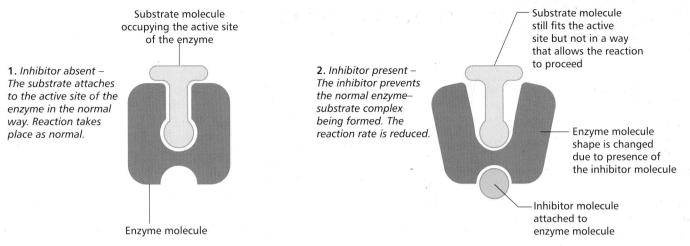

1. *Inhibitor absent –*
The substrate attaches
to the active site of the
enzyme in the normal
way. Reaction takes
place as normal.

Substrate molecule
occupying the active site
of the enzyme

Enzyme molecule

2. *Inhibitor present –*
The inhibitor prevents
the normal enzyme–
substrate complex
being formed. The
reaction rate is reduced.

Substrate molecule
still fits the active
site but not in a way
that allows the reaction
to proceed

Enzyme molecule
shape is changed
due to presence of
the inhibitor molecule

Inhibitor molecule
attached to
enzyme molecule

Fig 3.11 *Non-competitive inhibition*

3.4.3 Control of metabolic pathways

A metabolic pathway is a series of biochemical reactions in which one substance is broken down, or built up, into another. In the tiny space inside a single cell, many hundreds of different metabolic pathways take place. The pathways are not at all haphazard, but highly structured. The enzymes that control a pathway are often attached to the inner membrane of a cell organelle in a very precise sequence. This increases the chance of each enzyme coming into contact with its substrate and leads to greater efficiency. Inside each organelle, there are optimum conditions for the functioning of these enzymes. e.g. the pH may vary from organelle to organelle. To keep a steady level of a particular metabolite in a cell, the metabolite often acts as an inhibitor of an enzyme at the start of a reaction (Fig 3.13). In the example below, the end product inhibits enzyme A. The more end product there is, the greater the inhibition and the less end product is produced. A reduction in the amount of end product lessens the inhibition, so bringing the level of the metabolite back to normal. This is known as **end-product inhibition** and is a form of homeostasis. The enzyme that is inhibited therefore controls the rate of the reaction and is known as a **regulatory enzyme**. The type of inhibition involved is usually non-competitive.

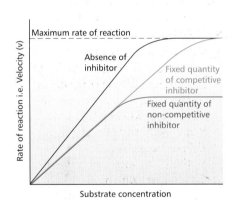

Fig 3.12 *Comparison of competitive and non-competitive inhibition on the rate of an enzyme-controlled reaction at different substrate concentrations*

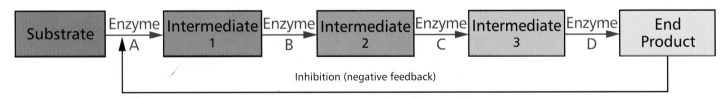

Fig 3.13 *Inhibition*

SUMMARY TEST 3.4

Inhibitors reduce the activity of enzymes. Where the inhibitor cannot be removed from the active site of the enzyme it is known as a **(1)** inhibitor. Inhibitors that can be removed from the active site are of two types. Competitive inhibitors are also known as **(2)** and compete with the substrate for active sites. Therefore if more substrate is added the effect of the inhibition is **(3)**. Non-competitive inhibitors become fixed to the enzyme at a point other than the active site. This point is called the **(4)**. If more substrate is added the inhibition is **(5)**.

1 A student investigated the activity of catalase by measuring the release of oxygen from hydrogen peroxide. The reaction occurs as follows:

$$2H_2O_2 \rightarrow 2H_2O + O_2$$

The student added 5cm³ of a catalase solution to 10cm³ of a 5% hydrogen peroxide solution and placed the mixture in the apparatus shown in figure 1. The total volume of gas collected was recorded every 20 seconds. The results are shown in figure 2.

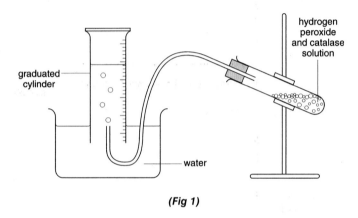

(Fig 1)

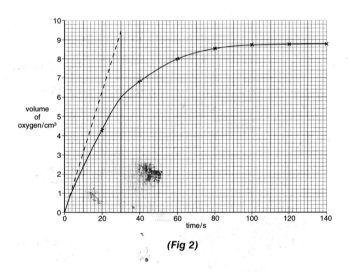

(Fig 2)

a One way of calculating the rate of reaction is to use the formula

$$\frac{\text{volume of oxygen collected}}{\text{time taken for collection}}$$

Use the formula to calculate the rate, **in cm³ min⁻¹**, over the first 30 seconds. *(2 marks)*

b In order to compare rates of reaction, the *initial* rate of reaction is used. This is the rate of reaction within the first few seconds, during which the reaction is proceeding at the maximum possible rate.

The dashed line on figure 2 shows the initial rate of oxygen production. Using this line, it can be calculated that the initial rate of production would be 19cm³ min⁻¹.

Explain why the initial rate of reaction is greater than the rate you calculated in **a**. *(2 marks)*

c Describe and explain the effect of factors, **other than substrate concentration**, on the rate of enzyme-controlled reactions.

Your answer should include reference to enzyme structure. *(10 marks)*

In this question, 1 mark is available for the quality of written communication.

QWC (1 mark)
(Total 15 marks)
OCR 2801 Jan 2003, B (BF), No.3

2 Maize grains contain an enzyme which can convert starch to maltose (a reducing sugar). 20 maize grains were soaked in water for 48 hours, after which 10 of the grains were killed by boiling. Extracts were prepared from both the living maize grains, extract **P**, and the dead maize grains, extract **Q**.

Extract **P** was added to an equal volume of starch solution in a test tube and maintained at 30 °C. Samples were taken at 30 second intervals and tested with iodine solution.

Extract **Q** was treated in exactly the same way.

The results are shown in the table below.

maize grain extract	time taken for blue-black colour to disappear/min
P (from living grains)	3.5
Q (from dead grains)	did not disappear

a Explain why

(i) the blue-black colour did not disappear with the extract from dead maize grains; *(2 marks)*

(ii) the extract from dead maize grains was included in the investigation. *(1 mark)*

b Explain why an enzyme which catalyses the conversion of starch to maltose is unable to catalyse the conversion of a protein into amino acids.

(2 marks)
(Total 5 marks)
OCR 2801 May 2002, B (BF), No.5

3 Enzymes are proteins with a tertiary structure.

a Explain the importance of the tertiary structure to the functioning of an enzyme. *(4 marks)*

Food spoilage is often caused by the enzyme action of microorganisms.

b Explain why

(i) some foods are preserved in vinegar; *(2 marks)*
(ii) foods are heated to very high temperatures before being canned. *(2 marks)*

During cheese production, rennet is used to solidify the protein in the milk. Rennet is a commercially prepared form of rennin, an enzyme found naturally in the stomachs of young mammals.

c Explain why the temperature of the milk needs to be kept between 30 °C and 40 °C during this stage of cheese production. *(2 marks)*

(Total 10 marks)

OCR 2801 Jan 2002, B (BF), No.6

4 The enzyme urease catalyses the following reaction:

urea + water → ammonium carbonate

Ammonium carbonate readily gives off ammonia.

The indictor bromo-thymol blue is yellow in neutral solution and blue in alkaline solution. In an investigation, urease was mixed with bromo-thymol blue. The mixture was divided equally between five test-tubes, labelled **A** to **E**. One test-tube was placed in each of five water baths at the temperatures shown in the table, until they reached the desired temperature. The same mass of urea was then added to each test-tube. The test-tubes were maintained at their temperatures and the time taken for the contents of each test-tube to turn blue was recorded.

The results of the investigation are shown in the table.

test-tube	temperature / °C	time taken for blue colour to appear / sec
A	0	89
B	15	21
C	35	5
D	45	17
E	55	33

a Suggest why the indicator changed colour after urea was added to the enzyme. *(1 mark)*

b With reference to the table,

(i) state what you can conclude about the optimum temperature of the enzyme; *(1 mark)*
(ii) explain the fact that the **time taken** for the blue colour to appear is greater at 15 °C than at 35 °C and is greater at 55 °C than at 35 °C. *(4 marks)*

The indicator was added to a separate sample of urea and to a separate sample of urease. In neither test-tube was a blue colour produced.

c Explain why the indicator was added to separate samples of both urea and urease. *(2 marks)*

d On the axes provided, sketch the curves you would expect if an enzyme-controlled reaction was carried out under optimum conditions with

(i) a fixed quantity of enzyme;

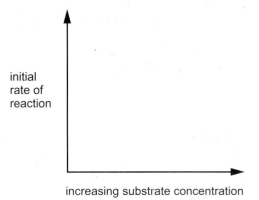

increasing substrate concentration *(2 marks)*

(ii) excess substrate.

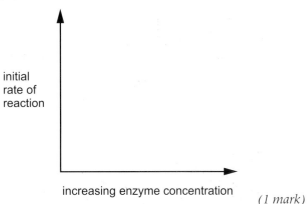

increasing enzyme concentration *(1 mark)*

The figure represents an enzyme, its substrate and a **non-competitive** inhibitor.

enzyme

substrate

non-competitive inhibitor

e With reference to the figure,

(i) label the active site; *(1 mark)*
(ii) explain how the inhibitor has its effect. *(3 marks)*

(Total 15 marks)

OCR 2801 June 2001, B (BF), No.4

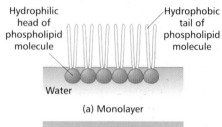

Hydrophilic head of phospholipid molecule

Hydrophobic tail of phospholipid molecule

Water

(a) Monolayer

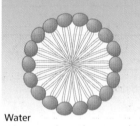

Water

(b) Spherical monolayer (micelle)

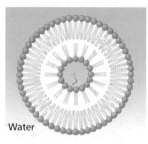

Water

(c) Bilayered vesicle

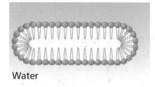

Water

(d) Bilayered sheet

Fig 4.1 *Different arrangements of phospholipid molecules in water*

Table 4.1 *Functions of membranes within cells*

- Control the entry and exit of materials in discrete organelles such as mitochondria and chloroplasts
- Isolate organelles so that specific metabolic reactions can take place within them
- Provide an internal transport system, e.g. endoplasmic reticulum
- Isolate enzymes that might damage the cell, e.g. lysosomes
- Provide surfaces on which reactions can occur, e.g. protein synthesis using ribosomes on rough endoplasmic reticulum

We saw in unit 1.7 that the cell surface membrane is the boundary between the cell cytoplasm and the environment and, as such, controls the movement of substances in and out of the cell. Before we look at how the cell surface membrane achieves this we need first to look in more detail at the molecules that form its structure – phospholipids, proteins, cholesterol, glycolipids and glycoproteins. The arrangement of these molecules within the cell surface membrane is detailed in figure 4.2, while table 4.2 summarises their functions. Membranes also occur within cells, surrounding organelles such as mitochondria and chloroplasts – their functions are listed in table 4.1.

4.1.1 Phospholipids

The molecular structure of a phospholipid is described in section 2.5.3. A phospholipid is made up of two parts:

- **a hydrophilic 'head'**, which is attracted to water but not to fat
- **a hydrophobic 'tail'**, which is repelled by water but mixes readily with fat.

This means that when phospholipid molecules are placed in water they take up positions that place the hydrophilic heads as close to the water as possible, and the hydrophobic tails as far away from the water as possible. The various formations are illustrated in figure 4.1.

The phospholipids in the cell surface membrane form a bilayer sheet, which is 7nm thick, because:

- one layer of phospholipids has its hydrophilic heads pointing inwards (attracted by the water in the cell cytoplasm)
- the other layer of phospholipids has its hydrophilic heads pointing outwards (attracted by the water which surrounds all cells)
- the hydrophobic tails of both phospholipid layers point into the centre of the membrane – protected, as it were, from the water on both sides.

Lipid-soluble material moves through the membrane via the phospholipid portion.

4.1.2 Proteins

The proteins of the cell surface membrane are arranged more randomly than the regular pattern of phospholipids. They are embedded in the phospholipid bilayer in two main ways:

- **Extrinsic (peripheral) proteins** occur either on the surface of the bilayer or only partly embedded in it, but never extend completely across it. They act either to give mechanical support to the membrane or, in conjunction with glycolipids, as cell receptors.
- **Intrinsic (integral) proteins** completely span the phospholipid bilayer from one side to the other. Some act as carriers to transport water-soluble material across the membrane while some are enzymes.

4.1.3 Cholesterol

Cholesterol molecules occur throughout the cell surface membrane of animal cells, where they are almost as numerous as phospholipid molecules. They add strength to animal cell surface membranes – a necessary feature in the absence of a cell wall. Cholesterol molecules are very hydrophobic and therefore have an important role in preventing leakage of water and dissolved ions out of the cell. They also pull together the fatty acid 'tails' of the phospholipid molecules, limiting their movement but without making the membrane as a whole too rigid.

4.1.4 Glycolipids

Glycolipids occur where a carbohydrate chain is associated with phospholipids in the cell surface membrane. The carbohydrate portion extends from the phospholipid bilayer into the watery environment outside the cell, where it acts as a recognition site for specific chemicals, e.g. the human ABO blood system operates as a result of glycolipids on the cell surface membrane.

4.1.5 Glycoproteins

Carbohydrate chains are attached to many extrinsic proteins on the outer surface of the cell membrane. These glycoproteins also act as recognition sites, more particularly for hormones and neurotransmitters. The term **glycocalyx** is applied collectively to the glycoproteins and the glycolipids.

IMPERMEABILITY OF CELL SURFACE MEMBRANE

The cell surface membrane is impermeable to most biological molecules because many are:

- not soluble in fat and therefore cannot pass through the phospholipid layer
- too large to pass through the pores in the membrane
- of the same charge as the charge on the protein channels and so, even if they are small enough to pass through, they are repelled
- electrically charged (i.e. are polar) and therefore have difficulty passing through the non-polar hydrophobic tails in the phospholipid bilayer.

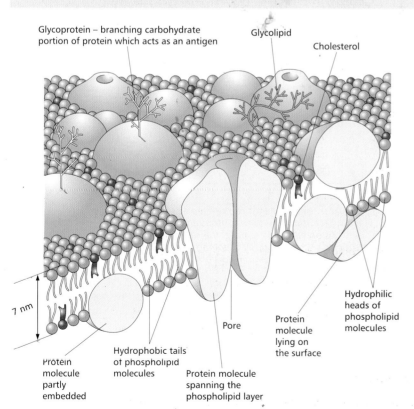

Glycoprotein – branching carbohydrate portion of protein which acts as an antigen

Glycolipid

Cholesterol

7 nm

Protein molecule partly embedded

Hydrophobic tails of phospholipid molecules

Pore

Protein molecule spanning the phospholipid layer

Protein molecule lying on the surface

Hydrophilic heads of phospholipid molecules

Fig 4.2 *Structure of the cell surface (plasma) membrane*

Table 4.2 *Summary of functions of the components of the cell surface membrane*

Proteins

- provide structural support
- act as carriers transporting water-soluble substances
- function as enzymes
- form ion channels for sodium, potassium, etc
- act as energy transducers
- form recognition sites by identifying cells
- help cells adhere together
- act as receptors, e.g. for hormones

Phospholipids

- allow lipid-soluble substances to enter and leave the cell
- prevent water-soluble substances entering and leaving the cell
- give the membrane fluidity

Cholesterol

- reduces lateral movement of phospholipids
- makes membrane less fluid at high temperatures
- prevents leakage of water and dissolved ions from the cell

Glycolipids

- act as recognition sites, e.g. ABO blood system
- help maintain stability of the membrane
- help cells attach to one another and so form tissues

Glycoproteins

- act as recognition sites for hormones and neurotransmitters
- help cells attach to one another and so form tissues
- act as antigens allowing cells to recognise one another, e.g. lymphocytes can recognise an organism's own cells.

SUMMARY TEST 4.1

The cell surface (plasma) membrane is made up of 5 main types of molecules. Phospholipid molecules from a **(1)** sheet in which their **(2)** heads point both inwards towards the cell cytoplasm and outwards towards the external environment. The **(3)** tails of the phospholipids point into the centre of the membrane. Within the phospholipid layer are both **(4)** proteins that span the complete membrane and **(5)** proteins that occur on the membrane surface or partly embedded in it. The remaining types of molecules within a membrane are **(6)**, **(7)** and **(8)**.

The movement of material into and out of cells occurs in a number of ways, some of which require energy **(active transport)** and some of which do not **(passive transport)**. Diffusion is an example of passive transport.

4.2.1 Explanation of diffusion

As all movement requires energy, it is possibly confusing to describe diffusion as passive transport. What is meant by passive, in this sense, is that the energy comes from the natural, inbuilt motion of particles, rather than from some outside source. To help understand diffusion and other passive forms of transport it is necessary to understand that

- all particles are constantly in motion due to the kinetic energy that they possess
- this motion is random, with no set pattern to the way the particles move around
- particles are constantly bouncing off one another as well as other objects, e.g. the sides of a vessel in which they are contained.

Given those facts, figure 4.3 shows in a series of diagrams how particles concentrated together in part of a closed vessel will, of their own accord, distribute themselves evenly throughout the vessel, due to diffusion. Diffusion is therefore defined as **'the net movement of molecules or ions from a region where they are more highly concentrated to one where their concentration is lower'**.

1. *If 10 particles occupying the left-hand side of a closed vessel are in random motion, they will collide with each other and the sides of the vessel. Some particles from the left-hand side move to the right, but initially there are no available particles to move in the opposite direction, so the movement is in one direction only. There is a large concentration gradient and diffusion is rapid.*

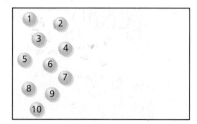

2. *After a short time the particles (still in random motion) have spread themselves more evenly. Particles can now move from right to left as well as from left to right. However, with a higher concentration of particles (7) on the left than on the right (3), there is a greater probability of a particle moving to the right than in the reverse direction. There is a smaller concentration gradient and diffusion is slower.*

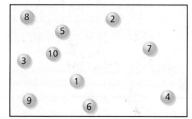

3. *Some time later, the particles will be evenly distributed throughout the vessel and the concentrations will be equal on each side. The system is in equilibrium. However, the particles are not static but remain in random motion. With equal concentrations on each side, the probability of a particle moving from left to right is equal to the probability of one moving in the opposite direction. There is no concentration gradient and no net diffusion.*

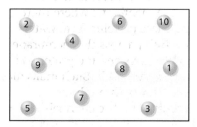

4. *At a later stage, the particles remain evenly distributed and will continue to do so. Although the number of particles on each side remains the same, individual particles are continuously changing position. This situation is called **dynamic equilibrium**.*

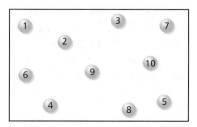

Fig 4.3 *Diffusion*

4.2.2 Rate of diffusion

A number of factors affect the rate at which molecules or ions diffuse. These include:

- **The concentration gradient** – the greater the difference in concentration between two regions of molecules or ions, the faster the rate of diffusion.
- **The area over which diffusion takes place** – the larger the area, the faster the rate of diffusion.
- **The distance over which diffusion occurs** – the shorter the distance, the faster the rate of diffusion.

The relationship between these three factors is expressed in **Fick's Law**, which states:

Diffusion is proportional to:

$$\frac{\text{surface area} \times \text{difference in concentration}}{\text{length of diffusion path}}$$

Although Fick's Law gives a good guide to the rate of diffusion, it is not wholly applicable to cells, because diffusion is also affected by

- **the nature of the cell membrane** – its composition and number of pores
- **the size and nature of the diffusing molecule** – for example:
  - small molecules diffuse faster than large ones
  - fat-soluble molecules such as glycerol diffuse faster than water-soluble ones
  - polar molecules diffuse faster than non-polar ones.

4.2.3 Facilitated diffusion

Facilitated diffusion is a passive process relying only on the kinetic energy of the diffusing molecules. Like diffusion, it occurs along a concentration gradient, but it differs in that it occurs at specific points on the membrane where there are special protein molecules. These proteins form water-filled channels **(protein channels)** across the membrane and therefore allow water-soluble ions and molecules such as glucose and amino acids to pass through. Such molecules would usually diffuse only very slowly through the phospholipid bilayer of the membrane. The channels are selective, each opening only in the presence of a specific molecule. When the particular molecule is not present, the channel remains closed. In this way, some control is kept over the entry and exit of substances. An alternative form of facilitated diffusion involves **carrier proteins** which also span the membrane. When a particular molecule specific to the protein is present it binds with the protein, causing it to change shape in such a way that the molecule is released to the inside of the membrane (Fig 4.4). Again, there is no use of external energy, and the molecules move from a region where they are highly concentrated to one of lower concentration, using only the kinetic energy of the molecules themselves.

SUMMARY TEST 4.2

Diffusion is the net movement of molecules or ions from where they are in a **(1)** concentration to a region where their concentration is **(2)**. The energy for this movement comes from the **(3)** energy of the molecules themselves and the process is therefore said to be a **(4)** one. If the area over which diffusion takes place is made smaller, its rate becomes **(5)**. If the concentration gradient is reduced, the rate becomes **(6)** and if the distance over which diffusion takes place is made shorter, its rate becomes **(7)**. Facilitated diffusion, which is faster than diffusion, may involve molecules known as **(8)** that span the cell membrane.

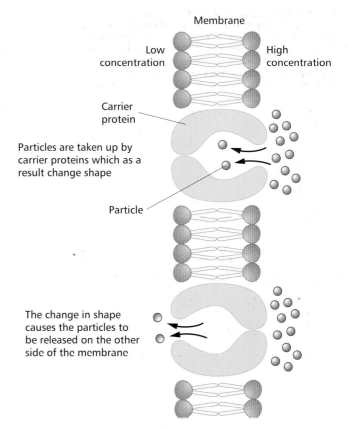

Fig 4.4 Facilitated diffusion

Osmosis

Osmosis is a special form of diffusion involving only water molecules. It is defined as **the passage of water from a region where it has a higher water potential to a region where it has a lower water potential through a partially permeable membrane.** Cell surface membranes and membranes surrounding organelles are **partially permeable**, i.e. they are permeable to water molecules and certain solute molecules but not to many other molecules. A **solute** is any substance that is dissolved in a **solvent**, e.g. water. The solute and the solvent together form a **solution**.

4.3.1 Explanation of osmosis

Consider the hypothetical situation in figure 4.5, in which a partially permeable membrane separates two solutions.

- The solution on the left has a low concentration of solute molecules while the solution on the right has a high concentration of solute molecules.
- Both the solute and water molecules are in random motion due to their inherent energy.
- The partially permeable membrane, however, only allows water molecules across it and not solute molecules.
- The water molecules diffuse from the left-hand side, which has the higher water potential, to the right-hand side, which has the lower water potential, i.e. along a water potential gradient.
- At this point, a dynamic equilibrium should be established and there is no **net** movement of water.

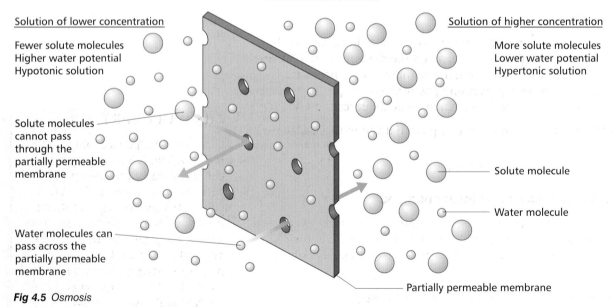

Solution of lower concentration

Fewer solute molecules
Higher water potential
Hypotonic solution

Solute molecules cannot pass through the partially permeable membrane

Water molecules can pass across the partially permeable membrane

Solution of higher concentration

More solute molecules
Lower water potential
Hypertonic solution

Solute molecule

Water molecule

Partially permeable membrane

Fig 4.5 *Osmosis*

4.3.2 Water potential

Water potential is represented by the Greek letter psi (Ψ), and is measured in units of pressure, usually kilopascals (kPa). It is the pressure created by water molecules. Under standard conditions of temperature and pressure (25°C and 100kPa), pure water is said to have a water potential of zero. It follows that:

- the addition of a solute to pure water will lower its water potential
- the water potential of a solution (water + solute) must always be less than zero, i.e. a negative value

- the more solute that is added (i.e. the more concentrated a solution), the lower (more negative) its water potential
- water will move by osmosis from a region of higher (less negative) water potential (e.g. −10kPa) to one of lower (more negative) water potential (e.g. −20kPa).

4.3.3 Understanding water potential

Water potential can be confusing, not least because its highest value, that of pure water, is zero, and so all other values are negative. The more negative the value, the lower the water potential. It may be easier to understand if you think of water potential as an overdraft at a bank. The bigger the overdraft, the more negative is the amount of money you have. The smaller the overdraft, the less negative is the amount of money you have. An account with a big overdraft (more negative value) has a greater need of money than an account with a small overdraft (less negative value). In the same way, a solution with a high water potential (less negative value) has less need of water than a solution with a lower water potential (more negative value). Water therefore moves from solutions with a high water potential (e.g. those containing fewer solute molecules) to ones with a low water potential (e.g. those containing more solute molecules) – but only when separated by a partially permeable membrane.

4.3.4 Solute potential

The presence of solute molecules in a solution lowers its water potential. The greater the concentration of solutes, the lower (more negative) is the water potential. This change in water potential as a consequence of the presence of solute molecules is called **solute potential** (Ψs). As the solute molecules always lower the water potential, the value of the solute potential is always negative.

4.3.5 Pressure potential

Take a situation such as that in figure 4.5 where water is entering a solution by osmosis. If a pressure were applied to the solution on the right of figure 4.5 then this would resist the entry of water. This pressure is known as the **pressure potential** and is given the symbol Ψp. Pressure potential is of importance in the water relations of plant cells and is dealt with in more detail in section 4.4.2.

The relationship between water potential, solute potential and pressure potential can be expressed in the equation;

$$\Psi = \Psi s + \Psi p$$

4.3.6 Hypotonic, isotonic and hypertonic solutions

Osmosis occurs not only when a solution is separated from pure water by a partially permeable membrane, but also when two solutions of different solute concentrations are similarly separated. The following terms are applied to this situation:

- **hypotonic solution** – a solution with a lower concentration of solutes (higher water potential), i.e. the more dilute of the two solutions
- **hypertonic solution** – a solution with a higher concentration of solutes (lower water potential), i.e. the more concentrated of the two solutions
- **isotonic solution** – a solution with the same concentration of solutes.

There is always net movement of water from the hypotonic solution to the hypertonic solution because there are more water molecules in the hypotonic solution. When the solutions are isotonic, the amount of water diffusing in one direction is exactly offset by that moving in the other direction, so there is no net movement of water.

SUMMARY TEST 4.3

Osmosis is the passage of (**1**) from a region where it is more highly concentrated to a region where its concentration is lower, through a (**2**) membrane. Water potential is the (**3**) created by water molecules and its value for pure water is (**4**). The addition of a solute to water makes its water potential (**5**). If solution A has a water potential of −10kPa and solution B has a water potential of −20kPa, water will move into solution (**6**). Solution B is said to be (**7**) to solution A because it is less dilute. Solution A is therefore (**8**) to solution B. If both solutions have the same solute concentration, they are said to be (**9**).

4.4 Osmosis and cells

REMEMBER

- Pure water has a water potential (Ψ) of zero.
- Solute makes the water potential lower (more negative).
- Water always moves from a higher (less negative) water potential to a lower (more negative) one.

In unit 4.3 we looked at what osmosis is and why it occurs. We now need to turn our attention to how it affects living cells. Due to the structural differences between plant and animal cells, osmosis affects each of them in a different way.

4.4.1 Osmosis and animal cells

Animal cells, such as a red blood cell, contain a variety of solutes dissolved in their watery cytoplasm. If a red blood cell is placed in pure water it will absorb water by osmosis because it has a lower **water potential** (i.e. it is hypertonic to the water). Cell surface membranes are very thin (7nm) and, although they are flexible, they cannot stretch to any great extent. The cell surface membrane will therefore break, bursting the cell and releasing its contents – an event known as **haemolysis**. To prevent this happening, animal cells are normally bathed in a liquid which has the same water potential as the cells, i.e. the two are isotonic. In our example, the blood plasma and red blood cells have the same water potential. If a red blood cell is placed in a solution with a lower water potential than its own, water leaves by osmosis and the cell shrinks, causing its shape to shrivel. The cell is said to be crenated (Fig 4.6).

Water potential (ψ) of external solution compared to cell solution	Higher (less negative)	Equal	Lower (more negative)
Net movement of water	Enters cell	Neither enters nor leaves	Leaves cell
State of cell	Swells and bursts	No change	Shrinks

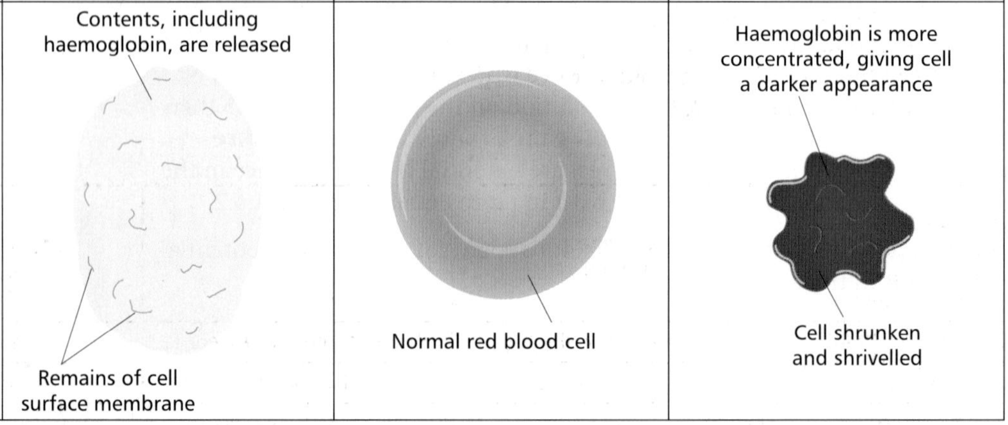

Contents, including haemoglobin, are released
Remains of cell surface membrane
Normal red blood cell
Haemoglobin is more concentrated, giving cell a darker appearance
Cell shrunken and shrivelled

Fig 4.6 Summary of osmosis in an animal cell, e.g. a red blood cell

4.4.2 Osmosis and plant cells

For the purposes of the following explanations, the plant cell can be divided into three parts:

- **the central vacuole** – containing a solution of salts, sugars and organic acids in water
- **the protoplast** – the outer cell surface membrane, cytoplasm and the inner vacuole membrane (tonoplast)
- **the cellulose cell wall** – a tough, inelastic covering that is permeable to even large molecules.

Like animal cells, plant cells also contain a variety of solutes, largely dissolved in the water of the large cell vacuole that each possesses. When placed in pure water they too absorb water by osmosis because of their lower (more negative) water potential. Unlike animal cells, however, they are unable to control the composition of the fluid around their cells. Indeed, plant cells are normally permanently bathed in almost pure water, which is constantly absorbed from the plant's roots (section 10.5.1). So why don't the cells burst? The answer lies in the cellulose cell wall which surrounds every plant cell. If this wall, or something similar, had not evolved, plants would not exist as we know them today. How then does it work?

- The arrangement of the β-glucose molecules in cellulose gives it considerable strength (unit 2.4).
- Water entering a plant cell by osmosis causes the protoplast to swell.
- The living portion of the cell, known as the protoplast (cell surface membrane, tonoplast and cytoplasm) pushes against the cellulose cell wall.
- Because the cell wall is capable of only very limited extension, a pressure builds up on it that resists the further entry of water. We saw in section 4.3.5 that such a pressure is called **pressure potential** and is given the symbol (Ψp).
- Because it prevents more water entering, the pressure potential increases the water potential of the cell.
- In this situation, the protoplast of the cell is kept pushed against the cell wall and the cell is said to be turgid.

If the same plant cell is placed in a solution with a lower water potential than its own, water leaves by osmosis. The volume of the cell decreases and so does the pressure potential. A stage is reached where the protoplast no longer presses on the cellulose cell wall. At this point the pressure potential is equal to zero and the cell is said to be at **incipient plasmolysis**. Further loss of water will cause the cell to shrink further and the protoplast to pull away from the cell wall. This condition is called **plasmolysis**. These events are summarised in figure 4.7.

SUMMARY TEST 4.4

A red blood cell will burst if placed in a solution which has a **(1)** water potential than itself. To prevent this, animal cells like red blood cells are normally bathed in a solution that has the same **(2)** as themselves. A plant cell placed in pure water takes in water by **(3)**, causing the living contents, called the **(4)**, to swell. This creates a pressure on the **(5)** that is known as the **(6)** potential and is represented by the symbol **(7)**. This pressure prevents water entering and therefore makes the water potential **(8)**. If a plant cell is placed in a solution with a greater solute potential than itself, the cell contents shrink away from the cell wall, a process called **(9)**.

Water potential (Ψ) of external solution compared to cell solution	Less negative (higher)	Equal	More negative (lower)
Net movement of water	Enters cell	Neither enters nor leaves cell	Leaves cell
Protoplast	Swells	No change	Shrinks
Condition of cell	Turgid	Incipient plasmolysis	Plasmolysed (flaccid)
	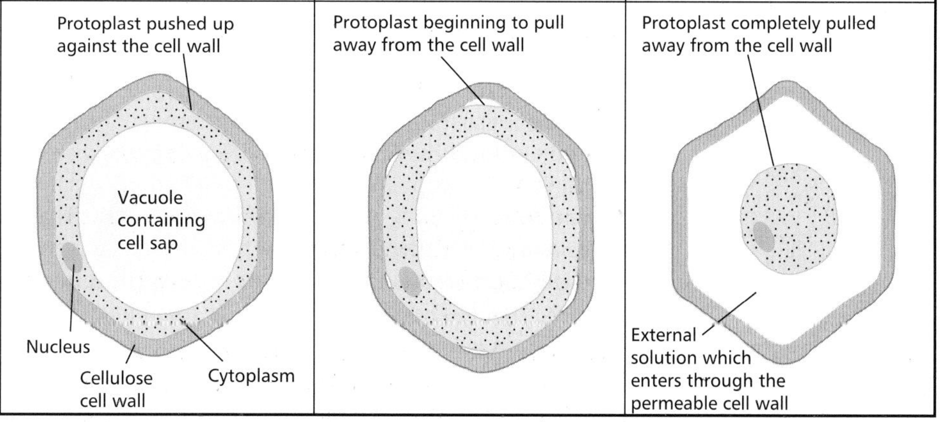 Protoplast pushed up against the cell wall — Vacuole containing cell sap — Nucleus — Cellulose cell wall — Cytoplasm	Protoplast beginning to pull away from the cell wall	Protoplast completely pulled away from the cell wall — External solution which enters through the permeable cell wall

Fig 4.7 Summary of osmosis in a plant cell

Active transport differs from passive forms of transport, such as diffusion and osmosis, in the following ways:

- metabolic energy in the form of **ATP** is needed
- materials are moved **against** a concentration gradient (i.e. from a lower to a higher concentration)
- carrier protein molecules which act as 'pumps' are involved
- the process is very selective, with specific substances being transported.

Materials may be transported, either as molecules or in bulk as larger particles, in a process called **cytosis**. There are two forms of cytosis (Fig 4.8):

- **endocytosis** – the bulk movement of material into the cell
- **exocytosis** – the bulk movement of the material out of the cell.

4.5.1 Mechanism of active transport

There are a number of possible explanations for how active transport occurs. One theory involves the proteins that span the **phospholipid** bilayer of a membrane. These proteins accept the molecules or ions to be transported on one side of the membrane and then change shape, using energy from ATP to carry the molecules or ions to the other side (Fig 4.9). Sometimes more than one substance may be moved in the same direction at the same time. Occasionally, one substance is moved into a cell at the same time as a different one is being removed from it. One example of this is the **sodium–potassium pump**. Here, sodium ions are actively removed from the cell while potassium ions are taken in from the surroundings. This process is essential to the creation of a nerve impulse.

4.5.2 Requirements for active transport

Certain conditions are necessary if a cell is to carry out active transport effectively. These include:

- the presence of numerous mitochondria
- a ready supply of ATP
- a high respiratory rate.

Clearly, any factor that affects respiratory rate will affect active transport, and therefore higher temperatures or an increased supply of oxygen will increase the rate of active transport. Lower temperatures, less oxygen or the presence of respiratory inhibitors, such as cyanide, will slow the rate of active transport.

4.5.3 Endocytosis

Endocytosis is the bulk movement of material into a cell by active means. It takes two forms:

- **Phagocytosis** involves the invagination of the cell to form a cup-shaped depression in which large particles or even whole organisms are contained (see photo). The depression is then pinched off to the inside of the cell, forming a **vacuole**. The process occurs in organisms such as *Amoeba*, in which it is used as a method of feeding. A few specialised cells in higher organisms also carry out phagocytosis. These cells are called **phagocytes** and include certain types of white blood cells that ingest harmful bacteria; in this way they fight infection or destroy worn-out cells, such as red blood cells. In both cases, lysosomes fuse with the vacuole formed during phagocytosis, releasing their enzymes into the vacuole. The particles inside are therefore broken down and any useful soluble products are absorbed.

Table 4.3 *Comparison of different forms of transport in cells*

Process	Occurs against a conc. gradient	Needs energy (ATP)	May use carrier molecules
Diffusion	No	No	No
Faciliated diffusion	No	No	Yes
Osmosis	No	No	No
Active transport	Yes	Yes	Yes

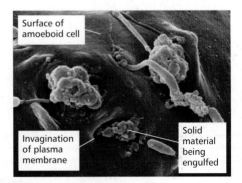

Phagocytosis

- **Pinocytosis** is very similar to phagocytosis, except that the vacuoles formed are smaller and are called **vesicles**. Pinocytosis is used more for the uptake of liquids or large molecules rather than solid particles, and is therefore also called 'cell drinking'.

4.5.4 Exocytosis

Exocytosis is the bulk movement of material out of the cell. It is the reverse of phagocytosis and pinocytosis: vacuoles and / or vesicles within the cell fuse with the cell surface membrane and their contents are expelled into the medium outside. Undigested material from the food vacuoles in *Amoeba* is removed in this way. In higher organisms, exocytosis is used to release hormones, e.g. insulin, from the cells that manufacture them.

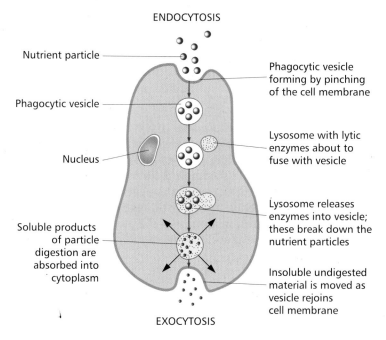

Fig 4.8 *Endocytosis and exocytosis*

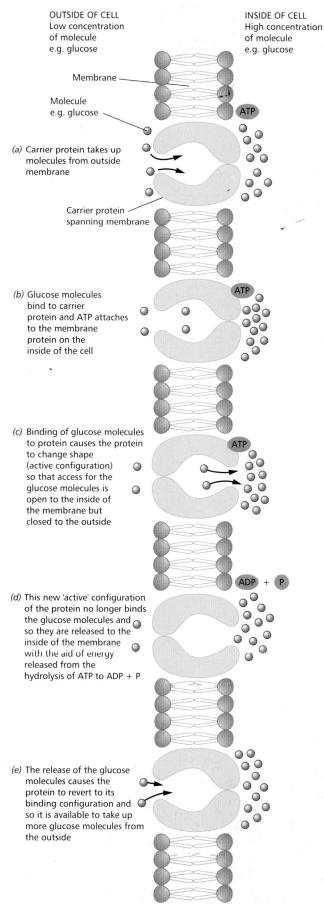

OUTSIDE OF CELL
Low concentration of molecule e.g. glucose

INSIDE OF CELL
High concentration of molecule e.g. glucose

Membrane

Molecule e.g. glucose

(a) Carrier protein takes up molecules from outside membrane

Carrier protein spanning membrane

(b) Glucose molecules bind to carrier protein and ATP attaches to the membrane protein on the inside of the cell

(c) Binding of glucose molecules to protein causes the protein to change shape (active configuration) so that access for the glucose molecules is open to the inside of the membrane but closed to the outside

(d) This new 'active' configuration of the protein no longer binds the glucose molecules and so they are released to the inside of the membrane with the aid of energy released from the hydrolysis of ATP to ADP + P

(e) The release of the glucose molecules causes the protein to revert to its binding configuration and so it is available to take up more glucose molecules from the outside

Fig 4.9 *Active transport*

SUMMARY TEST 4.5

Active transport occurs **(1)** a concentration gradient. It requires energy in the form of **(2)** and cells exhibiting active transport therefore have numerous **(3)** and a high **(4)** rate. The bulk movement of molecules and particles across the cell surface membrane is called **(5)**. Where this movement is into the cell, it is called **(6)**. Where this movement is out of the cell, it is called **(7)**. The invagination of a cell surface membrane to ingest particles and form a vacuole is known as **(8)**, whereas the invagination of the membrane to ingest liquids and form a vesicle is known as **(9)**.

Exchanges between organisms and their environment

Table 4.4 *How the surface area to volume ratio gets smaller as an object becomes larger*

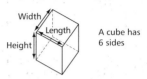

A cube has 6 sides

Length of edge of a cube / cm	Surface area of whole cube (area of one side × 6 sides) / cm²	Volume of cube (length × width × height) / cm³	Ratio of surface area to volume (surface area ÷ volume
1	1 × 6 = **6**	1 × 1 × 1 = **1**	$\frac{6}{1}$ = **6.0**
2	4 × 6 = **24**	2 × 2 × 2 = **8**	$\frac{24}{8}$ = **3.0**
3	9 × 6 = **54**	3 × 3 × 3 = **27**	$\frac{54}{27}$ = **2.0**
4	16 × 6 = **96**	4 × 4 × 4 = **64**	$\frac{96}{64}$ = **1.5**
5	25 × 6 = **150**	5 × 5 × 5 = **125**	$\frac{150}{125}$ = **1.2**
6	36 × 6 = **216**	6 × 6 × 6 = **216**	$\frac{216}{216}$ = **1.0**

For survival, organisms must transfer materials between themselves and their environment. Examples of things which need to be interchanged include:

- respiratory gases (oxygen and carbon dioxide).
- nutrients (glucose, fatty acids, amino acids, vitamins, minerals).
- excretory products (urea)
- heat.

This exchange can take place in two ways:

- passively (no energy is required) – by diffusion and **osmosis**
- actively (energy is required) – by **active transport, pinocytosis** and **phagocytosis**.

4.6.1 Surface area to volume ratio

Exchange takes place at the surface of an organism, but the materials absorbed are used by the cells that mostly make up its volume. For exchange to be effective, the surface area of the organism must therefore be large compared with its volume.

Small organisms like **protoctists** have a surface area that is large enough, compared with their volume, to allow efficient exchange across their body surface. However, as organisms become larger, their volume increases at a faster rate than their surface area (table 4.4), and so simple diffusion of materials across the surface can only meet the needs of the most sluggish organisms. Even if the surface could supply enough material, it would still take too long for it to reach the middle of the organism if diffusion alone was the method of transport. To overcome this problem, organisms have evolved one or more of the following features:

- **a flattened shape** so that no cell is ever far from the surface (e.g. flatworm)
- **a central region that is hollow** or filled with non-metabolising material at its centre (e.g. the gut at the centre of an earthworm)
- **specialised exchange surfaces** with large areas to increase the surface area to volume ratio (e.g. lungs in mammals, gills in fish).

4.6.2 Features of specialised exchange surfaces

To allow effective transfer of materials across them by diffusion or active transport, exchange surfaces have the following characteristics:

- **large surface area to volume ratio** – to speed up the rate of exchange
- **very thin** – to allow materials to cross rapidly
- **partially permeable** – to allow selected materials to cross without hindrance
- **movement of the environmental medium, e.g. air** – to maintain a diffusion gradient
- **movement of the internal medium, e.g. blood** – to maintain a diffusion gradient.

The relationship between certain of these factors is described in **Fick's Law**, which is expressed as:

Diffusion is proportional to:

$$\frac{\text{surface area} \times \text{difference in concentration}}{\text{distance over which diffusion occurs}}$$

Being thin, specialised exchange surfaces are easily damaged, and therefore are often located inside an organism for protection. Where an exchange surface is located inside the body, the organism needs to have a means of moving the external medium over the surface, e.g. a means of ventilating the lungs in a mammal.

4.6.3 Features of root hairs

Root hairs are the exchange surfaces in plants that are responsible for the absorption of water and mineral salts. Plants constantly lose water by the process of transpiration (unit 10.7). This loss can amount to up to 700 litres a day in a large tree – all of which must be replaced by water that is absorbed through the root hairs.

Each root hair is a tiny extension of a root epidermal cell. These root hairs remain functional for a few weeks before dying back to be replaced by others nearer the growing tip. They are efficient surfaces for the exchange of water and mineral ions because:

- they provide a large surface area because they are very long extensions and occur in their thousands on each of the branches of a root
- they have a thin surface, the cell surface membrane, across which materials can move
- they are permeable – the thin cellulose cell wall presents no barrier to the movement of water and ions, and the cell surface membrane allows water and minerals to cross it easily.

Root hairs absorb water passively by the process of osmosis (unit 4.3). The soil solution is a very low concentration of mineral ions in water and surrounds the particles that make up soil. The soil solution therefore has a high **water potential**. The root hairs, by comparison, have a relatively high concentration of ions, sugars and organic acids within their vacuoles and cytoplasm. Their cells have a lower water potential. Because the root hairs are in direct contact with the soil solution, water moves by osmosis from the higher (less negative) water potential of the soil solution to the lower (more negative) water potential within the root hair cells. Figure 4.10 illustrates this process. From the root hair cell, the water moves further into the root due to the osmotic gradient that exists between it and the xylem in the centre of the root (unit 10.5).

The absorption of mineral ions by root hairs is an altogether different situation. The concentration of ions inside the root hair cell is normally greater than that in the soil solution. The uptake of mineral ions is therefore against the concentration gradient and, as a result, requires active transport. This is achieved using special carrier proteins that use ATP to provide energy to transport particular ions from the soil solution, where they are in low concentrations, to the root hair cytoplasm and vacuole, where they are in higher concentrations. The details of this process are given in sections 4.5.1 and 4.5.2. On the rare occasions when a particular ion is in a greater concentration in the soil than in the root hair cell, the ion simply moves into the cell by the passive process of diffusion.

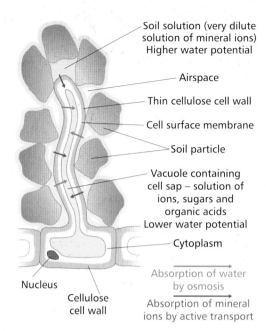

Fig 4.10 *Absorption of water and mineral ions by a root hair*

Labels:
Soil solution (very dilute solution of mineral ions) Higher water potential
Airspace
Thin cellulose cell wall
Cell surface membrane
Soil particle
Vacuole containing cell sap – solution of ions, sugars and organic acids Lower water potential
Cytoplasm
Absorption of water by osmosis
Absorption of mineral ions by active transport
Nucleus
Cellulose cell wall

SUMMARY TEST 4.6

The surface area to volume ratio of a cube with an edge length of 10cm is **(1)**. If the edge length is doubled to 20cm then the surface area to volume ratio is **(2)**. If the surface area over which diffusion is taking place is doubled, the rate of diffusion will **(3)** by a factor of **(4)**. If the distance across any diffusing surface is halved, then the rate of diffusion will **(5)**. In root hairs, water is absorbed by the process of **(6)** because the water potential of the soil solution is **(7)** than within the root hair. Mineral ions are absorbed by the process of **(8)**.

Gaseous exchange in mammals

All **aerobic** organisms require a constant supply of oxygen in order to produce a supply of energy in the form of **ATP** during respiration. The carbon dioxide produced in the process needs to be removed from the body.

The volume of oxygen that has to be absorbed and the volume of carbon dioxide that must be removed are large in mammals because:

- they are relatively large organisms with a large volume of living cells
- they maintain a high body temperature and therefore have high metabolic and respiratory rates.

As a result, mammals have evolved specialised surfaces, called **lungs**, to ensure efficient gaseous exchange between the air and their blood.

4.7.1 Mammalian lungs

Lungs are the site of gaseous exchange in mammals. All mammals evolved on land, even those like whales that have returned to water to live. As a result, their gaseous exchange surfaces evolved as internal organs because otherwise they would:

- collapse because air is not dense enough to support these delicate structures
- lose a great deal of water and therefore dry out.

The lungs have therefore evolved deep within the thorax of mammals and are connected to the outside air via a supported tube called the **trachea**. In this way, air can be warmed and moistened before it reaches the lungs, and so loss of heat and water are minimised. The lungs are supported and protected by a bony box called the **rib cage**. The ribs can be moved by the muscles between them, called the **intercostal muscles**. This enables the lungs to be ventilated by a tidal stream of air, thereby ensuring that the air within them is constantly replenished. The trachea divides into two smaller tubes, called **bronchi**, one of which enters each lung. On entering a lung, each bronchus divides again and again, forming a 'bronchial tree' whose smallest sub-divisions, those less than 1mm in internal diameter, are called **bronchioles**. These in turn, sub-divide into alveolar ducts which terminate in clusters of air-sacs called **alveoli**.

4.7.2 The alveoli as the site of gaseous exchange

The alveoli (see photo) are the actual site of gaseous exchange in mammals. They are tiny hollow sacs, each with a diameter of between $100\mu m$ and $300\mu m$. Each lung of an adult human contains around 300 million of them. They are adapted to their function of gaseous exchange in a number of ways:

- They have a very large surface area – in total around $70m^2$. This is about half the area of a tennis court.
- They are extremely thin – each alveolus is lined with squamous epithelium. (section 1.9.2). They are therefore only between $0.1\mu m$ and $0.5\mu m$ thick, which allows rapid diffusion across them.
- They are kept moist by fluid from the cytoplasm of the squamous epithelial cells. This fluid passes through the cell surface membrane onto the inner surface of the alveoli.
- They are in close contact with an extensive capillary network that quickly carries away the oxygen diffusing out of the alveoli and brings more carbon dioxide to the alveoli. This maintains a concentration gradient so that oxygen is continually absorbed by the blood and carbon dioxide is removed from the blood.

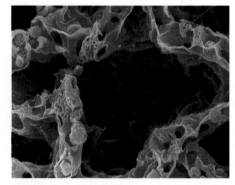

Alveolus in the lungs

- The passage of blood cells in these capillaries is slowed to allow more time for diffusion.
- The air within the alveoli is constantly changed during ventilation. This brings fresh supplies of oxygen and removes carbon dioxide, thus maintaining a concentration gradient of the two gases. Oxygen therefore constantly diffuses into the blood and carbon dioxide constantly diffuses from it.

Figure 4.11 illustrates the position of the lungs and the structure of alveoli. Further details of gaseous exchange in the alveoli are given in unit 8.5.

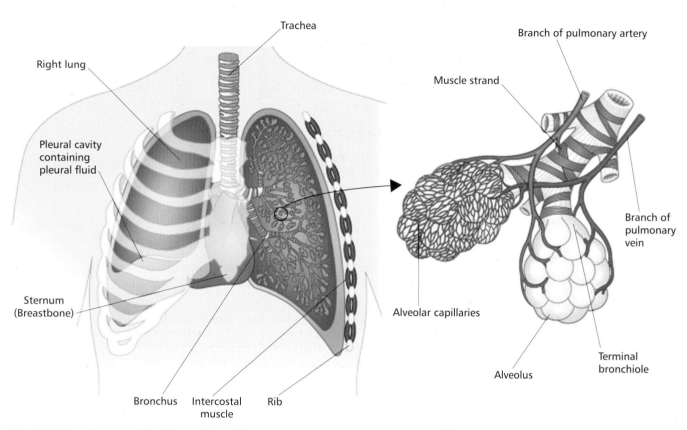

Fig 4.11 *Human lungs and alveoli*

SUMMARY TEST 4.7

Two reasons why mammals require a large and constant supply of oxygen are **(1)** and **(2)**. The main organs for gaseous exchange are the lungs, which are connected to the outside by a tube called the **(3)**. This branches into two **(4)**, one of which enters each lung. The actual site of gaseous exchange is in the alveoli, which have a diameter of **(5)** and have walls made of **(6)** which is very thin, being only **(7)** in thickness. The total number of alveoli for both lungs is around **(8)**, giving them a very large surface area of about **(9)**.

1 a The figure shows a mechanism for the transport of glucose across a plasma (cell surface) membrane using a carrier protein.

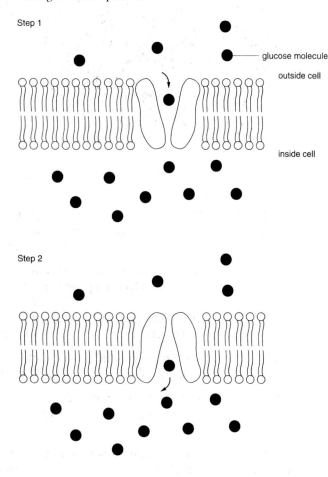

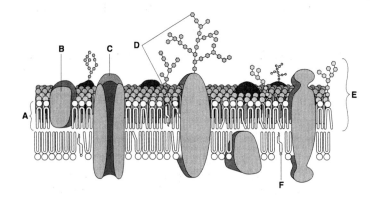

   (i) Name the process by which glucose is being transported across the membrane in the figure.
   (1 mark)

   (ii) State why glucose molecules cannot move through a phospholid bilayer. *(1 mark)*

  b The rate of glucose uptake into animal cells can vary.

   Suggest **two** changes in plasma (cell surface) membranes that could result in an increase in the rate of glucose uptake. *(2 marks)*
   (Total 4 marks)
   OCR 2801 Jun 2003, B (BF), No.6

2 The figure opposite represents the structure of the plasma (cell surface) membrane.

  a Name molecules **A** to **F**. *(6 marks)*

  b State the approximate width of the membrane. *(1 mark)*

Water moves across the plasma (cell surface) membrane by osmosis.

  c Complete the table below to

   (i) show the direction in which water will move across the plasma (cell surface) membranes of cells in different conditions;

   (ii) indicate whether or not the cell will burst.

   Place a (✔) or a cross (✗) in each box of the table as appropriate.

	initial net movement of water		
	in	out	cell bursts
leaf mesophyll cell immersed in distilled water			
red blood cell immersed in concentrated salt solution			
red blood cell immersed in distilled water			

(6 marks)

  d With reference to the figure, outline the ways in which substances, other than water, cross plasma (cell surface) membranes.

  (In this question, 1 mark is available for the quality of written communication.)
  (7 marks)
  (Total 20 marks)
  OCR 2801 Jan 2002, B (BF), No.2

3 The gaseous exchange surface of the mammalian lung possesses a number of features which allow efficient diffusion.

  a State how each of the following features assists in producing an efficient gaseous exchange system.

(i) The surface is composed of many alveoli. *(1 mark)*

(ii) The wall of each alveolus is 0.1μm thick. *(1 mark)*

(iii) A dense network of capillaries covers the outside of each alveolus. *(1 mark)*

(iv) The surface is well ventilated. *(1 mark)*

Root hairs absorb water and ions.

b State **two** features of root hairs which allow efficient absorption. *(2 marks)*

Substances cross the membranes of root hairs by a number of different methods, including osmosis and active transport.

c Describe the processes of osmosis and active transport, highlighting the differences between them. *(In this question, 1 mark is available for the quality of written communication.)*

(7 marks)

(Total 13 marks)

OCR 2801 June 2001, B (BF), No.5

4 Active transport and facilitated diffusion are two ways by which substances cross plasma (cell surface) membranes.

a State **one** difference and **one** similarity between active transport and facilitated diffusion.

(2 marks)

Vitamins C and D are essential for the correct functioning of the human body. They are, therefore, taken into cells but take different routes across the plasma (cell surface) membrane. Vitamin C is water-soluble while Vitamin D is lipid-soluble.

b Explain how the structure of the plasma (cell surface) membrane determines the route taken by each of these vitamins. *(4 marks)*

Excessive concentrations of salt in the blood and tissue fluid can cause serious damage to cells.

c Explain the effects of a high concentration of salt on animal cells. *(4 marks)*

(Total 10 marks)

OCR 2801 Jan 2001, B (BF), No.7

5 The diagram shows a cell surface membrane.

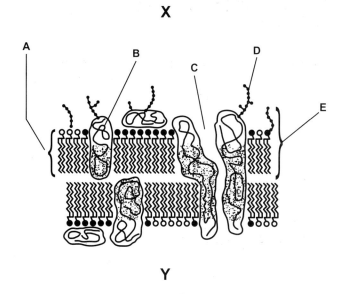

a (i) Name **A** to **E**. *(5 marks)*

(ii) State the width of a cell surface membrane. *(2 marks)*

b (i) On which side of the membrane, shown in the diagram, **X** or **Y**, is the cytoplasm of the cell? *(1 mark)*

(ii) Give a reason for your answer based on the evidence in the diagram. *(1 mark)*

c State the function of **three named** components of the cell surface membrane. *(6 marks)*

The properties of the components of cell surface membranes determine whether molecules can pass through membranes.

d Explain why cell surface membranes are impermeable to most biological molecules. *(4 marks)*

(Total 19 marks)

OCR 2801, 2000 Specimen paper, B (BF), No.1

5.1

Nucleotides and ribonucleic acid (RNA)

Adenosine monophosphate (adenylic acid)

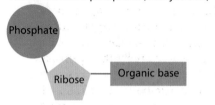

Fig 5.1 *Structure of a nucleotide*

Nucleotides are the basic units which make up a group of the most important chemicals in all organisms. These are the **nucleic acids**, of which the best known are:

- **ribonucleic acid (RNA)**
- **deoxyribonucleic acid (DNA)**.

5.1.1 Nucleotide structure

Individual nucleotides are made up of three components:

- **a pentose sugar**, of which there are two types: ribose and deoxyribose
- **a phosphate group**
- **an organic base**, of which five different forms are found in nucleic acids.

The five organic bases (Fig 5.2) are divided into two groups:

- **pyrimidines**, which are made up of a single six-sided ring, include **cytosine**, **thymine** and **uracil**
- **purines**, which are made up of a six-sided ring joined to a five-sided one. The two examples found in nucleic acids are **adenine** and **guanine**.

The pentose sugar, phosphate group and organic base are combined, as a result of **condensation reactions**, to give a **mononucleotide** (Fig 5.1). Two mononucleotides may, in turn, be combined as a result of a condensation reaction between the pentose sugar of one mononucleotide and the phosphate group of another. The new structure is called a **dinucleotide**. Continued linking of mononucleotides in this way forms a **polynucleotide**, such as RNA. While RNA and DNA perform essential functions in protein synthesis and heredity, they are by no means the only biologically important molecules containing nucleotides. A number of the others are listed in table 5.1.

NAME OF MOLECULE	REPRESENTATIVE SHAPE
Phosphate	
Pentose sugar	
Adenine (a purine)	 Adenine
Guanine (a purine)	 Guanine
Cytosine (a pyrimidine)	Cytosine
Thymine (a pyrimidine)	Thymine
Uracil (a pyrimidine)	Uracil

Fig 5.2 *Molecules found in nucleotides*

Table 5.1 *The functions of biologically important molecules containing nucleotides*

Molecule	Abbreviation	Function
Deoxyribonucleic acid	DNA	Contains the genetic information of cells
Ribonucleic acid	RNA	All three types play a vital role in protein synthesis
Adenosine monophosphate Adenosine diphosphate Adenosine triphosphate	AMP ADP ATP	Coenzymes important in making energy available to cells for metabolic activities, osmotic work, muscular contractions, etc
Nicotinamide adenine dinucleotide Flavine adenine dinucleotide	NAD FAD	Electron (hydrogen) carriers important in respiration in transferring hydrogen atoms from the Krebs cycle along the respiratory chain
Nicotinamide adenine dinucleotide phosphate	NADP	Electron (hydrogen) carrier important in photosynthesis for accepting electrons from the chlorophyll molecule and making them available for the photolysis of water
Coenzyme A	CoA	Coenzyme important in respiration in combining with pyruvate to form acetyl coenzyme A and transferring the acetyl group into the Krebs cycle

5.1.2 Ribonucleic acid (RNA) structure

Ribonucleic acid (RNA) is a polymer made up of repeating mononucleotide sub-units. It forms a single strand in which the pentose sugar is always **ribose** and the organic bases are adenine, guanine, cytosine and uracil (Fig 5.3). There are three types of RNA, all of which are important in protein synthesis:

- **ribosomal RNA (rRNA)**
- **transfer RNA (tRNA)**
- **messenger RNA (mRNA)**.

5.1.3 Ribosomal RNA (rRNA)

Ribosomal RNA (rRNA) is a large, complex molecule which is a major component of ribosomes, making up over half of their mass. It has a sequence of organic bases which is very similar in all organisms.

5.1.4 Transfer RNA (tRNA)

Transfer RNA (tRNA) is a relatively small molecule which is made up of around 80 nucleotides. It is manufactured by DNA and makes up 10–15% of the total RNA in a cell. Although there are a number of types of tRNA, they are very similar, each having a single-stranded chain folded into a clover-leaf shape, with one end of the chain extending beyond the other. This extended chain always has the organic base sequence of cytosine–cytosine–adenine; this is the part of the tRNA molecule to which amino acids can easily attach. There are at least 20 types of tRNA, each able to carry a different amino acid.

At the opposite end of the tRNA molecule is a sequence of three other organic bases, known as the **anticodon**. For each amino acid there is a different sequence of organic bases on the anticodon. During protein synthesis, this anticodon pairs with the complementary three organic bases that make up the triplet of bases on mRNA, known as the **codon**. The tRNA structure (Fig 5.4), with its end chain for attaching amino acids and its anticodon for pairing with the codon of the mRNA, is structually suited to its role of lining up amino acids on the mRNA template during protein synthesis.

5.1.5 Messenger RNA (mRNA)

Consisting of thousands of mononucleotides, mRNA is a long strand which is arranged in a single helix. Because it is manufactured when DNA forms a mirror-copy of part of one of its two strands, there is a great variety of different types of mRNA. Once formed, mRNA leaves the nucleus via pores in the nuclear envelope and enters the cytoplasm, where it associates with the ribosomes. There it acts as a template upon which proteins are built. Its structure is suited to this function because it possesses the correct sequence of many triplets of organic bases that code for specific polypeptides. It is also easily broken down and therefore exists only for as long as it is needed to manufacture a given protein.

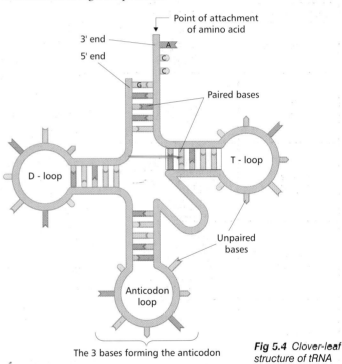

Fig 5.4 Clover-leaf structure of tRNA

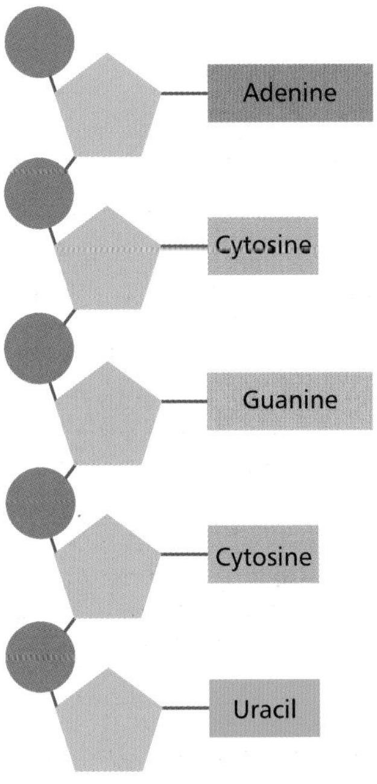

Fig 5.3 *Section of polynucleotide*

Adenine

Cytosine

Guanine

Cytosine

Uracil

SUMMARY TEST 5.1

Nucleotides are organic compounds that contain the elements carbon, hydrogen, oxygen, **(1)** and **(2)**. A mononucleotide contains a **(3)** sugar, which has **(4)** carbon atoms and has two forms: **(5)** and **(6)**. It also contains one of five organic bases, which fall into two groups. Those with a six-carbon ring only are called **(7)** and exist in three forms: thymine, **(8)** and **(9)**. Those of the second group, called **(10)**, have a six-sided ring joined to a **(11)**-sided ring; there are two such molecules: **(12)** and **(13)**. Ribonucleic acid, which never has the organic base **(14)**, exists in three forms. The form that has the same sequence of organic bases in all living organisms is called **(15)**; the form that has a sequence of three bases called an anticodon is **(16)**; and the remaining form, upon which proteins are formed, is **(17)**.

Deoxyribonucleic acid (DNA)

Deoxyribonucleic acid (DNA) is made up of two **nucleotide polymer** strands. In DNA, the pentose sugar is **deoxyribose** and the organic bases are adenine, guanine, cytosine and thymine. Each of the two polynucleotide strands is extremely long, and they are wound around one another to form a double helix. The differences between ribonucleic acid (RNA) and DNA are listed in table 5.2.

5.2.1 DNA structure

In 1953, James Watson and Francis Crick worked out the structure of DNA following pioneering work by Rosalind Franklin on X-ray diffraction patterns of DNA. This opened the door for many of the major developments in biology over the next half-century. In its simplified form, DNA can be thought of as a ladder in which the phosphate and deoxyribose molecules alternate to form the uprights and the organic bases pair together to form the rungs (Fig 5.5). The organic bases are of two types: the purines (adenine and guanine) are longer molecules than the pyrimidines (cytosine and thymine). It follows that, if the rungs of the DNA ladder are to be the same length, the base pairs must always be made up of one purine and one pyrimidine. In fact, the pairings are even more precise than this:

- Adenine always pairs with thymine by means of two **hydrogen bonds**.
- Guanine always pairs with cytosine by means of three hydrogen bonds.

It follows that the quantity of adenine and thymine in DNA is always the same, and so is that of guanine and cytosine. However, the ratio of adenine and thymine to guanine and cytosine varies from species to species.

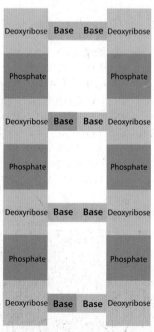

Simplified ladder

DNA structure may be likened to a ladder in which alternating phosphate and deoxyribose molecules make up the 'uprights' and pairs of organic bases comprise the 'rungs'.

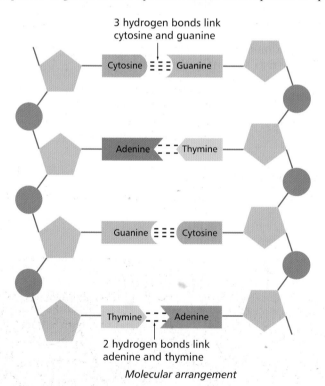

Molecular arrangement

Note the base pairings are always cytosine–guanine and adenine–thymine. This ensures a standard 'rung' length. Note also that the 'uprights' run in the opposite direction to each other (i.e. are antiparallel).

Fig 5.5 Basic structure of DNA

To appreciate the true structure of DNA, however, you have to imagine this ladder as being twisted, so that the uprights wind around one another to form a **double helix**. These uprights run in the opposite direction to each other and are therefore said to be **antiparallel**. They form the structural backbone of the DNA molecule. For each complete turn of this helix, there are 10 base pairs (Fig 5.6). In total, there are around 3.2 billion base pairs in the DNA of a typical mammalian cell. This vast number means that there is an almost infinite number of sequences of bases along the length of a DNA molecule and it is this variety that provides the immense genetic diversity within living organisms.

The DNA molecule is adapted to carry out its functions in a number of ways:

* It is very stable and can pass from generation to generation without change.
* Its two separate strands are, however, joined only with hydrogen bonds, allowing them to separate during replication (unit 5.3) and form **messenger RNA** during protein synthesis (unit 5.6).
* It is an extremely large molecule and it therefore carries an immense amount of genetic information.
* By having the base pairs within the helical cylinder of the sugar-phosphate backbone, the genetic information is protected to some extent from being corrupted by outside chemical and physical forces.

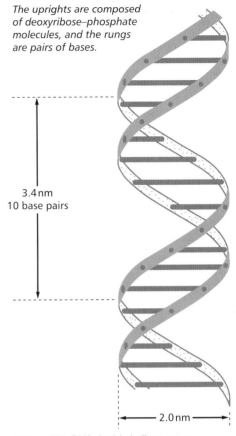

The uprights are composed of deoxyribose–phosphate molecules, and the rungs are pairs of bases.

3.4nm
10 base pairs

2.0nm

Fig 5.6 *The DNA double helix structure*

Table 5.2 *Differences between RNA and DNA*

RNA	DNA
Single polynucleotide chain	Double polynucleotide chain
Smaller molecular mass (20 000–2 000 000)	Larger molecular mass (100 000–150 000 000)
May have a single or double helix	Always a double helix
Pentose sugar is ribose	Pentose sugar is deoxyribose
Organic bases present are adenine, guanine, cytosine and uracil	Organic bases present are adenine, guanine, cytosine and thymine
Ratio of adenine to uracil and the ratio of cytosine to guanine varies	Ratio of adenine to thymine and the ratio of cytosine to guanine is one
Manufactured in the nucleus but found throughout the cell	Found mostly in the nucleus with some present in mitochondria and chloroplasts
Amount varies from cell to cell (and within a cell according to metabolic acivity)	Amount is constant for all cells of a species (except gametes and spores)
Chemically less stable	Chemically very stable
May be temporary – existing for short periods only	Permanent
Three basic forms: messenger, transfer and ribosomal RNA	Only one basic form, but with an almost infinite variety within that form

SUMMARY TEST 5.2

DNA is made up of a **(1)** sugar called **(2)**. This forms a ladder shape, with the sugar and **(3)** groups forming the uprights and the organic bases making up the rungs. These organic bases are always the same pairings: thymine is always paired with **(4)**, while guanine is always paired with **(5)**. The uprights run in opposite directions, i.e. they are **(6)** and are twisted around one another to form a **(7)** shape.

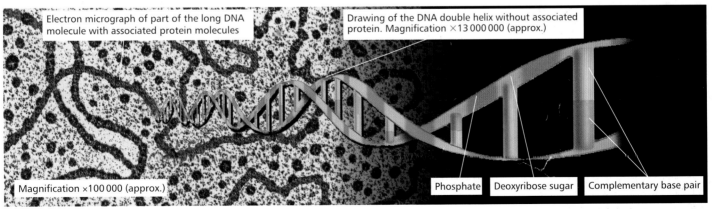

Electron micrograph of part of the long DNA molecule with associated protein molecules

Drawing of the DNA double helix without associated protein. Magnification ×13 000 000 (approx.)

Magnification ×100 000 (approx.)

Phosphate   Deoxyribose sugar   Complementary base pair

Deoxyribonucleic acid

DNA replication

We know that DNA is the hereditary material responsible for passing genetic information from cell to cell and generation to generation. We have only to look at identical twins to see just how perfectly the 3.2 million base pairs of DNA in the human genome can be copied. The process of DNA replication is clearly very precise. How then is it achieved?

5.3.1 Semi-conservative replication

The Watson–Crick model of DNA structure (unit 5.2) allows for a logical explanation of how DNA produces exact copies of itself. Basically, the **hydrogen bonds** linking the base pairs of DNA break and the double helix separates progressively, from one end, into its two strands. Each exposed strand then acts as a template to which complementary **nucleotides** are attracted. These nucleotides have been activated by having two phosphate molecules added to them to make them more reactive. These activated nucleotides zip together to form the 'missing' strand on each half, and two identical DNA molecules result (Fig 5.8). Each of the new DNA molecules contains one of the original DNA strands – i.e. half the original DNA has been saved and built into each of the new DNA molecules. The process is therefore termed **semi-conservative replication**. It takes place during interphase in the cell cycle (section 6.1.1).

The process outlined so far is, in practice, a complex one involving a series of enzymes. It can be summarised as follows:

- **Opening up the DNA double helix** – an enzyme (helicase) breaks the hydrogen bonds between the two parental strands of DNA, at a number of different points, called **replication forks** (Fig 5.7).

- **Unwinding the DNA** – where the DNA is separated it unwinds. To prevent a tangled mess, one strand of the parental DNA is temporarily broken by an enzyme (topoisomerase) and then rejoined once unwinding has occurred.

- **Assembling the leading strand** – one strand of DNA is made by **DNA polymerase** in a continuous process that occurs in the same direction (5′→3′) as the replication fork is moving.

- **Assembling the lagging strand** – the other strand of the DNA is antiparallel (section 5.2.1) and runs in the 3′→5′ direction. DNA polymerase can only work in the 5′→3′ direction. Therefore short sections of complementary DNA are made simultaneously in the 5′→3′ direction and these small sections are then linked together using the enzyme ligase.

- **Removing wrongly coded DNA** – errors that inevitably occur during replication could have disastrous effects on an organism and so an enzyme called **proof-reading endonuclease** cuts out any bases that are wrongly paired.

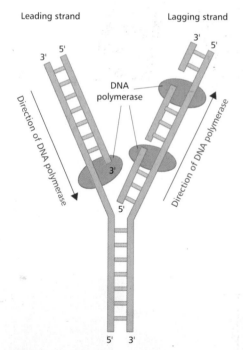

Leading strand   Lagging strand

DNA polymerase

Direction of DNA polymerase

Direction of DNA polymerase

DNA polymerase synthesises DNA which is complementary to both strands. It works in the 5′ – 3′ direction of the new strand. As the DNA molecule is antiparallel, the DNA polymerase makes new DNA on the leading strand as a continuous process in the direction of the replication fork. On the lagging strand it makes small sections simultaneously in the opposite direction. These sections are later joined by the enzyme DNA ligase.

Fig 5.7 *DNA replication fork showing the action of DNA polymerase*

1. A representative portion of DNA, which is about to undergo replication.

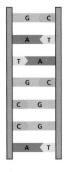

2. Helicase enzyme causes the two strands of the DNA to separate. Binding proteins then keep the two halves apart.

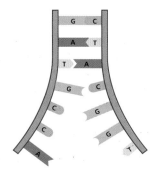

3. The helicase completes the splitting of the strand. Meanwhile, free nucleotides that have been activated by the addition of 2 phosphate molecules are attracted to their complementary bases.

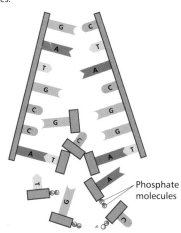

Phosphate molecules

4. Once the activated nucleotides are lined up, they are joined together by DNA polymerase (bottom three nucleotides) and the phosphate molecules are released. The remaining unpaired bases continue to attract their complementary nucleotides.

5. Finally, all the nucleotides are joined to form a complete polynucleotide chain using DNA polymerase and DNA ligase. In this way, two identical strands of DNA are formed. As each strand retains half of the original DNA material, this method of replication is called the semi-conservative method.

Fig 5.8 The semi-conservative replication of DNA

SUMMARY TEST 5.3

DNA replication involves the separation of the two **(1)** chains that make up the molecule. Firstly the double helix is opened up by an enzyme called **(2)**. The points where the helix splits are called **(3)**. To allow the helix to untwist without causing a tangled mass, one chain is broken by an enzyme called **(4)**. One new strand of DNA is made by the enzyme known as **(5)** in a continuous process. On the other strand the enzyme works in the opposite direction making small sections of **(6)** DNA simultaneously. The sections are then joined together by the enzyme **(7)**. Any incorrectly paired bases are removed using the enzyme **(8)**. As the new DNA contains half of the original DNA, the process is known as **(9)** replication.

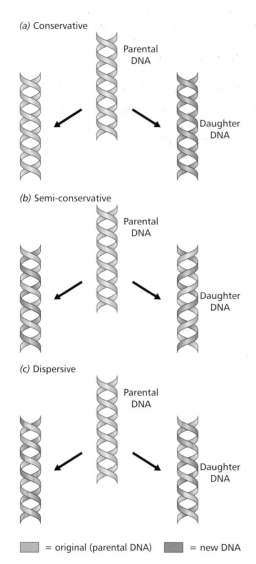

(a) Conservative

Parental DNA

Daughter DNA

(b) Semi-conservative

Parental DNA

Daughter DNA

(c) Dispersive

Parental DNA

Daughter DNA

■ = original (parental DNA)   ■ = new DNA

Fig 5.9 *Possible methods of DNA replication*

When James Watson and Francis Crick deduced the structure of DNA in 1953 with the help of Rosalind Franklin's diffraction studies, they remarked in their paper **'It has not escaped our notice that the specific pairing we have postulated immediately suggests a possible copying mechanism for the genetic material'.**

Their idea, namely the semi-conservative method, was, however, only one of three possible mechanisms that needed to be scientifically tested before a definite conclusion could be drawn.

5.4.1 Possible mechanisms of DNA replication

In 1953, the three most feasible explanations for how DNA might replicate were as follows.

- **The conservative model** suggested that the parental DNA remained intact and that a separate daughter DNA copy was built up from new molecules of deoxyribose, phosphate and organic bases. Of the two molecules present, one would be made of entirely new material while the other would be entirely original material (Fig 5.9).
- **The semi-conservative model** proposed that the DNA molecule split into two separate strands, each of which then replicated its mirror image (i.e. the missing half). Each of the two new molecules would therefore have one strand of new material and one strand of original material (Fig 5.9).
- **The dispersive model** predicted that parental DNA would be broken down and the **nucleotides** replicated before being dispersed randomly throughout the new molecules. The new molecules would contain both new and original material but this would be randomly distributed and not necessarily with equal amounts of old and new material in each molecule.

5.4.2 The experiments of Meselsohn and Stahl

In 1958, Matthew Meselsohn and Franklin Stahl of the California Institute of Technology evaluated the three proposed mechanisms of DNA replication. If we look at Fig 5.9, we can see that the distribution of the 'old' DNA after replication is different in each of the three cases. To find out which mechanism was correct was therefore easy, in theory at least – simply label the old DNA in some way and then look at how it was distributed after replication. Meselsohn and Stahl achieved this in a neat and elegant experiment using the isotope (section 2.1.3) heavy ^{15}N nitrogen. Their work can be summarised as follows.

- Cells of the gut bacterium *Escherichia coli* were grown on a medium containing a nitrogen source made up of the common form of nitrogen ^{14}N. This sample acted as the first control.
- Some bacterial cells were transferred to a medium where the nitrogen (in the form of ammonium chloride) was of the heavier isotope of nitrogen (^{15}N).
- The bacteria were grown, during which time the nitrogen they needed to make new DNA came from the growth medium and was therefore of the ^{15}N type. The resultant DNA was therefore heavier than that formed by the bacteria grown on the medium with ^{14}N. This was the second control.
- After many generations, the DNA of the cells was almost exclusively of the heavy type.
- Samples of the 'heavy DNA' bacteria were then transferred to a medium in which the nitrogen source (ammonium chloride) was of the light (^{14}N) type.
- The bacteria were grown just long enough for the cells to divide once (20–50 minutes, depending on temperature).

- A sample was removed and the DNA of the bacterial cells was extracted and placed in a solution of caesium chloride and centrifuged.
- Because the caesium atom is heavy, it sinks towards the bottom of the tube during **centrifugation** and a density gradient is established in the tube.
- DNA molecules sink in the gradient until they reach a level where their density equals that of the caesium chloride. They then 'float' at this level in the tube.
- The DNA containing ^{15}N is denser than that containing ^{14}N and therefore sinks to a lower level in the tube.
- After one generation in the ^{14}N medium, bacteria were found to have produced DNA with a density mid-way between that of the light (^{14}N) DNA and the heavy (^{15}N) DNA. This meant that one strand of each molecule of this DNA contained ^{14}N while the other strand contained ^{15}N.
- An extract of bacteria was taken after two generations and treated in the same way. Two bands were formed – one with the density of light DNA (indicating it contained only nitrogen of the ^{14}N type) and one of intermediate density (half ^{14}N and half ^{15}N).
- After three generations, there was three times as much DNA in the 'light' band than in the 'intermediate' band.

All these results are consistent with the semi-conservative theory of DNA replication. These experiments and the results are illustrated in figure 5.10.

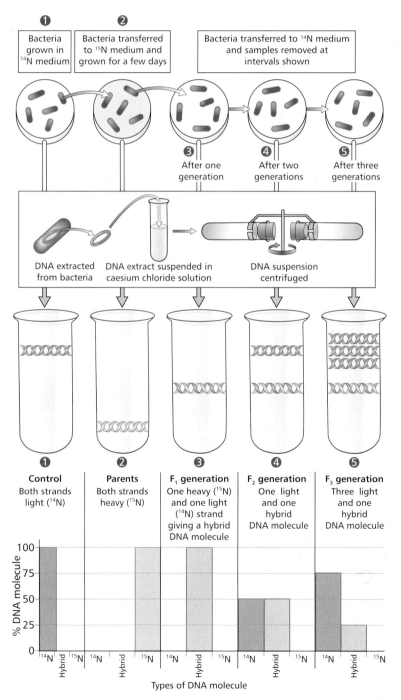

Fig 5.10 *Summary of Meselsohn–Stahl experiments on the semi-conservative replication of DNA*

The genetic code

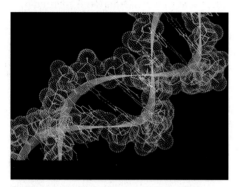

Computer representation of part of a DNA molecule

Table 5.3 The genetic code. The base sequences shown are those on mRNA

First position	Second position				Third position
	U	**C**	**A**	**G**	
U	Phe	Ser	Tyr	Cys	U
	Phe	Ser	Tyr	Cys	C
	Leu	Ser	Stop	Stop	A
	Leu	Ser	Stop	Trp	G
C	Leu	Pro	His	Arg	U
	Leu	Pro	His	Arg	C
	Leu	Pro	Gln	Arg	A
	Leu	Pro	Gln	Arg	G
A	Ile	Thr	Asn	Ser	U
	Ile	Thr	Asn	Ser	C
	Ile	Thr	Lys	Arg	A
	Met	Thr	Lys	Arg	G
G	Val	Ala	Asp	Gly	U
	Val	Ala	Asp	Gly	C
	Val	Ala	Glu	Gly	A
	Val	Ala	Glu	Gly	G

THE GENETIC CODE

A codon is made up of three nucleotide bases read in the sequence shown. For example, UGC codes for the amino acid Cys (cysteine). The first letter (U) is in the 'first position' column, the second letter (G) is in the 'second position' column and the third letter (C) is in the 'third position' column.

Once the structure of DNA had been discovered, and it had been established beyond doubt that it was the means by which genetic information was passed from generation to generation, scientists puzzled as to exactly how DNA determined the features of organisms.

5.5.1 DNA, proteins and enzymes

In 1902, a British doctor called Archibald Garrod studied a disorder called alkaptonuria in which one unusual symptom was that the urine of a sufferer turned black when exposed to the air. The chemical in the urine that was turning black was normally broken down by an enzyme before being excreted. Showing much insight, Garrod concluded that sufferers of the disorder were unable to make this enzyme, and that other inherited diseases could also be the result of enzyme deficiency. How right he was. We now know that it is the sequence of **nucleotides** on DNA that determines the amino acid sequence of a protein. We also know that enzymes are proteins and that enzymes control the production of materials within organisms. DNA therefore determines the characteristics of an organism through its ability to determine which proteins, especially enzymes, are produced.

The sequence of pairs of nucleotides, called **base pairs**, which determines the amino acid sequence of a polypeptide, is called a **gene**. It used to be thought that one gene determined one polypeptide. However recent evidence, much of it emerging from the Human Genome Project (section 11.5.1), has now questioned this view and it is thought that a single piece of DNA (gene) may code for up to 9 proteins. How exactly then, does a gene determine the sequence of amino acids in polypeptides it codes for?

5.5.2 The triplet code

Proteins show almost infinite variety. This variety likewise depends upon the sequence of amino acids in each protein. There are just 20 amino acids that regularly occur in proteins, and each must have its own code of bases on the DNA. As there are only four different bases (adenine, guanine, cytosine and thymine) present in DNA, if each coded for a different amino acid, only four different amino acids could be coded for. Using a pair of bases, 16 different codes are possible – which is still inadequate. A triplet of bases produces 64 codes, more than enough to satisfy the requirements of 20 amino acids. This is called the **triplet code**. All 64 codes are shown in table 5.3. Although there are only four nucleotide bases in DNA, the considerable length of a DNA molecule means that there is an almost unlimited variety of combinations of these bases. Even in a bacterium such as *Escherichia coli* there are 4 million base pairs in its single DNA molecule. Humans have 3.2 billion (3.2×10^9) base pairs, which provide a vast set of instructions for making proteins – so vast in fact that it is thought that less than 5% are actually used to make proteins. These are known as **exons**. The remainder, called **introns**, are not 'expressed' and are sometimes referred to as 'junk' DNA. The 3.2 billion bases that are 'expressed ' in humans were thought to be organised into up to 150 000 **genes** but recent evidence from the Human Genome Project suggests this may be as few as 25 000 genes.

5.5.3 Features of the triplet code

Further experiments, including **frame-shift** ones carried out by Watson and Crick, have revealed the following features of the triplet code. In each case, the codon referred to is the triplet of bases found on mRNA.

- A few amino acids have only a single triplet code, e.g. tryptophan is coded only by UGG.

- The remaining amino acids have between two and six codons each, e.g. leucine has six – UUA, UUG, CUU, CUC, CUA and CUG.

- The codon is always read in the 5'→3' direction.

- The code is a **degenerate code**, because most amino acids have more than one triplet code.

- The start of a sequence is always the codon AUG. This codes for the amino acid methionine. If this first methionine molecule does not form part of the final polypeptide, it is later removed.

- Three codons, UAA, UAG and UGA, do not code for any amino acid. These are called **stop codes**, or **nonsense codes**, and mark the end of a polypeptide chain (section 5.7.4). They act in much the same way as a full stop at the end of a sentence.

- The code is **non-overlapping**, i.e. each base in the sequence is read only once: six bases numbered 123456 are read as triplets 123 and 456, rather than as triplets 123, 234, 345, 456. Non-overlapping codes need more bases but are less likely to be affected by error. Some viruses, with limited amounts of DNA, use overlapping codes, but this is extremely rare.

- The code is **universal**, i.e. it is the same in all organisms – or almost so. There are differences in some protoctists and in the DNA of mitochondria and chloroplasts.

READING GENETIC CODE TABLES – TAKE CARE!

A **codon** is a sequence of three bases on a **messenger RNA** molecule that codes for a single amino acid (or is a stop codon). The genetic codes in table 5.3 on the opposite page are codons, i.e. they refer to the triplets of bases on mRNA. The codes on the coding strand of DNA (section 5.6.1) are therefore the same except that T (thymine) replaces U (uracil) in every case. Sometimes tables show the codes as they appear on the template strand of DNA. These codes are complementary to the codons on mRNA. For example, the codon UCA on mRNA is AGT on the template strand of DNA. When reading any table of genetic codes, you should therefore check carefully whether the codes refer to mRNA or DNA (template strand).

SUMMARY TEST 5.5

The sequence of nucleotides that determines the amino acid sequence of a polypeptide is called a **(1)**. Each amino acid in the sequence is coded for by a total of **(2)** nucleotide bases on a DNA molecule. The complementary sequence of these bases on a messenger RNA molecule is called a **(3)**. An immense variety of proteins can be made in an organism because the length of DNA in each cell is very large. In humans it totals **(4)** base pairs. Most amino acids have more than one code and the genetic code is therefore described as **(5)**. Each base sequence is read only once and the code is therefore said to be **(6)**. Three codes do not correspond to an amino acid. These are called **(7)** codes and show where one polypeptide ends and the next begins. Using a genetic code table, we can find that the sequence UAU on mRNA codes for the amino acid named **(8)**. The two base sequences on the template strand of DNA that could give rise to this same amino acid are **(9)** and **(10)**. The only mRNA code for the amino acid Met (methionine) is **(11)**.

Protein synthesis – transcription

Proteins, especially enzymes, are essential to all aspects of life. Every organism needs to make its own, sometimes unique, proteins. The biochemical machinery in the cytoplasm of each cell has the capacity to make any and every protein from just 20 amino acids. Exactly which proteins it manufactures depends upon the instructions that are provided, at any given time, by the DNA in the cell's nucleus. The process can be thought of as a bakery, where the basic equipment and ovens can manufacture any variety of bread or cake from relatively few basic ingredients. Which particular varieties are made depends on the recipe the baker uses on any particular day. By choosing different recipes at different times, rather than making everything all the time, the baker can meet seasonal demands, adapt to changing customer needs and avoid waste. DNA replication can be likened to the publication of many copies of a recipe book; taking a photocopy to use in the bakery is therefore **transcription**. Making the cakes, using the photocopied recipe, is **translation**. If the book is not removed from the library, many copies of the recipe can be made, and the same cakes can be produced in many places at the same time or over many years.

5.6.1 Production of messenger RNA (mRNA) – transcription

Transcription (Fig 5.11) is the process of making mRNA (section 5.1.5) using part of the DNA as a template. This mRNA then carries the information out of the nucleus to the ribosomes, in the cytoplasm, that are the site of protein synthesis. The process is as follows.

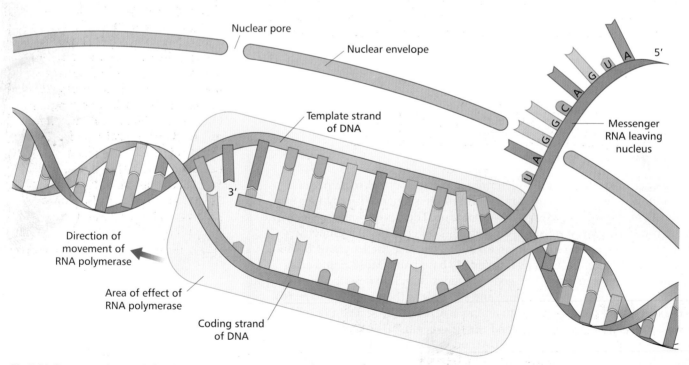

Fig 5.11 *Summary of transcription*

- The enzyme **helicase** acts on a specific region of the DNA molecule called a **cistron** to break the **hydrogen bonds** between the bases, causing the two strands to separate and expose the **nucleotide** bases in that region.
- The enzyme **RNA polymerase** moves along one of the two DNA strands known as the **template strand**, causing the bases on this strand to join with the individual complementary nucleotides from the pool which is present in the nucleus.
- In this way an exposed guanine base on the DNA is linked to the cytosine base of a free nucleotide. Similarly, cytosine links to guanine, and thymine joins to adenine. The exception is adenine, which links to uracil rather than thymine.
- As the RNA polymerase adds the nucleotides one at a time, to build a strand of mRNA, so the DNA strands rejoin behind it. As a result, only around 12 base pairs on the DNA are exposed at any one time.
- When the RNA polymerase reaches a particular sequence of bases on the DNA which it recognises as a 'stop' code, it detaches, and the production of mRNA is then complete.

5.6.2 Modification of messenger RNA (mRNA) – processing

Before leaving the nucleus, the mRNA produced during transcription is **modified** as follows:

- A guanine nucleotide is added to one end of the mRNA. This 'cap' is used to set off the process of translation when the mRNA reaches a ribosome.
- Around 100 adenine nucleotides are added to the other end of the mRNA. It is thought that this 'tail' may prevent the breakdown of the mRNA in the cytoplasm by nucleases because mRNA without a 'tail' is rapidly destroyed.
- Portions of mRNA called introns, which in eukaryotes have no functional value, are removed from the mRNA (Fig 5.12).

5.6.3 Translocation of messenger RNA (mRNA)

The mRNA molecules are too large to diffuse out of the nucleus and so, having been processed, they leave via a nuclear pore. Protected from enzymatic action by its adenine 'tail', the mRNA is attracted to the ribosomes to which it attaches itself.

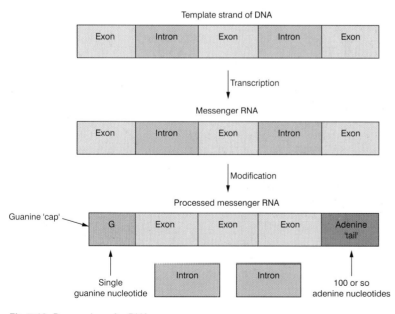

Fig 5.12 *Processing of mRNA*

SUMMARY TEST 5.6

DNA controls protein synthesis by the formation of a template known as (1), which is formed when a specific region of the DNA, called a (2), is opened up by the enzyme (3). Along the strand called the (4) strand, the enzyme (5) moves, causing the bases on the strand to link with their corresponding nucleotides, which in the case of adenine is the nucleotide containing the base (6). The template is then modified by the addition of a (7) nucleotide at one end and about 100 (8) nucleotides at the other. Non-functional portions of the template, called (9), are removed before it leaves the (10) and enters the cytoplasm.

Protein synthesis – translation

Translation is the process by which the messenger RNA from the nucleus of a cell forms a polypeptide, in accordance with the sequence of nucleotide bases along its length. The process begins with the activation of the amino acids which will make up the polypeptide.

5.7.1 Amino acid activation

The amino acids present in cells must first be **activated** before they can be assembled into a polypeptide. This occurs in two stages:

- The amino acid first forms an intermediate with **ATP**, which provides the energy for the next stage.
- The intermediate then combines with transfer RNA to form an amino acid–tRNA complex called **amino-acyl tRNA** (Fig 5.15). The reaction is controlled by the enzyme, amino-acyl tRNA synthetase.

Although the basic structure of tRNA is always the same (unit 5.1), a sequence of three bases on the anticodon loop varies. There are at least 60 variants, which correspond to a codon of three bases on the messenger RNA. At the other end of the tRNA molecule, there is always the sequence of bases adenine-cytosine-cytosine, and it is to this end that the amino acid attaches (Fig 5.13). Each amino acid therefore has its own tRNA molecule, with its own unique anticodon of bases.

5.7.2 Starting polypeptide construction

- A ribosome (Fig 5.14) becomes attached to one end of the mRNA molecule.
- The starting point on the mRNA is normally the triplet of bases **(codon)**, AUG.
- The amino-acyl tRNA molecule with the anticodon sequence of UAC moves to the ribosome and pairs up with the AUG sequence on the mRNA (Fig 5.17).
- As the tRNA which pairs with the AUG sequence on the mRNA always carries the amino acid methionine, polypeptides initially have methionine as the first amino acid.
- However, if this methionine molecule does not make up part of the finished polypeptide, it is removed at the end of the synthesis.

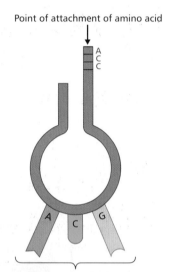

Point of attachment of amino acid

Anticodon – this sequence of ACG means that the amino acid cysteine will attach to the other end of this tRNA molecule. This anticodon will combine with the codon UGC on a mRNA molecule during the formation of a polypeptide. The mRNA codon UGC therefore translates into the amino acid cysteine.

Fig 5.13 Simplified structure of one type of tRNA

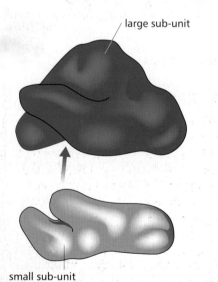

large sub-unit

small sub-unit

Fig 5.14 Structure of a ribosome. The smaller sub-unit fits into a depression on the surface of the larger one

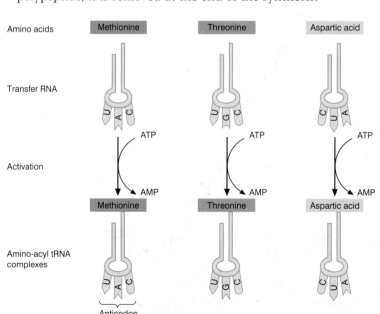

Fig 5.15 Amino acid activation

5.7.3 Making the polypeptide

The following explanation of how a polypeptide is made is illustrated in figure 5.17.

- The ribosome moves along the mRNA, bringing together two tRNA molecules at any one time, each pairing up with the corresponding two codons on the mRNA.
- By means of an enzyme (peptidyl transferase), the two amino acids on the tRNA are joined by a peptide bond.
- The ribosome moves on to the third codon in the sequence on the mRNA, thereby linking the amino acids on the second and third tRNA molecules.
- As this happens, the first tRNA is released from its amino acid (methionine) and is free to collect another methionine molecule from the amino acid pool in the cell.
- The process continues in this way, with up to 15 amino acids being linked each second, until a complete polypeptide chain is built up.
- Up to 50 ribosomes can pass immediately behind the first, so that many identical polypeptides can be assembled simultaneously (Fig 5.16). A group of ribosomes acting in this way is known as a **polysome**.

5.7.4 Finishing the polypeptide

The process described in section 5.7.3 continues until the ribosome reaches a **stop (nonsense) codon**. These are UGA, UAG and UAA, and do not attract a tRNA. At this point, therefore, the ribosome, mRNA and the last tRNA molecule all separate and the polypeptide chain is complete. It now needs to be assembled into the final protein.

5.7.5 Assembling the protein

What happens to the polypeptide next depends upon the protein being made, but usually involves the following:

- The polypeptide is made into its secondary structure by being either coiled into an α-helix or folded into a β-pleated sheet (section 2.6.4).
- The secondary structure is folded into its tertiary structure (section 2.7.1).
- Different polypeptide chains are linked to form the quaternary structure, along with associated (non-protein) prosthetic groups (section 2.7.2).

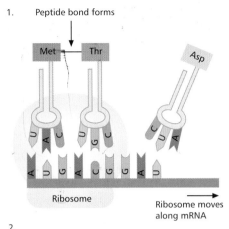

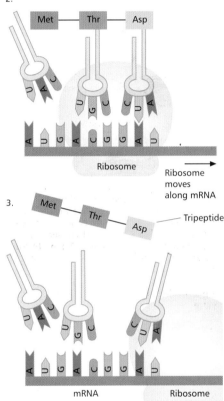

Fig 5.17 Translation

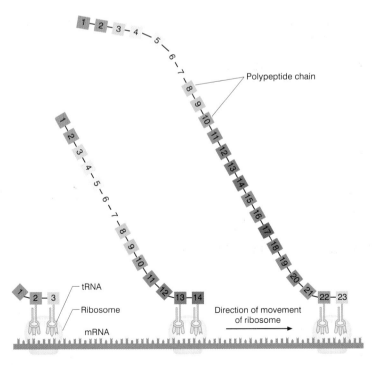

Fig 5.16 Polypeptide formation

SUMMARY TEST 5.7

The formation of a polypeptide from the sequence of bases on messenger RNA is known as (**1**). It begins with the addition of (**2**) to amino acids, in a process known as activation. These amino acids then combine with transfer RNA to form a complex called (**3**). To form the polypeptide, a (**4**) becomes attached to the mRNA molecule and the tRNA attaches to three bases, called a (**5**) on mRNA.

Gene technology is a general term covering those processes whereby genes are manipulated, altered or transferred from organism to organism. Also known as genetic engineering, it has enabled the characteristics of an organism to be changed, either by directly altering its genes or by transferring into it genes from another organism of the same, or a different, species. This process of recombining genes to make new combinations is known as recombinant DNA technology. The resulting organism is known as a genetically modified organism (GMO).

5.8.1 Recombinant DNA technology

Certain disorders and diseases of organisms are the result of individuals being unable to produce important substances because they have defective portions of DNA (genes). Two such substances in humans are:

- **Insulin** – the protein hormone which stimulates the conversion of glucose to glycogen, and so helps to maintain a constant level of sugar in the blood, and without which the debilitating, and potentially lethal, disorder known as **diabetes** occurs.
- **Human factor VIII (anti-haemophiliac globulin, or AHG)** – a protein essential to the clotting of blood, without which the potentially lethal condition known as haemophilia develops.

In both cases, it is possible to correct the deficiency using DNA technology. There are a number of stages to the process:

- **identification** of the required gene for the desired protein
- **isolation** of this gene from the rest of the DNA
- **multiplication** of the gene by the **polymerase chain reaction**
- **insertion** of the gene into a **vector** in order to transfer it to a host cell
- **introduction** of the gene into suitable host cells
- **identification** of host cells which have successfully taken up the gene
- **growth** of the population of host cells
- **production** of the desired protein by the host cells
- **separation** of the protein from the host cells
- **purification** of the protein for clinical use.

5.8.2 Identification and isolation of the gene

Before a gene can be transferred, it must be identified – no small feat given that the gene may consist of a sequence of a few hundred bases amongst the billions in human DNA. One method of identification uses an enzyme called **reverse transcriptase.** This enzyme, found in a group of viruses called **retroviruses**, reverses the process of transcription (unit 5.6), i.e. it makes DNA from RNA. The gene for insulin can be identified as follows:

- β cells of the islets of Langerhans are selected because these are known to produce insulin and so will contain the relevant mRNA.
- The mRNA carrying the code for insulin is extracted.
- Reverse transcriptase is used to make a single-stranded copy of DNA called **complementary DNA (cDNA)** from this RNA.
- The single-stranded copy is converted by an enzyme into the double-stranded form which contains the gene for insulin.
- Additional copies of the human DNA are made using a process called the **polymerase chain reaction**.

Table 5.4 *Advantages of using genetically engineered hormones rather than extracting them from organisms*

- Animals do not need to be slaughtered, e.g. calves do not need to be killed to extract insulin from their pancreases

- The hormones produced are more effective because they are exact copies of the human form rather than animal forms. There is therefore no immune response which might otherwise reject the hormone

- There is no risk of transferring infections such as hepatitis or HIV, as there is when hormones are extracted from human blood

- The hormones are cheaper to produce

• The DNA fragments have guanine **nucleotides** added to each of their ends to make them 'sticky' and hence easier to attach to plasmid DNA (section 5.8.3)

5.8.3 Insertion of DNA into a vector

With the relevant DNA isolated and copied many times, the next task is to join it into a carrying unit known as a **vector**. This vector is used to transport the DNA into the host cell. Most commonly used is a **plasmid**, which is a circular length of DNA found in bacteria. Plasmids almost always contain genes for **antibiotic resistance**, and it is at one of these antibiotic resistance sites that enzymes called **restriction endonucleases** are used to break the plasmid loop. The cut ends are made 'sticky' by adding cytosine nucleotides (cytosine nucleotides are used because they are complementary to the guanine nucleotides on the human DNA fragments). When the human DNA fragments are mixed with the opened plasmids removed from bacterial cells, the guanine and cytosine bases of the 'sticky' ends pair up and they become incorporated into the plasmid. The join is made permanent using the enzyme DNA ligase. These events are summarised in figure 5.18.

5.8.4 Introduction of DNA into host cells

With the DNA incorporated into at least some of the plasmids, these must then be re-introduced into bacterial cells. The process is called **transformation** and involves the plasmids and bacterial cells being mixed in a medium containing calcium chloride. The calcium chloride makes the bacterial cell wall permeable, allowing the plasmids to pass through into the cytoplasm. However, not all the bacterial cells will possess the DNA fragments, and the task now is to identify which bacterial cells contain the gene (DNA fragment). This identification can be carried out by treating bacteria with the appropriate antibiotic, i.e. the one for which the plasmid has a resistance gene. Those bacteria that have taken up the plasmid will have the antibiotic resistance gene and will survive. Those that have failed to take up the plasmid will be killed by the antibiotic.

5.8.5 Large-scale culturing

The bacteria which have the human gene for insulin incorporated into them are known as **genetically modified (GM) bacteria**. These bacteria are multiplied by growing them in an industrial fermenter. They now secrete insulin into the medium in which they grow alongside the other secretions that they previously produced. After they have been grown for some time, the insulin is then separated from the GM bacteria and other secretions, purified and used to treat diabetes.

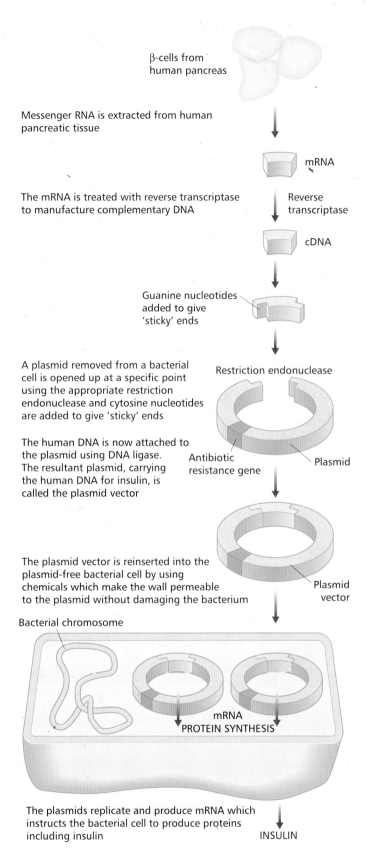

β-cells from human pancreas

Messenger RNA is extracted from human pancreatic tissue

mRNA

The mRNA is treated with reverse transcriptase to manufacture complementary DNA — Reverse transcriptase

cDNA

Guanine nucleotides added to give 'sticky' ends

A plasmid removed from a bacterial cell is opened up at a specific point using the appropriate restriction endonuclease and cytosine nucleotides are added to give 'sticky' ends

Restriction endonuclease

The human DNA is now attached to the plasmid using DNA ligase. The resultant plasmid, carrying the human DNA for insulin, is called the plasmid vector

Antibiotic resistance gene — Plasmid

The plasmid vector is reinserted into the plasmid-free bacterial cell by using chemicals which make the wall permeable to the plasmid without damaging the bacterium

Plasmid vector

Bacterial chromosome

mRNA
PROTEIN SYNTHESIS

The plasmids replicate and produce mRNA which instructs the bacterial cell to produce proteins including insulin

INSULIN

Fig 5.18 *Summary of the use of gene technology in the production of insulin*

1 A DNA molecule is made up of two polynucleotide strands which are twisted into a double helix, as shown in the figure.

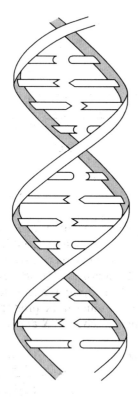

DNA is involved in transcription, which is part of protein synthesis, and replication.

Complete the table, putting a tick (✔) in the column(s), where relevant, to indicate whether each statement refers to transcription, replication or both. The first one has been done for you.

	transcription	replication
free nucleotides bond to the DNA strand	✔	✔
two new DNA molecules are produced		
only the part of the DNA molecule containing the gene unwinds		
hydrogen bonds are broken between the two DNA strands		
cytosine nucleotides bond to guanine on the DNA strand		
uracil nucleotides bond to adenine on the DNA strand		

(5 marks)
(Total 5 marks)
OCR 2801 Jan 2003, B (BF), No.4

2 One of the uses of genetic engineering is in the synthesis of human insulin. Before this technology had been perfected, insulin was obtained from animals, such as pigs.

a Describe how

(i) the isolated human insulin gene is inserted into a bacterial plasmid; *(4 marks)*
(ii) the bacteria which take up the modified plasmids can be identified. *(2 marks)*

b Suggest why it is considered preferable to use genetically engineered sources of human insulin rather than insulin obtained from pigs. *(1 mark)*

c Describe the process of protein synthesis in cells.
In your answer, include the roles of messenger RNA, transfer RNA and ribosomes.
(In this question, 1 mark is available for quality of written communication.)

(9 marks)
QWC (1 mark)
(Total 17 marks)
OCR 2801 May 2002, B (BF), No.6

3 Figure 1 represents a nucleotide which forms part of a DNA molecule.

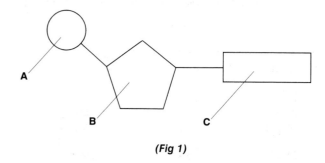

(Fig 1)

a (i) Name **A** to **C**. *(3 marks)*
(ii) State which part of the nucleotide contains nitrogen. *(1 mark)*

During research into the mechanism of DNA replication, bacteria were grown for many generations in a medium containing only the 'heavy' isotope of nitrogen, ^{15}N. This resulted in all the DNA molecules containing only ^{15}N. This is illustrated in figure 2.

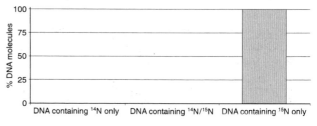

(Fig 2)

These bacteria were then grown in a medium containing only 'light' nitrogen, ^{14}N. After the time taken for the DNA to replicate once, the DNA was analysed. The results are shown in figure 3.

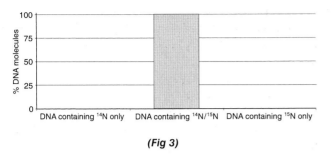

(Fig 3)

b Explain how these data support the semi-conservative hypothesis of DNA replication. *(3 marks)*

The bacteria continued to grow in the 'light' nitrogen, ^{14}N, medium until the DNA had replicated once more. The DNA molecules were analysed. The results are shown in figure 4.

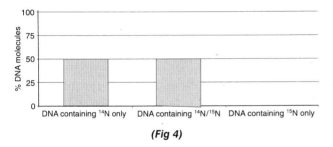

(Fig 4)

Figure 5 shows simple diagrams of DNA molecules, indicating the nitrogen content of each.

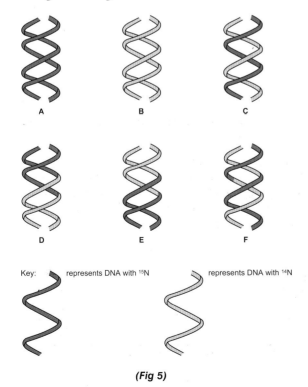

(Fig 5)

c With reference to figure 5, select the letter or letters which best represent the bacterial DNA. *(3 marks)*

The bacteria continued to grow in the 'light' nitrogen, ^{14}N, medium until the DNA had replicated once more. The DNA molecules were analysed.

d Complete the bar chart below to indicate the expected results of the composition of these DNA molecules.

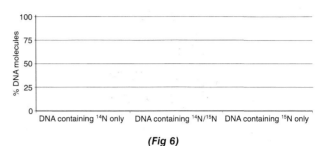

(Fig 6)

(3 marks)
(Total 13 marks)
OCR 2801 Jan 2002, B (BF), No.7

4 Humans produce insulin from certain cells in the pancreas. The insulin gene is isolated from a human pancreas cell and then inserted into a plasmid. The DNA responsible for the synthesis of insulin is then inserted into a bacterium. The figure, which is **not** drawn to scale, shows how insulin can be produced in this way. Different enzymes function at **X** and **Y**.

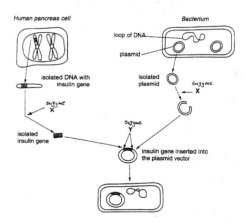

a State a general term for the technique shown in the figure. *(1 mark)*

b Outline the roles of the enzymes that function at **X** and **Y**. *(3 marks)*

c Explain why the plasmid is described as a *vector*. *(2 marks)*

d Outline the role of the bacterium in the process once the vector has been inserted into the host cell. *(4 marks)*
(Total 10 marks)
OCR 2801, 2000 Specimen paper, B (BF), No.5

6 Nuclear division

6.1 Chromosomes and the cell cycle

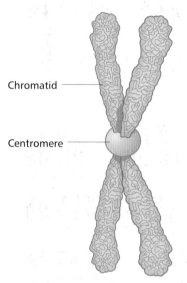

Fig 6.1 *Structure of a chromosome*

(labels: Chromatid, Centromere)

The cells that make up organisms are always derived from existing cells, by the process of division, which occurs in two main stages:

- **Nuclear division**, is the process by which the nucleus divides. There are two types of nuclear division:
  - **Mitosis**, results in two daughter nuclei having the same number of chromosomes as the parent nucleus. The nuclei formed are normally genetically identical to the parent one.
  - **Meiosis**, results in four daughter nuclei having half the number of chromosomes as the parent nucleus. The nuclei formed have a genetic composition different from the parent one.
- **Cell division (cytokinesis)** follows nuclear division, and is the process by which the whole cell divides.

6.1.1 The cell cycle

Cells do not divide continuously, but undergo a regular cycle of division separated by periods of cell growth. This is known as the **cell cycle** and has three stages:

- **Interphase** – occupies most of the cell cycle, and is sometimes known as the resting phase, because no division takes place. In one sense, this could hardly be further from the truth, as interphase is a period of intense chemical activity, divided into three parts:
  - **First growth (G_1) phase**, when the proteins from which cell organelles are synthesised are produced.
  - **Synthesis (S) phase**, when DNA is replicated.
  - **Second growth (G_2) phase**, when organelles grow and divide and energy stores are increased.
- **Nuclear division**, when the nucleus divides into two (mitosis) or four (meiosis).
- **Cell division (cytokinesis)**, when the cell divides into two (mitosis) or four (meiosis).

Nuclear division, or mitosis, typically occupies 5–10% of the total cycle. The cycle may take as little as 20 minutes in a bacterial cell, although it typically takes 8–24 hours.

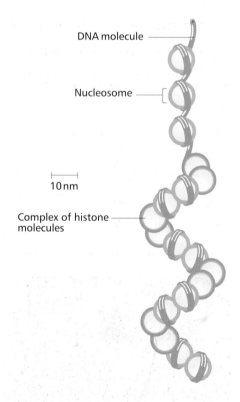

Fig 6.2 *Detailed structure of a chromosome*

(labels: DNA molecule, Nucleosome, 10 nm, Complex of histone molecules)

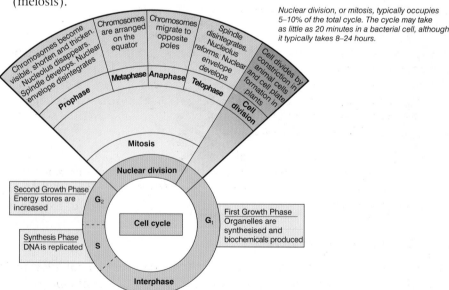

Fig 6.3 *The cell cycle*

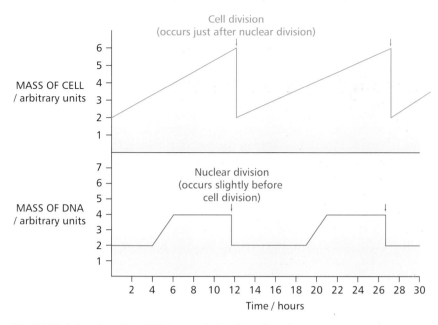

Fig 6.4 Variations in cell and DNA mass during the cell cycle

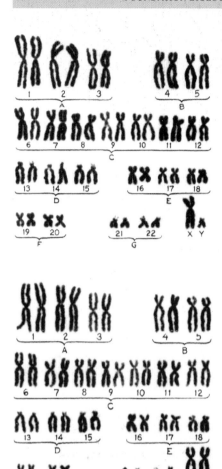

Human karyotypes (male above and female below)

Table 6.1 Chromosome numbers of various organisms

Species	Chromosome number
Crocus	6
Locust	24
Tomato	24
Cat	38
Human	46
Potato	48
Dog	78
Some protozoa	300

6.1.2 Chromosome structure

Chromosomes have a characteristic shape, occur in pairs, and carry the hereditary material of the cell. Chromosomes are only visible as discrete structures when a cell is dividing. The rest of the time, they consist of widely spread areas of darkly staining material called **chromatin**. When they are visible, chromosomes appear as long, thin threads around 50 μm long. They are made up of two strands called **chromatids**, joined at a point called the **centromere** (Fig 6.1).

Chromosomes are made up mainly of three materials:

- **proteins** (70%), mostly in the form of **histones**, scaffold proteins and **polymerases**
- **deoxyribonucleic acid** – DNA (15%)
- **ribonucleic acid** – RNA (10%).

To fit in, the considerable length of DNA found in each cell (around 2 metres in humans) is highly coiled and folded. This DNA is held in position by proteins called **histones**, which together form a complex known as **chromatin**. The chromatin has a beaded appearance due to the presence of **nucleosomes** (Fig 6.2). These make up a portion of DNA which is 146 base pairs in length and wrapped around eight histone molecules.

For convenience, and to make them easier to study, photographs of chromosomes are cut out and pasted into a logical format where they are arranged in their pairs and given numbers to identify them. These arrangements are called **karyotypes**.

6.1.3 Chromosome number

Although the number of chromosomes is always the same for normal individuals of a species, it varies from one species to another (table 6.1). The number of chromosomes is no indication of the level of organisation, complexity or evolutionary status of a species. It has no significance.

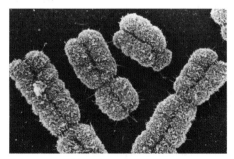

False-colour SEM of a group of human chromosomes

By mitosis, the nucleus of a cell divides in such a way that the two resulting nuclei both have the same number and type of **chromosomes** as the parent nucleus. Except in the rare event of a mutation, the genetic make-up of the two daughter nuclei is also identical to that of the parent nucleus. Mitosis is always preceded in the cell cycle by a period during which the cell is not dividing. This period is called **interphase** (section 6.1.1). Although mitosis is a continuous process, it can be divided into four stages for convenience:

- **Prophase** – chromosomes become visible and the nuclear envelope disappears.
- **Metaphase** – chromosomes arrange themselves at the centre (equator) of the cell.
- **Anaphase** – **chromatids** migrate to opposite poles.
- **Telophase** – the nuclear envelope reforms.

The complex process is illustrated in the photographs and in figure 6.5.

6.2.1 Prophase

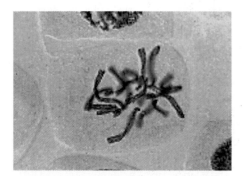

In prophase, the chromosomes first become visible, to begin with as long thin threads, which later shorten and thicken. Animal cells contain two cylindrical organelles called **centrioles** (section 1.7.3), each of which moves to opposite ends (called **poles**) of the cell. From each of the centrioles, **microtubules** develop which span the cell from pole to pole. Collectively, these microtubules are called the **spindle apparatus**. As plant cells lack centrioles but do develop a spindle apparatus, centrioles are clearly not essential to microtubule formation. The nucleolus disappears and the nuclear envelope breaks down, leaving the chromosomes free in the cytoplasm of the cell. These chromosomes are drawn towards the equator of the cell.

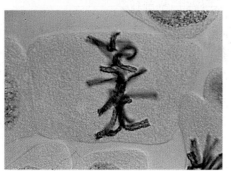

6.2.2 Metaphase

By metaphase, the DNA has replicated and the chromosomes are made up of two **chromatids** joined by the **centromere** (section 6.1.2). It is to this centromere that some microtubules from the poles are attached, and the chromosomes are pulled along the spindle apparatus and arrange themselves across the equator of the cell.

6.2.3 Anaphase

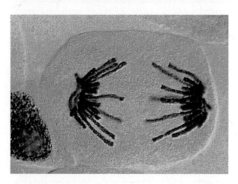

In anaphase, the centromeres divide into two and the microtubules joined to each contract, causing the individual chromatids that make up the chromosome to separate and move to opposite poles of the cell. Now called **daughter chromosomes**, the chromatids move rapidly to their respective poles. The energy for the process is provided by mitochondria, which gather around the spindle fibres. If cells are treated with chemicals that destroy the spindle, the chromosomes remain at the equator, unable to reach the poles.

6.2.4 Telophase

In this stage, the daughter chromatids reach their respective poles and become longer and thinner, finally disappearing altogether, leaving only widely spread **chromatin**. The spindle fibres disintegrate, and the nuclear envelope and nucleolus re-form.

The main stages of mitosis: prophase, metaphase, anaphase and telophase

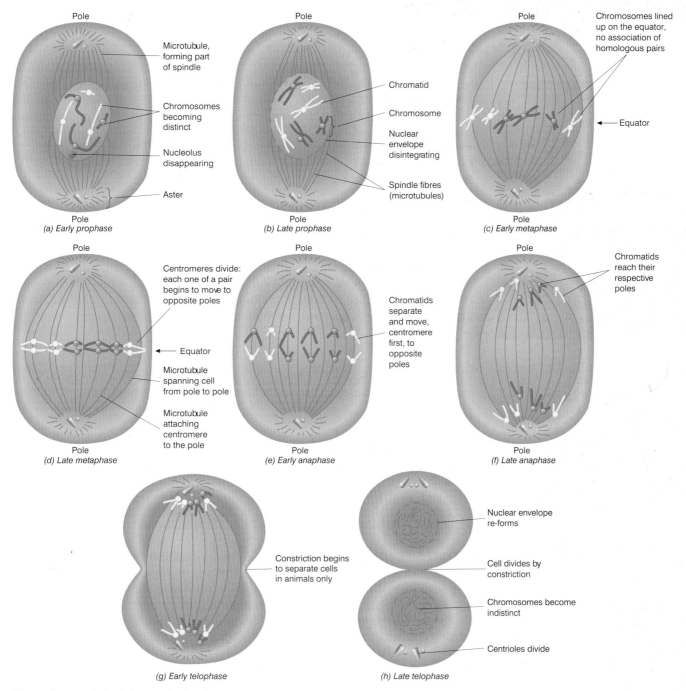

Pole

Microtubule, forming part of spindle

Chromosomes becoming distinct

Nucleolus disappearing

Aster

Pole
(a) Early prophase

Pole

Chromatid

Chromosome

Nuclear envelope disintegrating

Spindle fibres (microtubules)

Pole
(b) Late prophase

Pole

Chromosomes lined up on the equator, no association of homologous pairs

Equator

Pole
(c) Early metaphase

Pole

Centromeres divide: each one of a pair begins to move to opposite poles

Equator

Microtubule spanning cell from pole to pole

Microtubule attaching centromere to the pole

Pole
(d) Late metaphase

Pole

Chromatids separate and move, centromere first, to opposite poles

Pole
(e) Early anaphase

Pole

Chromatids reach their respective poles

Pole
(f) Late anaphase

Constriction begins to separate cells in animals only

(g) Early telophase

Nuclear envelope re-forms

Cell divides by constriction

Chromosomes become indistinct

Centrioles divide

(h) Late telophase

Fig 6.5 *Stages of mitosis in an animal cell*

SUMMARY TEST 6.2

The stage when a cell is not dividing is called (1). The first stage of mitosis is called (2). During this stage in animal cells, two cylindrical structures called (3) move to the opposite ends, also known as (4), of the cell. Thin threads, called microtubules, develop, which span the cell from end to end and together form the (5). Towards the end of this stage, the (6) breaks down and the (7) disappears. During the second stage, called (8), the chromosomes arrange themselves across the centre of the cell, which is also called the (9). By the third stage, called (10), the (11) of each chromosome divides into two and the microtubules attached to each pull the individual (12) to opposite ends of the cell. In the final stage, known as (13), the daughter chromosomes disappear, a nucleolus reforms and the cell divides into two.

The importance of mitosis and meiosis

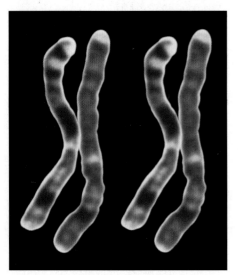

Homologous chromosomes

In the process of sexual reproduction, offspring are formed as a result of the fusion of gametes. If each gamete contained the full number of **chromosomes** (e.g. 46 in humans), when two gametes fused the resulting organism would have twice as many chromosomes as each of the parents. It therefore follows that, at some stage in the life cycle, the number of chromosomes must be halved. This halving of the chromosome number is brought about by **meiosis**, which is a feature of the life cycle of all sexually reproducing organisms.

6.3.1 Haploid and diploid cells

Each species has a fixed number of chromosomes in each of its cells (section 6.1.3). This is usually made up of two of each type of chromosome (one from the mother and the other from the father), i.e. there are two chromosomes that determine the same characteristics. Any cell that contains this double set of chromosomes is called **diploid**, and this is indicated by the symbol **2n**. When a diploid cell divides by **mitosis**, it produces another diploid cell:

$$\text{diploid (2n) cell} \xrightarrow{\text{mitosis}} \text{diploid (2n) cell}$$

However, when a diploid cell divides by meiosis, it produces four nuclei, each with half the number of chromosomes. This is known as a **haploid** number, which is indicated by the symbol **n**.

$$\text{diploid (2n) cell} \xrightarrow{\text{meiosis}} \text{haploid (n) cell}$$

In most organisms, it is the gametes (sperm and ova) that are the haploid cells.

6.3.2 Homologous chromosomes

All diploid cells of organisms have two sets of chromosomes, one set provided by each parent. There are therefore always two sets of genetic information for each characteristic of an individual. Any two chromosomes which determine the same characteristics are termed a **homologous pair**. 'Determining the same characteristics' is not the same as being identical. For instance, a homologous pair of chromosomes may each possess information on eye colour and hair texture, but one chromosome may carry the code for blue eyes and curly hair while the other carries the code for brown eyes and straight hair. During meiosis, the halving of the number of chromosomes is not done randomly. Instead, each daughter cell receives one of each homologous pair of chromosomes. In this way, each cell has one set of information on each characteristic of the organism. When these haploid cells (usually gametes) combine, the diploid state, together with its paired homologous chromosomes, is restored.

6.3.3 Life cycles

All life cycles are, in effect, an alternation of a haploid stage and diploid stage. Although there are other life cycles, the main types are those of animals and plants. The animal life cycle (Fig 6.6) has the following features.

- Gametes are usually produced by meiosis (an exception is male bees, whose gametes are produced by mitosis).
- Gametes fuse almost immediately they are formed, and so the interval between the end of meiosis and fertilisation is short.
- The diploid phase is long and dominant.
- The haploid phase is short.

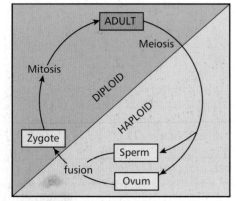

Fig 6.6 Life cycle of most animals

The plant life cycle (Fig 6.7) has the following features.

- Gametes are produced by mitosis.
- Spores (pollen grains and embryo sac) are produced by meiosis.
- The interval between meiosis and fertilisation is longer than in animals.
- The haploid stage can sometimes be long (some pollen grains remain viable for many years).

6.3.4 The importance of mitosis

Mitosis is important in organisms because it produces daughter cells that are genetically identical to the parent cells. Why then is it so essential to make exact copies of existing cells? There are three main reasons:

- **Growth** – when two haploid cells (e.g. a sperm and an ovum) fuse together to form a diploid cell (e.g. a zygote) this diploid cell has all the genetic information needed to form the new organism. If the new organism is to resemble its parents, all the cells that grow from this original cell must possess this identical set of genetic information. Mitosis ensures that this is the case. The cell firstly divides to give a group of identical cells. Although all cells have a complete set of genetic information, only part of it is expressed in any one cell. Depending on which part is expressed, cells change (differentiate) to give groups of specialised cells, e.g. muscle or epithelium in animals, xylem or phloem in plants. These different cell types also divide by mitosis to give tissues made up of identical cells which perform a particular function. This is essential as the tissue can only function efficiently if all its cells have the same structure and perform the same function.
- **Repair** – if cells are damaged or die it is important that the new cells produced have an identical structure and function to the ones that have been lost. If they were not exact copies the tissue would not function as effectively as before. Mitosis is therefore the means by which new cells replace damaged or dead ones.
- **Asexual reproduction** – produces offspring that are identical (assuming no mutations) to their parents and to each other. This has one main advantage. Because they have been able to survive, grow and produce offspring, the parents must be well enough suited to the environmental conditions in which they currently live. By producing identical offspring they can be sure that these too will thrive as long as conditions do not change. Asexual reproduction is a relatively rapid form of reproduction (there is no delay while searching for a mate) and so large numbers can be quickly built up and the local area colonised. This is of particular advantage to plants, which can gain a competitive advantage, especially for light, by this means. Asexual reproduction by mitosis is therefore more common in plants, where a variety of forms of **vegetative propagation** are used. Mitosis is also the basis of natural and artificial cloning.

6.3.5 The importance of meiosis

Meiosis is important because:

- It halves the number of chromosomes in the gametes, so that when two haploid gametes fuse, the diploid number is restored.
- It leads to increased genetic variation.

Variation is achieved by meiosis because:

- It produces haploid gametes which, when they fuse at fertilisation, combine the different genetic material of the two parents.
- During meiosis, pairs of homologous chromosomes arrange themselves randomly and so they are re-sorted in the daughter cells. This is known as **independent assortment**. This is similar to shuffling a pack of cards and dealing them into two hands. The combinations will be different each time.
- During meiosis, **chromatids** break at points called **chiasmata** and portions of them are 'swapped' with equivalent portions of homologous chromosomes. This separates linked alleles and creates new genetic combinations.

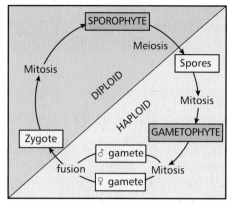

Fig 6.7 *Life cycle of most plants*

SUMMARY TEST 6.3

To avoid chromosome numbers doubling with each generation, a type of cell division called **(1)** halves the number of chromosomes. Most cells have two sets of chromosomes and are therefore said to be **(2)**. These chromosomes occur in pairs, each member of which determines the same characteristics. These are known as **(3)** pairs. When the number of chromosomes in a cell is halved, only one of these pairs is present in the daughter cell, which now has a **(4)** number of chromosomes. In most organisms, the cells which have a single set of chromosomes are the **(5)**. Mitosis produces cells which are genetically identical, unless a **(6)** occurs. This makes mitosis important for three main reasons, namely **(7)**, **(8)** and **(9)**.

The word cancer (Latin for 'crab'), was first used by Hippocrates in 400BC. He saw a similarity between the swollen veins radiating from a breast tumour and the legs of a crab. Cancer is a common and destructive disease. Of all those born in around the year 2000, one-third will suffer from cancer at some time during their lives. Worldwide there are estimated to be 10 million new cases each year and some 6 million deaths from the disease. That being said, cancer is to some extent avoidable and, if diagnosed early enough, successfully treatable.

6.4.1 What is cancer?

Cancer is a group of diseases (around 200 in total) caused by a growth disorder of cells. It is the result of damage to the genes that regulate **mitosis** and the cell cycle, which leads to unrestrained growth of cells. As a consequence, a group of abnormal cells, called a **tumour**, develops and constantly expands in size. Cancers can develop in any organ of the body, but are most commonly found in the lungs, prostate (male), breast and ovaries (female), large intestine, stomach, oesophagus and pancreas.

6.4.2 Cancer and the genetic control of cell division

DNA analysis of tumours has shown that, in general, cancer cells are derived from a single mutant cell. The initial mutation causes uncontrolled mitosis in this cell. Later, a further mutation in one of the descendent cells leads to other changes that cause subsequent cells to be different from normal in growth and appearance. In the process, a number of factors that regulate cell division are altered.

- **Oncogenes** are mutated forms of the proto-oncogenes that regulate normal cell division. The normal control mechanism is shown in figure 6.8. Oncogenes can trigger uncontrolled cell division in a number of ways including:
  - permanently activating the receptor protein (Fig 6.8), even without the presence of growth factors
  - producing excessive amounts of the growth factor.
- **Tumour suppressor genes**, such as p53, code for proteins that stop the cell cycle and also destroy mutated cells. A mutated p53 gene fails to do this and so damaged DNA continues to divide, passing its mutations to daughter cells.
- **Telomeres** are non-coding regions of DNA at the end of chromosomes. They are essential for DNA replication but, every time a cell divides, they get shorter. This means that normal cells cannot divide indefinitely. In cancer cells the telomeres do not shorten, however many times the cells divide, and they are therefore considered **immortal.**

6.4.3 Types of tumour

Most mutated cells die. However, any that survive are capable of making clones of themselves and forming tumours. Not all tumours are harmful **(malignant)**; some are harmless **(benign)**. These tumours are compared in table 6.2.

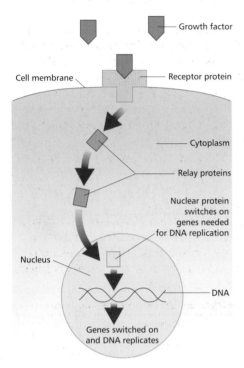

Fig 6.8 *Normal cell receiving signals from growth factors that tell it when to divide*

(labels on figure: Growth factor; Cell membrane; Receptor protein; Cytoplasm; Relay proteins; Nuclear protein switches on genes needed for DNA replication; Nucleus; DNA; Genes switched on and DNA replicates)

Table 6.2 *A comparison of benign and malignant tumours*

Benign tumours	Malignant tumours
Remain within the tissue from which they arise	Tend to spread to other regions of the body (metastasis)
Grow very slowly	Grow rapidly
Have cells which are often well differentiated (specialised)	Have cells that become de-differentiated (unspecialised)
Cells tend to stay together, surrounded by a capsule of dense tissue	Cells are not surrounded by a capsule
Not usually life-threatening, but can disrupt functioning of a vital organ	May be life-threatening, as abnormal tumour tissue replaces normal tissue

6.4.4 Primary and secondary tumours

- **Primary tumours** are masses of abnormal tissue in their place of origin. As shown in figure 6.9, they can be large and develop their own blood supply. Since they are localised, treatment of these tumours is often very successful.
- **Secondary tumours** are more difficult to treat. As shown in figure 6.9, these arise from cells that break away from the primary tumour and migrate via blood or lymphatic vessels. This method of spreading is called **metastasis**. Cells that do not metastasise may spread by growing rapidly, extending the tumour into nearby tissues.

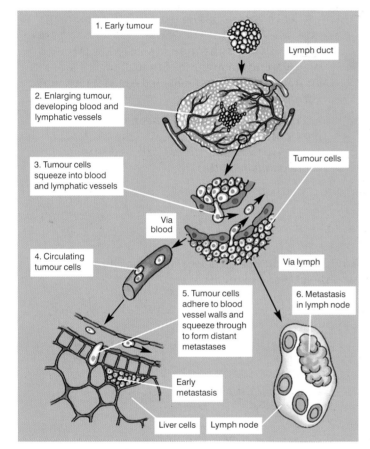

Fig 6.9 *Early primary tumour and its development and spread as a secondary tumour*

6.4.5 Causes of cancer

Cancer is not a single disease and, likewise, it does not have one single cause. Currently, the following factors are known to play a part.

- **Genetic factors** – more than a dozen forms of cancer are known to be directly inherited, e.g. 5–10% of breast cancer. Many more develop in those who inherit a susceptibility to certain factors that cause cancer. Cancers with known genetic risk factors include forms of skin cancer and breast cancer.
- **Carcinogens** – these are chemicals that affect genetic activity in some way, causing abnormal cell division. Some carcinogens are **mutagens**, causing changes in the cell's DNA structure (unit 5.2). Carcinogens include:
  - chemicals such as the polycyclic hydrocarbons found in soot and cigarette smoke (section 14.3.2)
  - short wavelength radiation, such as X-rays, gamma rays and some ultra-violet rays (UVA and UVB). Table 6.3 lists some chemical carcinogens and the cancers they may cause.
- **Age** – some cancers, e.g. leukaemia, are found primarily in young people. Others, such as colon cancer, are found in older adults. Cancers in later life may result partly from the accumulated effects of cell damage. The rate of cancer deaths increases sharply after the age of 40 years, and even more steeply after the age of 60 years, suggesting that several independent mutations must accumulate to give rise to cancer.
- **Environment** – exposure to certain types of radiation or chemicals can cause cancer to develop. For example, breathing in asbestos fibres can cause lung cancer, and sunlight can cause skin cancer (Table 6.3).
- **Viruses** – some cancers have now been shown to have a viral origin. This makes sense because viruses interfere with the genetic make-up of infected cells. For example, the papilloma virus, which causes genital warts, has been linked to some cases of cervical cancer.

Table 6.3 *Chemical carcinogens and the cancers they may cause*

Chemical	Cancer
Diesel exhaust	Lung
Mineral oils	Skin
Pesticides	Lung
Cigarette tar	Lung
Asbestos	Mesothelioma; lung; bladder
Hair dyes	Bladder
Polychlorinated biphenyls (PCBs)	Liver; skin
Soot	Skin
Arsenic	Lung; skin
Formaldehyde	Nose

SUMMARY TEST 6.4

Any chemical or other form of energy that causes abnormal cell division through altered genetic activity is called a **(1)** and the group of abnormal cells produced is called a **(2)**. These groups of abnormal cells may grow rapidly and be unspecialised and life-threatening, in which case they are said to be **(3)**. Cells may break away and spread to other parts of the body, a process called **(4)**. A susceptibility to some forms of cancer is inherited. Two examples are forms of **(5)** cancer and **(6)** cancer. Viruses may cause cancer – the papilloma virus, for example, is thought to cause certain forms of **(7)** cancer. Forms of short wavelength radiation, such as **(8)** or **(9)**, may also cause cancer, while certain environmental factors also contribute, such as **(10)** fibres, which are implicated in lung cancer.

1 a The figure is a series of three diagrams showing some of the events of mitosis in an animal cell. Each diagram represents a stage in mitosis. They are **not** in the correct order.

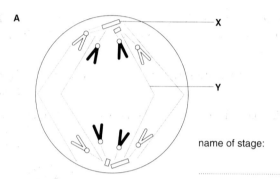

A

name of stage:

....

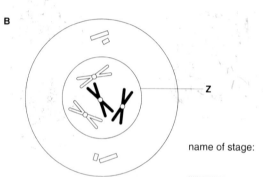

B

name of stage:

....

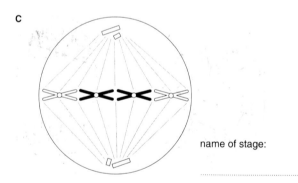

C

name of stage:

....

(i) Using the letters **A** to **C**, put the stages in the correct order. *(1 mark)*

(ii) For each diagram, **A**, **B** and **C**, write the **name** of the stage in the space provided on the figure.

(3 marks)

(iii) Name structures **X** to **Z**. *(3 marks)*

b Interphase is often described as the 'resting stage' of the mitotic cell cycle, but this can be misleading.

Explain why interphase is not a 'resting stage' in the cell cycle. *(2 marks)*

(Total 9 marks)

OCR 2801 Jun 2003, B (BF), No.2

2 Some plant cells divide by mitotic cell division.

a Explain the role of mitotic cell division in the life of plants. *(4 marks)*

Cancer in humans results from uncontrolled mitotic cell division.

b Explain why

(i) it is advisable to apply high factor sun screen when going out into the sun in the summer;

(1 mark)

(ii) it may be less harmful to smoke low tar cigarettes than those with high tar; *(1 mark)*

(iii) radiographers stand behind a lead screen while taking X-rays. *(1 mark)*

c Describe the behaviour of a **chromosome** during a mitotic cell cycle.

(In this question, 1 mark is available for the quality of written communication.)

(8 marks)

(Total 15 marks)

OCR 2801 Jan 2002, B (BF), No.5

3 Figure 1 indicates the appearance of a chromosome at early prophase of mitosis.

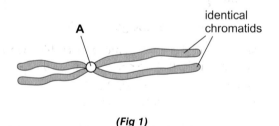

A

identical
chromatids

(Fig 1)

a With reference to figure 1,

(i) name the structure labelled **A**; *(1 mark)*

(ii) explain why the two chromatids are identical. *(2 marks)*

Figure 2 represents the nucleus of an animal cell (2n = 6) at early prophase of mitosis.

b On figure 2, shade **one** pair of homologous chromosomes. *(1 mark)*

c Draw an annotated diagram to indicate what happens in this cell at anaphase of mitosis.

(4 marks)

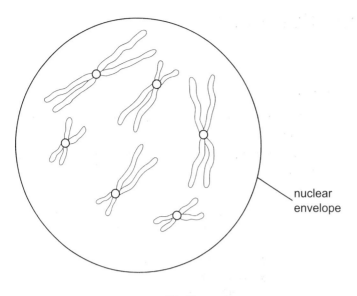

(Fig 2)

d (i) State the number of chromosomes which would be found in a **haploid** cell in this animal.

(1 mark)

(ii) Explain why haploid cells need to be produced during a life cycle which includes sexual reproduction.

(3 marks)

(Total 12 marks)

OCR 2801 June 2001, B (BF), No.6

4 Some cells divide by mitosis.

a (i) Describe the products of mitosis. *(2 marks)*

(ii) State **two** processes in which mitosis occurs in humans. *(2 marks)*

The diploid number of the gorilla is 48.

b State the number of chromosomes which would be found in the following cells of the gorilla.

brain cell …   epithelial cell …
sperm …   muscle cell …   *(4 marks)*

The figure lists six of the events that take place during mitotic cell division. Each event is identified by a letter

A	chromatids separate
B	centromeres divide
C	chromosomes become visible
D	nuclear membrane disintegrates
E	chromosomes align at equator
F	cytoplasm divides (cytokinesis)

c List the letters shown in the figure in the order in which the events occur during a mitotic cell division. The first one (c) has been entered for you. *(5 marks)*

Cancer is caused by uncontrolled mitotic cell division.

d State **three** factors which can increase the chances of cancerous growth. *(3 marks)*

(Total 16 marks)

OCR 2801 Jan 2001, B (BF), No.1

5 The figure shows six stages of mitosis, labelled **A** to **F**, in a plant root tip, as seen under high power of a light microscope.

a (i) Name the stages of mitosis shown in the figure. *(6 marks)*

b Explain the importance of mitosis to living organisms. *(3 marks)*

(Total 9 marks)

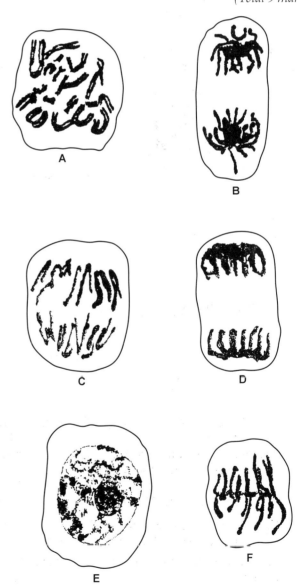

OCR 2801, 2000 Specimen paper, B (BF), No.7

7.1 Ecosystems

Bluebells in a woodland habitat

Ecology is the study of the inter-relationships between organisms and their environment. The term **environment** refers to the conditions that surround an organism. These include both non-living (**abiotic**) components and living (**biotic**) components. Ecology is therefore a complex area of study which incorporates not only most aspects of biology but also elements of physics, chemistry, geography and geology. It is, in effect, the study of the life-supporting layer of land, air and water that surrounds the Earth, called the **biosphere**.

7.1.1 Ecosystems

An ecosystem is made up of all the interacting biotic (living) and abiotic (non-living) elements in a specific area. Ecosystems are more or less self-contained functional units. Within an ecosystem there are two major processes to consider:

- the flow of energy through the system
- the cycling of nutrients within the system.

In theory, the biosphere can be considered as a single ecosystem because energy flows through it and nutrients are recycled within it. In practice, there are much smaller units that are more or less self-contained in terms of energy and nutrients. A fresh-water pond, for example, has its own community of plants to collect the necessary sunlight energy to supply the organisms within it. Nutrients such as nitrates and phosphates are recycled within the pond, with little or no loss or gain between it and other ecosystems. Another example of an ecosystem is an oak woodland (Fig 7.1). Within each ecosystem, there are a number of groups of organisms known as a community.

7.1.2 Community

A **community** is defined as **all the populations** of different organisms living and interacting in a **particular place** at the **same time**. Within an oak woodland, a community might include a large range of organisms, such as oak trees, hazel shrubs, bluebells, nettles, sparrowhawks, blue tits, ladybirds, aphids, woodlice, earthworms, fungi and bacteria. Each species is made up of many groups of individuals, called populations.

7.1.3 Populations

A **population** is a group of organisms of the **same species** that occupy the **same place** at the **same time** and that have the chance to **interbreed** with one another. In the oak woodland there are populations of nettles, worms, woodpeckers, beetles etc. The boundaries of a population are difficult to define, except perhaps within a small pond. In our oak woodland, for example, all the mature sparrowhawks can breed with one another and so form a single population. However, the woodlice on a decaying log at one side of the wood can, in theory, breed with those on a log a kilometre or more away at the other side of the wood. In practice, the sheer distance makes interbreeding unlikely and therefore they can be considered as separate populations. Where exactly the boundary lies between these two populations is, however, unclear.

SUMMARY TEST 7.1

The study of the inter-relationships between organisms and their environment is called **(1)**. The layer of land, air and water that surrounds the Earth is called the **(2)**. An ecosystem is a more or less self-contained functional unit made up of all the living or **(3)** elements and non-living or **(4)** elements in a specific area. Within each ecosystem are groups of organisms, called a **(5)**, which live and interact in a particular place at the same time. A group of interbreeding organisms occupying the same place at the same time is called a **(6)**, and the place where they live is known as a **(7)**.

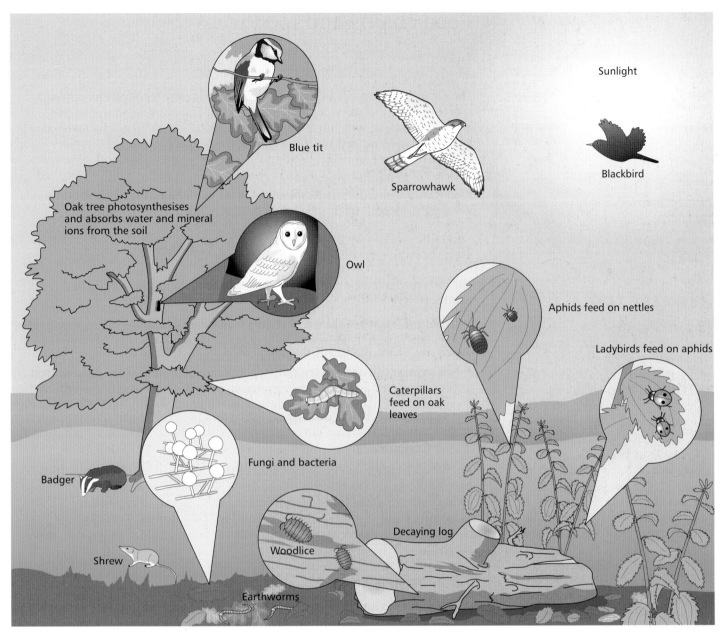

Fig 7.1 *Part of oak woodland ecosystem*

7.1.4 Habitat

A **habitat** is the place where an organism lives. Within an ecosystem there are many habitats. For example, in our oak woodland, the leaf canopy of the trees may be a habitat for blue tits while a decaying log is the habitat for woodlice. A stream flowing through the woodland provides a very different habitat, within which aquatic plants and water beetles live. For a water vole, the stream and its banks form its habitat. Within each habitat there are smaller units, each with its own microclimate. These are called **microhabitats**. The mud at the bottom of the stream may be the microhabitat for a blood worm while a crevice on the bark of an oak tree may be the microhabitat for a lichen.

7.1.5 Ecological niche

This is all of the ranges of environmental conditions and resources required for an organism to survive, reproduce and maintain a viable population. It is also sometimes referred to as the ecological role of a species within its community. Some species may appear very similar, but their nesting habits or other aspects of their behaviour will be different, or they may show different levels of tolerance to, e.g. a pollutant or a shortage of oxygen or nitrates. Any differences in niche, however small, limit competition between species. No two species occupy exactly the same niche.

Food chains and food webs

Table 7.1 *Net primary production in different ecosystems*

Ecosystem	Mean NPP / kJ m^{-2} yr^{-1}
Deserts	260
Oceans	4700
Temperate grasslands	15 000
Intensive agriculture	30 000
Tropical rainforest	40 000

The organisms found in an oak woodland, or any other **ecosystem**, rely on a source of energy to carry out all their activities. The ultimate source of this energy is sunlight, which is converted to chemical energy by plants (producers) and then passed as food from one animal (consumer) to another.

7.2.1 Producers

Producers are photosynthetic organisms that manufacture organic substances using light energy, water and carbon dioxide:

$$6CO_2 + 6H_2O \xrightarrow[\text{chlorophyll}]{\text{light energy}} C_6H_{12}O_6 + 6O_2$$

The rate at which they produce this organic food is referred to as their **productivity**.

- **Gross primary productivity (GPP)** is the total production of organic food in a given area and in a given time. It depends on the types of plant growing there, their density and the climate.
- **Net primary productivity (NPP)** is the rate of production of organic food after allowing for that lost via respiration by the plant – in other words, the production of material that might be eaten by consumers. Table 7.1 shows the mean NPP for a number of different ecosystems.

7.2.2 Consumers

Animals can only gain energy and nutrients by eating other organisms. Those that eat plants are **herbivores**, those eating animals are **carnivores**, and those eating both are **omnivores**. Herbivores are also referred to as **primary consumers** and they are eaten by **secondary consumers**. **Tertiary consumers** and **quaternary consumers** are usually predators but they may also be scavengers or **parasites**.

7.2.3 Decomposers

When producers and consumers die, some energy is 'locked up' in the complex organic molecules of which they are made. This energy is used by a group of organisms that break down these complex materials into simple components again. In doing so, they also release valuable minerals and elements in a form that can be absorbed by plants and so contribute to recycling. The majority of this work is carried out by saprobiontic fungi and bacteria called **decomposers** and, to a lesser extent, by certain animals, such as earthworms, called **detritivores**.

7.2.4 Food chains

The term **food chain** describes a feeding relationship in which plants are eaten by herbivores, which are in turn eaten by carnivores (see table 7.2). Each stage in this chain is referred to as a **trophic level**. The first trophic level is represented by producers, the second by herbivores, and all subsequent ones by carnivores. The shortest food chains usually have three levels:

$$\text{grass} \rightarrow \text{sheep} \rightarrow \text{human}$$

and the longest usually no more than four or five:

nettle	→	aphid	→	ladybird	→	tit	→	sparrowhawk
producer		primary consumer		secondary consumer		tertiary consumer		quaternary consumer

SUMMARY TEST 7.2

The Sun's energy is passed from one feeding level to another through the ecosystem. Each feeding stage in this chain is called a **(1)**. As green plants manufacture complex organic molecules from simple ones during the process of **(2)**, they are known as **(3)**. The total amount of organic food made in a given time for a given area is the **(4)**. Some of this food is used by the plant for the process of **(5)**. The remainder, known as the **(6)**, is available for consumption by other organisms called consumers. **(7)** are consumers that feed directly off green plants, whereas **(8)** feed off other animals. In this way energy is passed along a hierarchy of feeding levels. This is referred to as a **(9)**. When organisms die, some energy remains locked up in the chemicals of which they are made. This energy may be used by a group of organisms called **(10)**.

Ladybird feeding on aphids

The arrows on food chain diagrams represent the **direction of energy flow**. Many herbivores feed on dead plant material and are part of a **decomposer** chain.

dead leaves → earthworms → shrews → badgers

7.2.5 Food webs

In reality, most animals do not rely upon a single food source and, within a single habitat, many food chains will be linked together to form a **food web**. For example, on the edge of an oak woodland, the food chain shown in section 7.2.4 can be combined with others to form the web shown in figure 7.2.

Table 7.2 Food chains in other habitats

Level	Grassland	Pond	Sea-shore
Quaternary consumer	Stoat	Pike	Sea gull
Tertiary consumer	Grass snake	Stickleback	Crab
Secondary consumer	Toad	Leech	Whelk
Primary consumer	Caterpillar	Water snail	Limpet
Producer	Grass	Pondweed	Seaweed

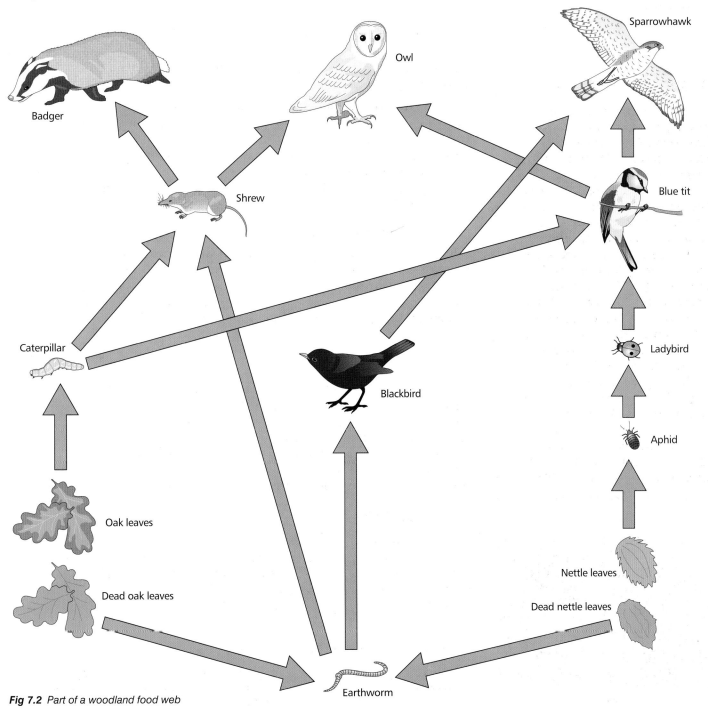

Fig 7.2 Part of a woodland food web

103

The Sun is the source of energy for **ecosystems**. Only around 1% of this light energy is captured by green plants and hence made available to successive organisms in the food chain. These in turn pass on only a fraction of the available energy at each stage.

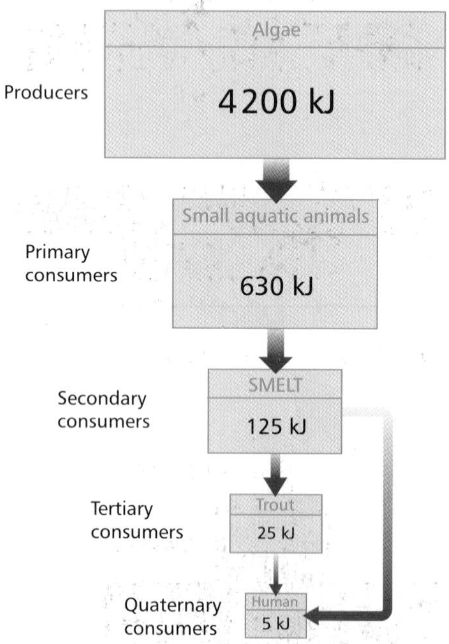

7.3.1 Energy losses in food chains

Plants normally convert between 1% and 3% of the Sun's energy available to them into organic matter. Losses occur in a number of ways.

- Over 90% of solar energy is reflected back into space by clouds and dust or absorbed by the atmosphere and re-radiated.
- Not all wavelengths of light can be absorbed and used for photosynthesis.
- Light may not fall on a chlorophyll molecule.
- Low carbon dioxide levels may limit the rate of photosynthesis.

Plants then lose 20–50% of their gross primary production (section 7.2.1) via respiration, leaving little to be stored as potential food for herbivores. Even then, only about 10% of the **net primary production** of plants is used by **herbivores** for growth. This low percentage is the result of the following:

- Some of the plant is not eaten.
- Some parts are eaten but cannot be digested (e.g. they are lost in faeces).
- Some of the energy is lost in excretory materials (e.g. urine).
- Some energy losses occur in respiration and heat loss to the environment. These losses are high in mammals and birds because of their constant body temperature. Much energy is needed to maintain their body temperature when heat is constantly being lost to the environment.

Carnivores are slightly more efficient, transferring about 20% of the energy available from their prey into their own bodies. It is the relative inefficiency of energy transfer between **trophic levels** that explains why:

- Most food chains have only four or five trophic levels because insufficient energy is available to support a breeding **population** at trophic levels higher than these.
- The biomass of organisms is less at higher trophic levels.
- The total amount of energy stored is less at each level as one moves up a food chain.

Fig 7.3 Food chain in Cayuga Lake, New York State. Figures illustrate the relative amount of energy available at each stage in the food chain

Producers — Algae — **4 200 kJ**

Primary consumers — Small aquatic animals — 630 kJ

Secondary consumers — SMELT — 125 kJ

Tertiary consumers — Trout — 25 kJ

Quaternary consumers — Human — 5 kJ

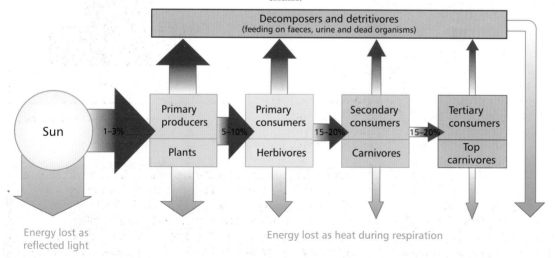

Decomposers and detritivores (feeding on faeces, urine and dead organisms)

Sun — 1–3% — Primary producers / Plants — 5–10% — Primary consumers / Herbivores — 15–20% — Secondary consumers / Carnivores — 15–20% — Tertiary consumers / Top carnivores

Energy lost as reflected light

Energy lost as heat during respiration

Fig 7.4 Energy flow through different trophic levels of a food chain. The arrows are not to scale and give only an idea of the proportion of energy transferred at each stage. Likewise the figures for % energy transfer between trophic levels are only a rough average as they vary considerably between different plants, animals and habitats

It is therefore more energy-efficient for humans to eat food from lower trophic levels, but tastes often prevent this happening. This is illustrated in figure 7.3, where we can see that the eating of small fish, called smelt, would yield five times the biomass of the more popular trout.

Energy flow along food chains is summarised in Fig 7.4. It is possible to construct **ecological pyramids** representing the numbers, biomass or stored energy of organisms at different trophic levels in a food chain.

7.3.2 Pyramids of number

Usually the numbers of organisms at lower trophic levels are greater than the numbers at higher levels. This can be shown by drawing bars with lengths proportional to the numbers present at each trophic level (Fig 7.5). However, figures 7.5 b) and c) indicate that there can be significant drawbacks to using these pyramids to describe a food chain, as the following examples show.

* No account is taken of size – one tree is equated to one aphid and each parasite has the same numerical value as its larger host.
* The number of individuals can be so great that it is impossible to represent them accurately on the same scale as other species in the food chain. For example, one tree may have millions of greenfly living off it.
* No account is taken of juveniles and other immature forms of a species, whose diet and energy requirements may differ from those of the adult.

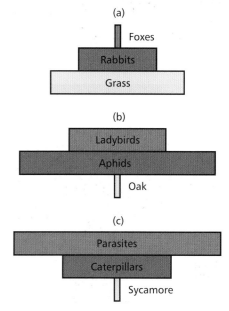

Fig 7.5 *Pyramids of numbers*

7.3.3 Pyramids of biomass

Biomass is the total mass of the plants and / or animals in a particular place. It is normally measured over a fixed period of time. The term is sometimes used to refer to all living organisms on Earth, or a major part of the Earth, such as the oceans. It may also refer to plant or animal material that is exploited as fuel or raw material for industry. A more reliable, quantitative description of a food chain is provided when, instead of counting the organisms at each level, their biomass is measured. The fresh mass is quite easy to assess, but the presence of varying amounts of water makes it unreliable. The use of dry mass measurement overcomes this problem but, because the organisms must be killed, it is usually only made on a small sample, and this sample may not be representative. In both pyramids of numbers and pyramids of biomass, only the organisms present at a particular time are shown; seasonal differences are not apparent. This is particularly significant when the biomass of some marine ecosystems is measured: over the course of a whole year, the mass of **phytoplankton** (plants) must exceed that of **zooplankton** (animals), but at certain times of the year this is not seen. For example, in early spring around the British Isles, the standing crop of zooplankton is greater than that of phytoplankton (Fig 7.6).

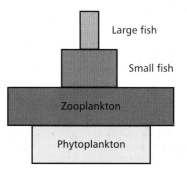

Fig 7.6 *Pyramid of biomass for a marine system*

7.3.4 Pyramids of energy

Collecting the data for pyramids of energy (Fig 7.7) can be difficult and complex, but the result is a true representation of the energy flow through a food web, with no anomalies. Data are collected in a given area (e.g. 1 square metre) for a set period of time, usually a year. The results are much more reliable than those for biomass, because two organisms of the same dry mass may store different amounts of energy. For example, 1 gram of fat stores twice as much energy as 1 gram of carbohydrate. The energy flow in these pyramids is usually shown as $kJ\,m^{-2}\,yr^{-1}$.

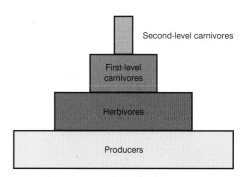

Fig 7.7 *Pyramid of energy, based on oak trees*

SUMMARY TEST 7.3

The Sun is the ultimate source of energy for all living organisms on Earth. Between 1% and **(1)**% of this solar energy is converted to carbohydrate by plants. Of this amount, around **(2)**% is used by plants during respiration. Of the remainder, known as **(3)**, only around 10% is consumed by **(4)**. Carnivores convert around 20% of the energy in their prey, for their own use. This inefficiency in transferring energy from one **(5)** to the next along a food chain is the reason why each food chain is short and the numbers of organisms at each level usually reduce, producing a pyramid of numbers. It is more reliable, however, to use the total mass of organisms, rather than numbers, to give a pyramid of **(6)**. More reliable still is a pyramid of **(7)**.

The nitrogen cycle

Legume with root nodules

The flow of energy through living systems is **linear** (unit 7.3), but the flow of matter is **cyclical**. There is a limited amount of nitrogen available to living organisms. If it is not to run out, it must be used over and over again. Most nutrient cycles have two components:

- **geochemical** – including rocks and deposits in oceans and the atmosphere
- **biological** – including **producers**, **consumers** and **decomposers** that, in some way, help to convert one form of the nutrient into another.

7.4.1 Cycling of nitrogen

All living organisms require a source of nitrogen from which to manufacture proteins, nucleic acids and other nitrogen-containing compounds, such as **ATP** and NAD. Although 78% of the atmosphere is nitrogen, there are very few organisms that can use nitrogen gas directly. This is because the nitrogen molecule has two atoms linked by a triple **covalent bond**, which makes it extremely stable and unreactive. Plants take up most of the nitrogen they need in the form of nitrate ions (NO_3^-), from the soil. These are absorbed, using **active transport**, by the root hairs (section 4.6.3). Animals obtain nitrogen-containing compounds by eating and digesting plants.

Nitrate ions are very soluble, and easily leach through the soil, beyond the reach of plant roots. One way of re-building the nitrate levels is to add fertilisers, but this is also achieved through the natural recycling of nitrogen-containing compounds. When plants and animals die, the process of decomposition begins a series of steps by which microorganisms replenish the nitrate levels in the soil. Within the nitrogen cycle (Fig 7.8), five main stages can be recognised:

- nitrogen fixation
- assimilation
- ammonification
- nitrification
- denitrification.

7.4.2 Nitrogen fixation

This is the process by which nitrogen gas is converted into nitrogen-containing compounds. There are three ways of doing this, all of them requiring energy.

- **Lightning** allows nitrogen and oxygen to combine, producing oxides of nitrogen. These are washed into the soil by rain and absorbed by plant roots in the form of nitrates.
- **Industrial processes**, such as the Haber process, use high temperatures and pressures to combine nitrogen and hydrogen to produce ammonia. Much of this is added to the soil as nitrogen-containing fertilisers.
- **Fixation by microorganisms** is carried out by many bacteria and cyanobacteria living freely in the soil, and by some that live in nodules on the roots of leguminous plants.
  - **Free-living nitrogen-fixers** include bacteria that reduce gaseous nitrogen to ammonia, which they then use to manufacture amino acids. Nitrogen-rich compounds are released from them when they die and decay.
  - **Mutualistic nitrogen-fixers** include *Rhizobium*, which lives in nodules on the roots of plants such as peas and beans. Nodules are swellings on the roots of these plants (see photo) in which *Rhizobium* uses an enzyme called **nitrogenase** to convert nitrogen gas (N_2) into ammonium ions (NH_4^+) using hydrogen ions (H^+) and ATP. The process requires **anaerobic** conditions and so the plant produces a pigment similar to **haemoglobin**, called leghaemoglobin, which absorbs any oxygen in the root nodule. The plant combines the ammonium ions produced with carbohydrate to form amino

acids for making proteins. In return, *Rhizobium* receives energy and a place to live. Because both partners to this relationship benefit, it is called **mutualism**.

7.4.3 Assimilation

In plants which have a mutualistic relationship with *Rhizobium*, some nitrogen is assimilated in the form of ammonium ions from the nodules. All plants, however, can absorb nitrates from the soil via their root hairs (section 4.6.3). These are then reduced to nitrite ions (NO_2^-) and then ammonium ions (NH_4^+) for incorporation into amino acids, and hence proteins. Animals assimilate their nitrogen in the form of protein, which forms part of the plants or animals that they eat.

7.4.4 Ammonification

Ammonification is the production of ammonia from organic ammonium-containing compounds. In nature, these include urea (from the breakdown of excess amino acids) and proteins, nucleic acids and vitamins (found in faeces and dead organisms). Decomposers, mainly fungi and bacteria, feed on these materials, releasing ammonia which forms ammonium compounds in the soil.

7.4.5 Nitrification

The conversion of ammonium ions to nitrates involves oxidation reactions, which release energy for the nitrifying bacteria which carry out this process. This conversion occurs in two stages:

- **oxidation of ammonium ions to nitrites (NO_2^-)** by nitrifying bacteria, such as *Nitrosomonas*, which live freely in well-aerated soil:

$$2NH_3 \quad + \quad 3O_2 \quad \rightarrow \quad 2NO_2^- \quad + \quad 2H^+ \quad + 2H_2O$$
ammonia　　　oxygen　　　nitrite ions　　hydrogen ions　water

- **oxidation of nitrites to nitrates (NO_3^+)** by other free-living nitrifying bacteria, such as *Nitrobacter*.

$$2NO_2^- \quad + \quad O_2 \quad \rightarrow \quad 2NO_3^-$$
nitrite ions　　oxygen　　　nitrate ions

The oxygen requirements of nitrifying bacteria mean that it is important for farmers to keep soil structure light and well aerated by ploughing. Good drainage also prevents the air spaces from being filled with water, which would displace air, and hence oxygen, from the soil.

7.4.6 Denitrification

When soils become waterlogged, and therefore short of oxygen, a different type of microbial flora flourishes. Fewer nitrifying and free nitrogen-fixing bacteria are found, and there is an increase in anaerobic denitrifying bacteria, which reduce soil nitrates into gaseous nitrogen. The stages in reduction are as follows:

$$NO_3^- \quad \rightarrow \quad NO_2^- \quad \rightarrow \quad N_2O \quad \rightarrow \quad N_2$$
nitrate　　　　nitrite　　　dinitrogen oxide　　nitrogen

This reduces the availability of nitrogen-containing compounds for plants.

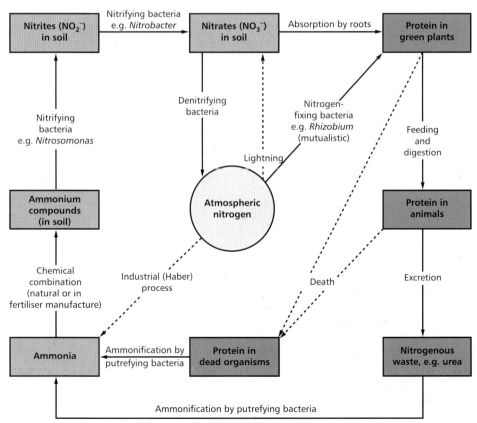

Fig 7.8 *The nitrogen cycle*

1 The figure represents **part** of a food web of a lake.

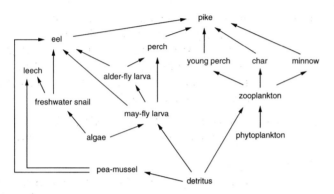

a Name **two** organisms from the food web shown in the figure that feed only as secondary consumers.

(1 mark)

b Explain what is meant by each of the following terms. In each case, give a suitable example from the information provided above.

  (i) *producer* *(2 marks)*
  (ii) *trophic level* *(2 marks)*
  (iii) *ecosystem* *(2 marks)*

c The table shows the efficiency of energy conversion of two trophic levels in the ecosystem.

	percentage conversion efficiency
producer	5.3
primary consumer	14.8

Suggest reasons for these differences in efficiency of energy conversion. *(2 marks)*

(Total 9 marks)

OCR 2801 Jan 2003, B (BF), No.6

2 The figure represents the transfer of energy through a woodland ecosystem.

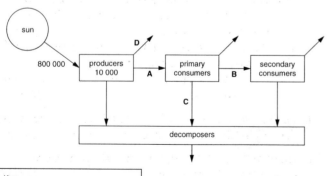

Of the 800 000kJ of energy which strikes the producers, only 10 000kJ of energy is converted by the producers in photosynthesis.

a (i) Calculate the percentage of energy striking the producers that is converted in photosynthesis.

(2 marks)

  (ii) Explain what happens to the energy **striking the producers** that is not converted in photosynthesis. *(2 marks)*

b (i) State the approximate percentage energy transfer between the producers and primary consumers at **A**. *(1 mark)*

  (ii) State **two** ways in which energy is transferred from primary consumers to decomposers at **C**. *(2 marks)*

  (iii) State how energy is lost from the producers at **D**. *(1 mark)*

  (iv) Suggest why the percentage energy transfer between producers and primary consumers at **A** is less than that between the primary consumers and secondary consumers at **B**. *(2 marks)*

(Total 10 marks)

OCR 2801 May 2002, B (BF), No.4

3 The figure represents part of the cycling of nitrogen within an ecosystem.

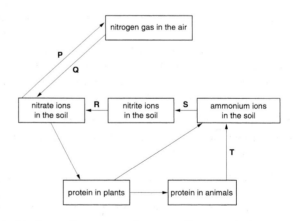

a (i) State the name of process **P**. *(1 mark)*
  (ii) State which type of organism carries out process **P**. *(1 mark)*
  (iii) State a **natural** process represented by **Q**. *(1 mark)*
  (iv) Name the organisms which carry out processes **R** and **S**. *(2 marks)*

b Outline how protein in animals is converted to ammonium ions in the soil, as indicated by **T** on the figure. *(3 marks)* c Complete the following passage by inserting the appropriate term in each space. The first one has been inserted for you.

The grey squirrel lives in *deciduous woodland* (its ..HABITAT..). It feeds on nuts, acorns and toadstools and is, therefore, a primary ... . As it can also eat animal material, such as eggs or young birds, it occupies more than one ... ... (position in the food chain). Surveys have been carried out to establish the *number of squirrels* (the ... of squirrels) in oak woodland on the island of Anglesey, North Wales. It has been discovered that the number of grey squirrels has been increasing dramatically. The grey squirrel was introduced into Britain from North America and has outcompeted the native red squirrel so effectively that the numbers of red squirrels are now very low. As a result of their interaction with *the organisms of the other species in the habitat* (the ...), the stability of the ... (the environment in which they live and with which they interact) has been affected.

(5 marks)
(Total 13 marks)
OCR 2801 Jan 2002, B (BF), No.3

4 Organisms, both plants and animals, are able to exist together in the same environment. They interact, both with members of their own species and with other species.

a State the word or phrase which best describes each of the following.

(i) The place where an organism lives. *(1 mark)*
(ii) A number of **different** species interacting in a particular place. *(1 mark)*
(iii) All members of the **same** species in a particular place. *(1 mark)*
(iv) The ecological role of an organism. *(1 mark)*
(v) The first trophic level in **all** food chains. *(1 mark)*
(vi) A natural unit of living and non-living parts, interacting to produce a stable system. *(1 mark)*

Some farmers spread waste from sewage works on their fields. This waste contains nitrogen compounds such as protein, urea and ammonia.

b (i) Describe how the nitrogen compounds present in the sewage waste are converted naturally into a form which can be taken up by the plants in the fields. *(4 marks)*
(ii) Suggest a possible **disadvantage** of spreading this waste on the fields. *(1 mark)*
(Total 11 marks)
OCR 2801 Jun 2001, B (BF), No.2

5 a Explain the meaning of the terms *producer* and *trophic level*. *(4 marks)*

The table shows the estimated energy content for four trophic levels of a grassland ecosystem.

	energy content / kJ m^{-2}
producers	5600
herbivores	125
omnivores	15
carnivores	10

b (i) Calculate the percentage energy decrease between the producers and herbivores. (Show your working and give your answer to the nearest whole number.) *(2 marks)*
(ii) Explain why the energy content of the herbivores is less than that of the producers. *(4 marks)*

c Suggest **two** factors which might **reduce** the productivity of the producers. *(2 marks)*

d Suggest
(i) why it can be a problem placing omnivores into a trophic level; *(1 mark)*
(ii) a reason for the difference in energy content between the omnivores and the carnivores in this ecosystem. *(1 mark)*
(Total 14 marks)
OCR 2801 Jan 2001, B (BF), No.3

6 The figure shows the flow of energy through the trees in a forest ecosystem. The numbers represent inputs and outputs of energy in kilojoules per m^2 per year.

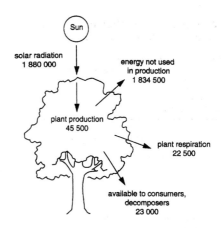

a (i) On the figure, draw a ring around the number which indicates the energy entering the system via photosynthesis. *(1 mark)*
(ii) The total energy available to the plants in the ecosystem is 1 880 000 kJ per m^2 per year. Calculate the efficiency of photosynthesis. Show your working. *(2 marks)*

b Suggest **four** reasons why so much solar energy is **not** used in production in the forest ecosystem. *(4 marks)*

c In what form will energy from plant respiration escape from the ecosystem? *(1 mark)*
(Total 8 marks)
OCR 2801, 2000 Specimen paper, B (BF), No.6

PART TWO

Transport

8

The mammalian transport system

8.1

The need for transport systems in organisms

Giant redwoods need to transport water to a height of 100m without the expenditure of energy

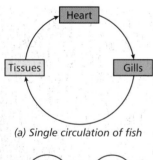

(a) Single circulation of fish

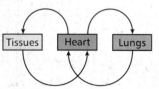

(b) Double circulation of other vertebrates

Fig 8.1 *Single and double circulatory systems*

Transport over short distances, e.g. between adjacent cells, is adequately achieved by processes such as diffusion (unit 4.2), osmosis (unit 4.3) or active transport (unit 4.5). The situation in larger organisms is, however, altogether different.

8.1.1 Why large organisms need a transport system?

All organisms need to exchange materials between themselves and their environment (unit 4.6). In small organisms, this exchange takes place satisfactorily over the surface of the body. With increasing size, however, the surface area to volume ratio decreases to a point where the needs of the organism cannot be met (section 4.6.1). A specialist exchange surface is therefore needed to absorb nutrients and respiratory gases and to remove excretory products. These exchange surfaces are located in specific regions of the organism. A transport system is required to take materials from the cells that need them or produce them to or from the exchange surfaces that absorb or remove them. However, not only do materials have to be transported between these organs and the environment, they also need to be transported between different parts of the organism. As organisms have evolved into larger and more complex structures, the tissues and organs of which they are made have become more specialised and dependent upon one another, making a transport system all the more essential. Whether or not there is a specialised transport medium, and whether or not it is circulated by a pump, depends on two factors – the surface area to volume ratio and how active the organism is. The lower the surface area to volume ratio, and the more active the organism, the greater is the need for a specialised transport system with a pump. Table 8.1 shows the types of transport system in a range of organisms.

Table 8.1 *Transport systems related to surface area to volume ratio and the degree of activity*

Organism	Surface area to volume ratio	Degree of activity	Method of transport
Protoctista, e.g. Amoeba	High	Medium – moves to obtain food	Diffusion
Cnidaria, e.g. sea anemone	Moderate	Low – little or no movement	Diffusion
Arthropoda, e.g. housefly	Low	High – very active flying insect	Blood circulated by pumps (hearts)
Chordata, e.g. trout	Very low	High – very active; swims	Blood pumped by heart in a single circulation
Chordata, e.g. human	Very low	High – very active; walks and runs	Blood pumped by heart in a double circulation
Spermatophyta, e.g. oak tree	High	Low – no movement	Through xylem and phloem without a pump

8.1.2 Features of transport systems

Whether plant or animal, any large organism encounters the same problems in transporting materials within itself. Not surprisingly, therefore, the transport systems of organisms have many common features:

- A suitable medium in which to carry materials, e.g. blood. This is ordinarily a liquid based on water because water readily dissolves substances and can be moved around relatively easily.
- A form of mass flow transport in which the transport medium is moved around in bulk over large distances.
- A closed system of tubular vessels that contains the transport medium and forms a branching network to distribute it, along specific routes, to all parts of the organism.
- A mechanism for moving the transport medium within vessels. This requires a pressure difference between one part of the system and another. It is achieved in two main ways:
  - Animals use muscular contraction either of the body muscles or of a specialised pumping organ such as the heart (unit 9.1).
  - Plants do not possess muscles and so rely on passive natural physical processes such as the evaporation of water or differences in solute concentrations.
- A mechanism to maintain the mass flow movement in one direction, e.g. valves.
- A means of controlling the flow of the transport medium to suit the changing needs of different parts of the organism.

8.1.3 Transport systems in animals

Animals have circulatory systems that move the transport medium, the blood, around the body. Smaller animals, such as insects, use an open blood system where blood flows freely over the cells and tissues. Large organisms have a closed blood system where blood is confined to vessels. A muscular pump called the heart circulates the blood around the body. A greater diversity of substances is transported in animals compared to plants. These include nutrients such as glucose, amino acids, minerals and vitamins, respiratory gases such as oxygen and carbon dioxide, as well as hormones and waste metabolites such as urea. The blood system of animals also carries white blood cells that are involved in providing immunity and protecting against disease. The ability to clot, and so prevent leakage, which results from blood being under pressure, is another feature. Mammals and birds have a double circulatory system (Fig 8.1) in which the blood, having had its pressure reduced as it is forced over the lung capillaries, is returned to the heart to boost its pressure before being circulated to the rest of the body. This assists the rapid delivery of material which is necessary as birds and mammals have a higher rate of metabolism, due to their constant high body temperature.

8.1.4 Transport systems in plants

The range of materials transported by plants is more limited than that moved by animals. Plants, for example, have no need to transport respiratory gases in bulk because:

- Most gases they make or use are produced or required by leaves, which have a large surface area for the capture of light, and these gases can therefore diffuse directly in and out of them through stomata.
- Plants do not move from place to place and their energy requirements are therefore low, which means a reduced need for **ATP** and hence respiratory gases.
- In light, the oxygen that plants require for respiration can mostly be supplied from photosynthesis, while the carbon dioxide they produce during respiration is used up in photosynthesis.
- No photosynthetic or respiratory tissue is far from the surface, and so diffusion can meet the needs of cells. The furthest from the surface is the central part of the trunk of trees, and this is composed of dead, non-respiring tissue.

Plants need mainly to transport water and mineral salts from the roots to aerial parts and to transport the products of photosynthesis – sugars, from the regions of manufacture (sources) to regions where they are utilised or stored (sinks). Plants do not possess contractile cells and therefore depend on passive modes of transport. Evaporation through the stomata in the leaves creates a water potential gradient which draws water across the leaf from the xylem (section 10.6.1). The xylem forms an uninterrupted column of water and so it is moved up this column in much the same way as water can be moved up through a straw (section 10.6.2), i.e. a lower pressure is created at the top so that water moves upwards along a pressure gradient. Water likewise moves across the root due to a **water potential** gradient. Only in passing from the root tissues into the xylem is energy expended (section 10.5.5). The movement of sugars is less clearly understood, but would seem to combine elements of active and passive transport through a different tissue, the phloem (unit 10.3).

The veins of a leaf form a network that allows material to be transported to and from its cells

Arteries, veins and capillaries

To permit both rapid transport of blood and control of its distribution, larger, more active organisms have evolved a **closed circulatory system** in which blood is retained within **vessels**. In mammals, the blood loses pressure as it passes through the capillaries of the lungs. It is therefore returned to the heart to boost its pressure and ensure its rapid circulation to the rest of the body. Blood passes twice through the heart in each complete circuit of the body. This is known as a **double circulatory system**. There are three types of blood vessel in a closed circulatory system:

- **Arteries** carry blood away from the heart. Smaller arteries are called **arterioles**.
- **Veins** carry blood towards the heart. Smaller veins are called **venules**.
- **Capillaries** are smaller vessels that link arteries to veins.

8.2.1 Artery structure related to function

Figure 8.2 illustrates the structure of an artery. While the four main layers are the same for both arteries and veins, the proportions of each differ. These different proportions are related to the different functions of each vessel. Arteries function in transporting blood rapidly, under high pressure from the heart to the tissues. Their structure is related to this function in the following ways:

- **The elastic layer is relatively thick** for two main reasons. Firstly, when blood is forced by the heart into the arteries, it creates a pulse of very high pressure. The elastic wall allows the arteries to expand rather than burst under this pressure. As the pressure gets less when blood passes along the arteries, so the relative thickness of the elastic wall also gets less the further an artery is from the heart. Secondly, it is important that blood pressure in arteries is kept high if blood is to reach the extremities of the body. When the elastic wall is stretched by the pressure within it, it springs back in the same way as a stretched elastic band. This **recoil action** creates another surge of pressure that carries blood forwards in a series of pulses.
- **The collagen fibres provide a tough outer layer** that prevents the artery bursting under the pressure of blood within it.
- **The overall thickness of the wall is large**. This again resists the bursting under pressure.
- **There are no valves** except in the arteries leaving the heart because blood is under constant high pressure and therefore does not tend to flow backwards.

Fig 8.2 *Comparison of arteries, veins and capillaries*

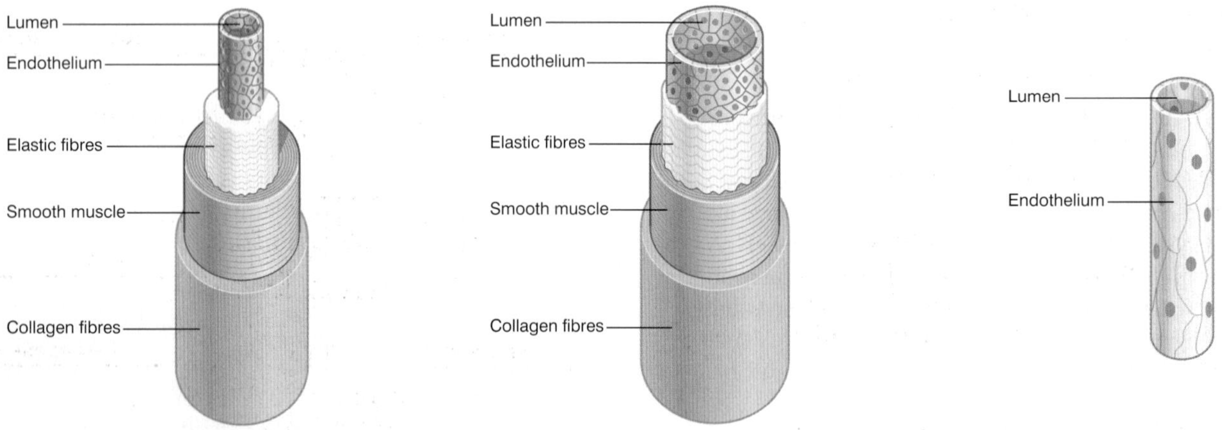

Artery	Vein	Capillary
Thick muscular wall	Thin muscular wall	No muscle
Much elastic tissue	Little elastic tissue	No elastic tissue
Small lumen relative to diameter	Large lumen relative to diameter	Large lumen relative to diameter
Capable of constriction	Not capable of constriction	Not capable of constriction
Not permeable	Not permeable	Permeable
Valves in aorta and pulmonary artery only	Valves throughout all veins	No valves
Transports blood from the heart	Transports blood to heart	Links arteries to veins
Oxygenated blood except in pulmonary artery	Deoxygenated blood except in pulmonary vein	Blood changes from oxygenated to deoxygenated
Blood under high pressure (10–16 kPa)	Blood under low pressure (1kPa)	Blood pressure reducing (4–1kPa)
Blood moves in pulses	No pulses	No pulses
Blood flows rapidly	Blood flows slowly	Blood flow slowing

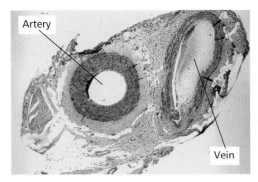

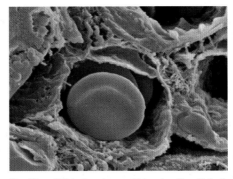

Blood vessels – artery and vein

Capillary SEM showing red blood cells within it

8.2.2 Vein structure related to function

Figure 8.2 illustrates the structure of a vein. Veins function in transporting blood slowly, under low pressure, from the tissues to the heart. Their structure is related to this function in the following ways:

- **The elastic layer is relatively thin** because the low pressure (around 1kPa) of blood within the veins will not cause them to burst and pressure is too low to create a recoil action.
- **The muscular wall is relatively thin** because veins carry blood away from tissues and therefore their constriction and dilation cannot control the flow of blood to the tissues.
- **The collagen fibres provide a tough outer layer** in order to prevent the veins bursting – more from external pressures such as physical forces (they are nearer the skin surface than arteries) than from the blood pressure within them.
- **The overall thickness of the wall is small** because there is no need for a thick wall as the pressure within the veins is too low to create any risk of bursting. It also allows them to be flattened easily, aiding the flow of blood within them (see below).
- **There are semi-lunar valves throughout** to ensure that blood does not flow backwards, which it might otherwise tend to do because the pressure is so low. When **striated muscles** contract, veins are compressed, pressurising the blood within them. The valves ensure that this pressure directs the blood in one direction only – towards the heart.

8.2.3 Capillary structure related to function

Figure 8.2 shows the structure of a capillary. The function of capillaries is to exchange materials like oxygen, carbon dioxide and glucose between the blood and the cells of the body. Their structure is related to their function as follows:

- **Their walls consist only of endothelium** making them extremely thin. This allows for rapid diffusion of materials between the blood and the cells, due to the short distance over which diffusion takes place.
- **They are numerous and highly branched** thus providing a large surface area for diffusion.

- **They have a narrow diameter** and so permeate tissues, which means that no cell is far from a capillary.
- **Their lumen is so narrow** – around 7μm in diameter – that red blood cells are squeezed flat against the side of a capillary. This brings them even closer (as little as 1μm) to the cells to which they supply oxygen. This again reduces the diffusion distance between the air in the alveoli and the red blood cell.
- **There are spaces (fenestrations) between endothelial cells** which allow white blood cells to escape in order to combat infections within tissues. The degree to which material can escape from capillaries varies from tissue to tissue, being greatest in the kidney and least in the brain.

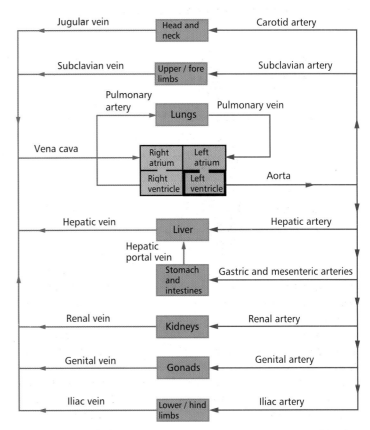

Fig 8.3 Plan of the mammalian circulatory system

Surface view

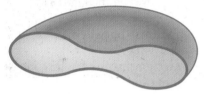

8μm

Transverse section

Fig 8.4 *Red blood cell*

Blood is the medium by which materials are transported between different parts of the body. Humans have between 4dm³ and 6dm³ of blood. It is made up of a liquid – the **plasma** (55%) and two types of cells (45%) – **red cells** and **white cells**.

8.3.1 The plasma

Blood plasma is 90% water and 10% chemicals, which are either dissolved or suspended in it. The function of the plasma is to transport these chemicals, along with heat, from where they are produced or absorbed, to the cells that use or excrete them. These chemicals include:

- **nutrients**, e.g. glucose, amino acids and vitamins
- **waste products**, e.g. urea
- **mineral salts**, e.g. calcium, iron
- **hormones**, e.g. insulin, adrenaline
- **plasma proteins**, e.g. fibrinogen, prothrombin, albumin
- **respiratory gases**, e.g. oxygen, carbon dioxide.

Between them, these chemicals make the plasma slightly alkaline – around pH 7.4. The removal from the plasma of the proteins that are involved in clotting (prothrombin and fibrinogen) results in a liquid called **serum**, which does not clot.

8.3.2 Red blood cells

Red blood cells, or **erythrocytes**, are bi-concave discs, i.e. they are like a doughnut with a hole that does not quite go all the way through (Fig 8.4). Around 7–8μm in diameter, there are 5 million in each mm³ of blood and each lives for around 120 days. This means that, in adults, to maintain their numbers, the bone marrow of certain bones (cranium, sternum, vertebrae and ribs) needs to make over 2 million red blood cells each second in humans. Red blood cells are unusual in having no nucleus, mitochondria, rough endoplasmic reticulum or Golgi apparatus when mature – a feature which, although it gives them a shorter life-span, makes them more efficient in their role of transporting oxygen because, without the nucleus and other organelles:

- They are much thinner in the middle and so form a bi-concave shape which gives them a larger surface area to volume ratio.
- They can more easily change shape, allowing them to be flattened against the capillary walls, thereby reducing the distance across which diffusion takes place, and so speeding up the process (section 4.2.2).
- Without the nucleus and mitochondria, there is more room for the pigment **haemoglobin** which carries oxygen.

It is the pigment **haemoglobin** that gives red blood cells their characteristic colour. The structure of haemoglobin is illustrated in section 2.8.2 and its role in oxygen transport is described in unit 8.6.

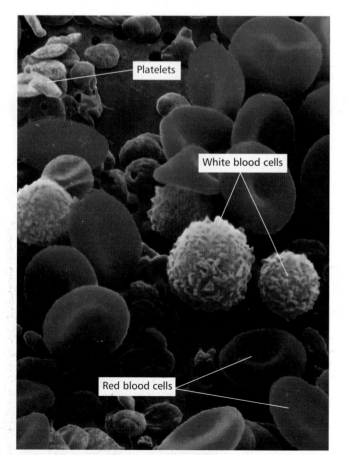

Platelets

White blood cells

Red blood cells

SEM of blood showing red cells (red), white cells (yellow) and platelets (pink)

8.3.3 White blood cells

White blood cells, or **leucocytes**, exist in a variety of forms, as shown in figure 8.5. They all contain a nucleus and are either spherical or irregular in shape. Most are larger than red blood cells. Unlike red blood cells, leucocytes can pass out of blood vessels into the fluid that surrounds the cells of tissues. Made in the marrow of the limb bones, the white blood cells function to protect the body against infection. They can be divided into two groups, based upon how they carry out this function.

- **Phagocytes**, such as neutrophils and monocytes, remove microorganisms, other foreign material and dead cells by the process of phagocytosis (section 16.1.2). Phagocytosis is non-specific and occurs whatever the infection.
- **Lymphocytes** – act against microorganisms, with some lymphocytes secreting **antibodies** – a range of **glycoproteins** that between them immobilise and kill microorganisms and make them ready for phagocytes to engulf. More about this process can be found in section 16.2.2. Each type of lymphocyte acts against one particular pathogen, i.e. they are specific. They can provide long-term immunity to future infections.

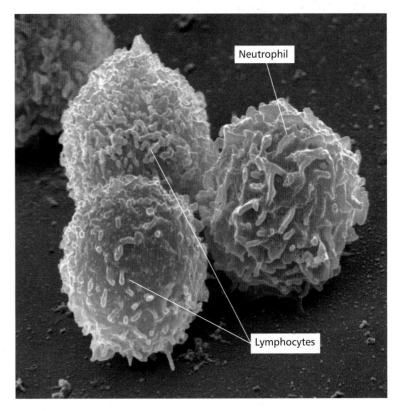

SEM of white blood cells

PHAGOCYTES	LYMPHOCYTES
Engulf bacteria by phagocytosis	Produce glycoproteins such as antibodies that kill/immobilise microorganisms and prepare them for attack by phagocytes

MONOCYTE

Indented (kidney-shaped) nucleus

Non-granular cytoplasm

NEUTROPHIL

Multilobed nucleus

Granular cytoplasm

Compact nucleus

Non-granular cytoplasm

Fig 8.5 *Three of the many types of white blood cell*

SUMMARY TEST 8.3

Blood is made up of a watery liquid called the **(1)**, in which lie a variety of cells. The most numerous are the **(2)**, which are bi-concave discs about **(3)** in diameter. They live for about **(4)** and contain a red pigment called **(5)** that carries **(6)**. The second type of cells are the **(7)**, which have a variety of forms. Those that engulf bacteria are called **(8)**, an example of which is a cell type called **(9)**. Those that secrete chemicals called **(10)** are known as **(11)**.

Tissue fluid and lymph

Blood supplies nutrients to the tissues of the body via tiny vessels called capillaries. Small though they are, these cannot serve every single cell directly, and therefore the final stage of the nutrient's journey is made in solution in a liquid that bathes the tissues – **tissue fluid**.

8.4.1 Tissue fluid

Tissue fluid is formed from the plasma of the blood. It is a watery liquid which contains glucose, amino acids, fatty acids, salts and oxygen, all of which it supplies to the tissues. In return, it receives carbon dioxide and other waste materials from the tissues. Tissue fluid is therefore the means by which materials are exchanged between blood and cells and, as such, it bathes all cells of the body.

8.4.2 Formation of tissue fluid

Blood pumped by the heart passes along arteries, then the narrower arterioles and, finally, the even narrower capillaries. This creates a pressure, called **hydrostatic pressure**, of around 4.8kPa at the arterial end of the capillaries, which tends to force liquid out of the blood. The outward pressure is, however, opposed by two other forces:

- hydrostatic pressure of the tissue fluid outside the capillaries, which prevents outward movement of liquid
- **osmotic** forces due to the plasma proteins, which tend to pull water back into the capillaries.

The combined effect of all these forces is to create an overall pressure of 1.7kPa, which pushes tissue fluid out of the capillaries. This pressure is only enough to force small molecules out of the capillaries, leaving all cells and proteins in the blood. This type of filtration under pressure is called **ultrafiltration**. The loss of the tissue fluid reduces the pressure in the capillaries and so, by the time the blood has reached the venous end of the network, its hydrostatic pressure is less than that of the tissue fluid outside it. Along with the osmotic forces due to the proteins in the blood plasma that pull water back into the capillary, there is also an overall negative pressure of 1.5kPa drawing tissue fluid back into the capillaries. This fluid has lost much of its oxygen and nutrients by diffusion to the cells it bathed, but has gained carbon dioxide and excretory products in return. These events are summarised in figure 8.7. Not all the tissue fluid can return to the capillaries; the remainder is carried back via the lymphatic system.

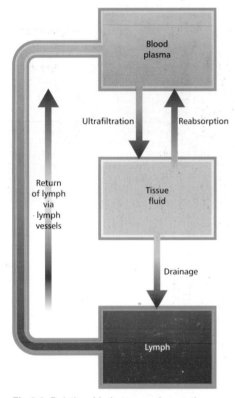

Fig 8.6 *Relationship between plasma, tissue fluid and lymph*

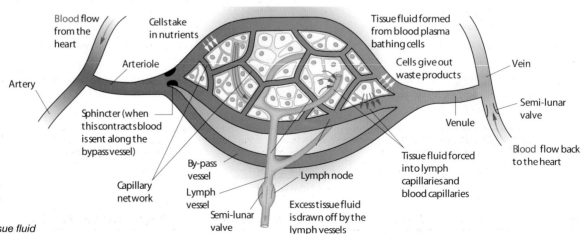

Fig 8.7 *Formation of tissue fluid*

8.4.3 Lymph and the lymphatic system

Lymph is a milky liquid made up of material from three sources:

- **tissue fluid** that has not been reabsorbed at the venous end of the capillary network
- **fatty substances** absorbed by lacteals in the ileum
- **lymphocytes** (section 8.3.3) which have either been produced by the lymph nodes or have migrated from capillaries to fight infection.

Lymph is carried in the **lymphatic system**, which is made up of capillaries resembling blood capillaries, which merge into larger vessels that form a network around the body (Fig 8.8). The lymph vessels drain their contents back into the blood stream via two ducts:

- **the right lymphatic duct** drains the thorax, right side of the head and right arm into the right subclavian vein near the heart
- **the thoracic duct** draws the rest of the body's lymph into the left subclavian vein.

At points along the lymph vessels are a series of **lymph nodes**, which produce and store lymphocytes. Lymph nodes filter from the blood any bacteria and other foreign material, which are then engulfed by lymphocytes. This causes the nodes to swell with dead cells, and is the reason for the tenderness often felt in the groin, armpits and neck during an infection.

Lymph is moved along lymph vessels in three ways:

- **hydrostatic pressure** of the tissue fluid leaving the capillaries
- **contraction of body muscles** squeezes the lymph vessels. Valves ensure that the fluid inside them moves away from the tissues in the direction of the heart
- **enlargement of the thorax during breathing in** reduces pressure in the thorax, drawing lymph into this region and away from the tissues.

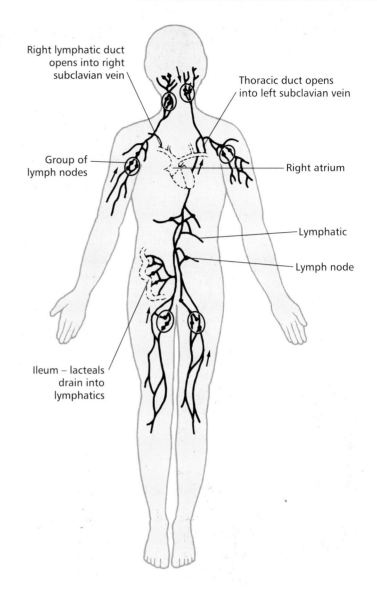

Right lymphatic duct opens into right subclavian vein

Thoracic duct opens into left subclavian vein

Group of lymph nodes

Right atrium

Lymphatic

Lymph node

Ileum – lacteals drain into lymphatics

Fig 8.8 Human lymphatic system

Table 8.2 Comparison of various body fluids

	Blood	Plasma	Tissue fluid	Lymph
Cells	Erythrocytes, leucocytes and **platelets** (cell debris)	None	None	Lymphocytes
Location	Within heart, arteries, veins and capillaries	Within heart, arteries, veins and capillaries	Outside vessels, bathing cells	Within lymph vessels
Moved by	Pumping of heart, muscle contraction and breathing action	Pumping of heart, muscle contraction and breathing action	Hydrostatic and osmotic forces	Hydrostatic forces, muscle contraction and breathing action
Direction of flow	Circulates around the body – to and from heart	Circulates around the body – to and from heart	Out of arterial end and into venous end of capillaries	Towards the heart from the tissues
Derived from	Bone marrow, thymus gland and lymph nodes	Water and substances absorbed by alimentary canal	Plasma	Tissue fluid, fatty material absorbed by ileum and lymph nodes
Function	Transport and defence	Transport over long distances	Transport over short distances	Transport and defence

Obtaining oxygen – gaseous exchange in the alveoli

The sites of gaseous exchange in mammals are the alveoli, minute air-sacs some 100–300μm in diameter and situated in the lungs. The lungs are paired organs found in the thorax. More details of their structure and the method of moving air in and out of them are given in unit 4.7.

8.5.1 Role of the alveoli in gaseous exchange

The 300 million alveoli in each lung of a human have a total surface area of around 70m². Their structure is shown in figures 8.9 and 8.10. Each alveolus is lined mostly with squamous epithelial cells only 0.1–0.5μm thick; their structure is described in section 1.9.2. Around each alveolus is a network of pulmonary capillaries, so narrow (7–10μm) that red blood cells are flattened against the thin capillary walls in order to squeeze through. These capillaries have walls comprising only a single layer of endothelial cells. A review of Fick's Law (section 4.2.2) tells us that diffusion of gases between the alveoli and the blood will be very rapid because:

- the red blood cells are slowed as they pass through the pulmonary capillaries, allowing more time for diffusion
- the distance between alveolar air and red blood cells is reduced as the red blood cells are flattened against the capillary walls
- the walls of both alveoli and the capillaries are very thin and therefore the distance over which diffusion takes place is very short
- the alveoli and blood capillaries have a very large total surface area
- breathing movements constantly ventilate the lungs, and the action of the heart constantly circulates blood around the alveoli. Together, these ensure that the steep concentration gradient of the gases to be exchanged is maintained.

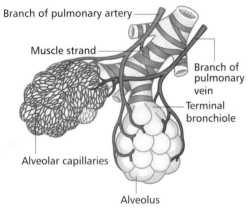

Branch of pulmonary artery

Muscle strand

Branch of pulmonary vein

Terminal bronchiole

Alveolar capillaries

Alveolus

Fig 8.9 *Alveoli*

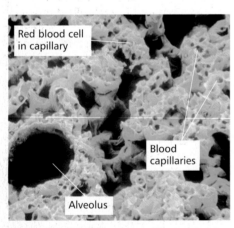

Red blood cell in capillary

Blood capillaries

Alveolus

False-colour SEM of a section of human lung tissue showing alveoli (large dark areas) surrounded by capillaries (small dark areas). Red blood cells can be seen in several of the capillaries

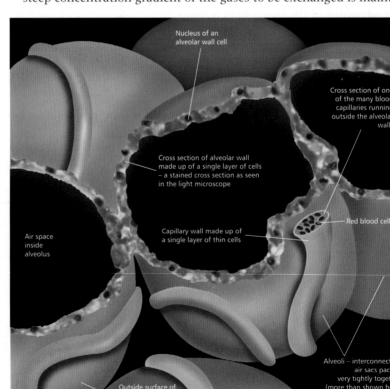

Nucleus of an alveolar wall cell

Cross section of one of the many blood capillaries running outside the alveolar walls

Cross section of alveolar wall made up of a single layer of cells – a stained cross section as seen in the light microscope

Red blood cell

Air space inside alveolus

Capillary wall made up of a single layer of thin cells

Outside surface of alveolar wall – the plasma membrane of alveolar cells

Alveoli – interconnecting air sacs packed very tightly together (more than shown here)

Fig 8.10 *External appearance of a group of alveoli (×300 approx.)*

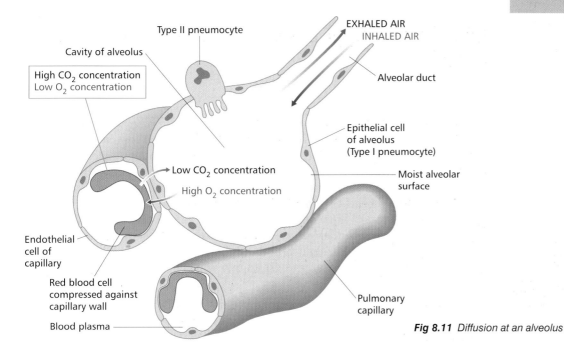

Fig 8.11 *Diffusion at an alveolus*

8.5.2 Lung surfactants and their function

The walls of the alveoli are made up of two types of cell (Fig 8.12):

- **type I pneumocytes** – these are the squamous epithelial cells referred to in section 8.5.1.
- **type II pneumocytes** – a special type of epithelial cell which produces a chemical called a surfactant.

Surfactants are chemicals that reduce the surface tension of substances, most commonly water. They are best known for their role in detergents. As you are probably aware, detergents such as washing-up liquid make things less sticky and more slippery. The surfactants in the lungs perform exactly the same function, only in this case it is to prevent the surfaces of the alveoli sticking to each other, rather than preventing grease sticking to a plate. Without the phospholipid-protein surfactants of the lung, the moist alveolar surfaces would stick together, making it difficult, if not impossible, to inflate the lungs.

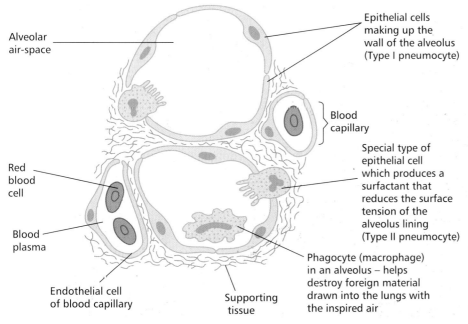

Fig 8.12 *Arrangement of cells and tissues in human alveoli*

SUMMARY TEST 8.5

Gaseous exchange occurs in the alveoli, with the gas called **(1)** moving into the blood and the gas called **(2)** moving in the opposite direction. The diameter of an alveolus is **(3)** and it is surrounded by squamous epithelial cells that are only **(4)** thick and so allow rapid **(5)** of gases across them. Each alveolus is surrounded by a network of **(6)** that are around **(7)** in diameter, causing **(8)** within them to be flattened against their surface, thus improving the rate of exchange of gases between themselves and the alveoli. The alveoli also contain cells called **(9)** which produce a chemical, **(10)**, that reduces the surface tension of their walls and therefore makes them less **(11)**. Within the alveoli can be found **(12)** cells that help to destroy foreign material brought into the lungs during breathing.

119

The amount of a gas, such as oxygen, that is present in a mixture of gases is measured by the pressure it contributes to the total pressure of the gas mixture. This is known as the **partial pressure** of the gas and, in the case of oxygen, is written as pO_2. It is also known as **oxygen tension** and is measured in the usual unit for pressure, namely **kilopascals (kPa)**. Normal atmospheric pressure is 100kPa. As oxygen makes up 21% of the atmosphere, its partial pressure is normally 21kPa.

USE OF ABBREVIATION – Hb

Hb is commonly used as an abbreviation for haemoglobin. However, in the context of oxygen transport, it is used to represent a single haem group. As a haemoglobin molecule has four haem groups, each of which carries an oxygen molecule (O_2), oxyhaemoglobin is shown as Hb_4O_8.

As organisms evolved, they became more complex and, in many cases, much larger. Metabolic rates increased and with them the demand for oxygen increased. Specialised gaseous exchanges surface, such as lungs, developed to meet this need. This solved the problem only if there was a mechanism to transport the oxygen from these surfaces to the cells requiring it. Even with blood vessels and a heart to pump the blood around them, the transport of oxygen would be totally inadequate if the gas were simply dissolved in the plasma. Only the evolution of specialised molecules capable of carrying large quantities of oxygen could adequately supply the tissues. These molecules are called **respiratory pigments**, the best known of which is **haemoglobin**.

8.6.1 Haemoglobin

Haemoglobin is a red pigment that has a relative molecular mass of 68 000. As such, it could be lost from the body during ultrafiltration in the kidneys. It is therefore contained within the red blood cells that carry it around the body, separated from the plasma. Its structure is described in section 2.8.2. One oxygen molecule can combine with each of its four haem groups to form oxyhaemoglobin.

$$4Hb \quad + \quad 4O_2 \quad \rightleftharpoons \quad Hb_4O_8$$
$$\text{haemoglobin} \qquad \text{oxygen} \qquad\qquad \text{oxyhaemoglobin}$$

To be efficient, haemoglobin must:

- readily associate with oxygen at the gaseous exchange surface
- readily dissociate from oxygen at those tissues requiring it.

These two requirements may appear contradictory, but are achieved by the remarkable property of haemoglobin changing its affinity for oxygen under different conditions (Table 8.3).

Table 8.3 Affinity of haemoglobin for oxygen under different conditions

Region of body	Oxygen tension (concentration)	Carbon dioxide tension (concentration)	Affinity of haemoglobin for oxygen	Result
Gaseous exchange surface	High	Low	High	Oxygen is absorbed
Respiring tissues	Low	High	Low	Oxygen is released

Haemoglobin has an affinity for carbon monoxide some 250 times greater than for oxygen – a feature which makes this gas potentially lethal, especially because, once it is attached to haemoglobin to form carboxyhaemoglobin, the carbon monoxide molecule remains permanently, and so prevents oxygen molecules being loaded. As carbon monoxide is found in both vehicle exhaust fumes and cigarette smoke, both reduce the oxygen-carrying capacity of the blood.

8.6.2 Oxygen dissociation curves

When haemoglobin is exposed to different partial pressures of oxygen (see box in margin), it does not absorb the oxygen evenly. At very low concentrations of oxygen, the four polypeptides of the haemoglobin molecule are closely united,

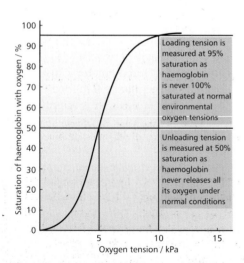

Fig 8.13 Oxygen dissociation curve for adult human haemoglobin

and so it is difficult to absorb the first oxygen molecule. However, once loaded, this oxygen molecule causes the polypeptides to bind more loosely, and so the remaining three oxygen molecules are loaded very easily. The graph of this relationship is known as the **oxygen dissociation curve**; it is illustrated in figure 8.13. You will notice from this that a very small decrease in the partial pressure of oxygen leads to a lot of oxygen becoming dissociated from haemoglobin. The graph tails off at very high oxygen concentrations simply because the haemoglobin is almost saturated with oxygen. There are many different oxygen dissociation curves because:

- there are a number of different respiratory pigments
- haemoglobin exists in a number of different forms
- the characteristics of each pigment change under different conditions.

The many different oxygen dissociation curves are better understood if two facts are always kept in mind:

- the more to the left the curve is, the more readily the pigment associates with oxygen but the less easily it dissociates from it
- the more to the right the curve is, the less readily the pigment associates with oxygen but the more easily it dissociates from it.

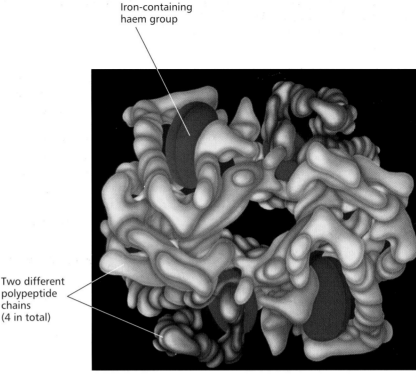

Iron-containing haem group

Two different polypeptide chains (4 in total)

Computer graphic representation of a haemoglobin molecule showing two pairs of polypeptide chains (orange and blue) associated with the haem groups (red)

EXTENSION – MYOGLOBIN

Myoglobin is a different respiratory pigment whose main role is to store oxygen, rather than transport it. Found in the muscles of all vertebrates, it is used to provide an emergency store of oxygen at times when the rate at which muscles are using oxygen exceeds the rate of supply by the blood. Myoglobin has a higher affinity for oxygen than haemoglobin and so its dissociation curve is to the left of that of haemoglobin (Fig 8.14). This ensures that oxygen for muscle action is taken from haemoglobin rather than from myoglobin – the latter is used only when the haemoglobin supply is exhausted. It also ensures that myoglobin is rapidly reloaded with oxygen after exercise has ended, when supply again exceeds demand. The muscles of diving mammals, such as whales and seals, contain a lot of myoglobin. This acts as a store of oxygen to sustain them during long periods of submersion.

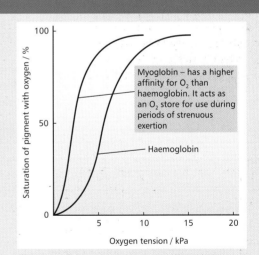

Myoglobin – has a higher affinity for O_2 than haemoglobin. It acts as an O_2 store for use during periods of strenuous exertion

Haemoglobin

Fig 8.14 *Comparison of oxygen dissociation curves of human haemoglobin and myoglobin*

SUMMARY TEST 8.6

Haemoglobin in humans is an example of a **(1)**. One molecule has a relative molecular mass of **(2)** and possesses a total of **(3)** haem groups, each of which possesses a single atom of **(4)**. The 'globin' part of the molecule comprises four **(5)** that are of two different types: **(6)** and **(7)**. Each haemoglobin molecule can carry a total of **(8)** molecules of oxygen. The amount of oxygen present in a mixture of gases is known as the oxygen **(9)** and the graph of the relationship between this and the amount of oxygen taken up by haemoglobin is called the **(10)**.

Table 8.4 *Changes in atmospheric pressure with altitude*

Altitude / m	Atmospheric pressure / kPa	Partial pressure of oxygen / kPa
0	100	21
3500	65.5	13.8
8500	31.9	6.7

Low atmospheric pressure at high altitude makes it difficult to load haemoglobin with oxygen. Climbers frequently breathe oxygen from cylinders to compensate

In unit 8.6, we saw how oxygen is transported under normal circumstances. There are, however, some particular circumstances in which adaptations have to be made to ensure efficient transport of oxygen. Two examples are high altitude and during fetal development.

8.7.1 Oxygen transport at high altitude

Most people are aware of the problems associated with high altitude and, in particular, the need to use oxygen when climbing in areas such as the Himalayas. This results in the popular misconception that there is less oxygen at high altitude. In fact, the proportion of oxygen in the atmosphere is the same (21%) everywhere in the world. What changes at high altitude is the air pressure, which decreases as we ascend from sea-level. Put simply, the higher we go, the less air there is above us weighing down on our bodies. As a result, the air pressure on the top of Mount Everest (altitude 8848m) is less than one-third that at sea-level (see table 8.4). Such low pressures make breathing, and the loading of **haemoglobin** with oxygen, difficult.

Sudden exposure to the reduced pressure at high altitude could result in death within 10 minutes. This is because, if haemoglobin cannot be loaded with oxygen, the level of oxygen in the blood and tissues falls, resulting in a condition called **hypoxia**. However, if ascent is slow, i.e. over many days or weeks, the body can adjust by trying to maximise, in other ways, the amount of oxygen delivered to the tissues. This process is called **acclimatisation**.

One main way of acclimatising is for the number of red blood cells in circulation to increase. This is achieved in the following ways:

- After a few days at high altitude, water is absorbed from the circulation, concentrating the red blood cells and thickening the blood.
- After 1–2 weeks, the kidneys increase the production of a hormone called **erythropoietin**, which stimulates the formation and release of more red blood cells from the bone marrow.

The result of these changes is that, after 2–3 weeks at high altitude, the proportion of the total blood that is made up of red blood cells increases from an average of 45% to an average of 60%.

EXTENSION – OTHER CHANGES DURING ACCLIMATISATION

Apart from an increase in red blood cell count other changes that maximise the amount of oxygen delivered to the tissues are:

- **an increase in cardiac output** in the first few days, i.e. the heart beats more rapidly and pumps more blood at each beat
- **hyperventilation** – increasing the rate and depth of breathing, so delivering more air, and therefore oxygen, to the alveoli. Until acclimatisation is complete, this creates a conflict, because hyperventilation also removes larger than normal volumes of carbon dioxide. This results in reduced ventilation and so the overall pattern of breathing can be irregular
- **an increase in the rate of exchange between alveoli and lung capillaries** by increasing the lung volume and the flow of blood through the capillaries

- **an increase in the concentration of haemoglobin in each red blood cell** by as much as 20%
- **an increase in the level of myoglobin (unit 8.6) in the muscles**.

Native highlander adaptations

There is little evidence that people who spend all their lives at high altitude are genetically adapted to do so. Although they may have a greater lung volume, and consequently be more barrel-chested than their lowland counterparts, in other respects they are merely better acclimatised. Both natives and visitors will show some **hyperventilation**, and have a higher red blood cell count and greater concentration of haemoglobin. For these last two reasons, athletes find it advantageous to train at high altitude. These features are lost in both groups if they move to sea-level.

EXTENSION – ACUTE MOUNTAIN SICKNESS

More than half the people trekking to 5000m will experience the altitude-related illness called **acute mountain sickness (AMS)**. Its many symptoms include:

- headaches and dizziness
- nausea and vomiting
- insomnia
- general lethargy
- dry, irritating cough
- breathlessness.

If ignored, these symptoms may worsen, leading to the following:

- **Mental impairment** – at moderate altitude, many people are shown to have slower reactions than at sea-level. As they climb higher, they tend to lack concentration and they will find it difficult to calculate and make judgements. Still higher and they may suffer hallucinations and have a dangerous sense of well-being.

- **Redistribution of body fluids** – severe hypoxia leads to an increased production of ADH (antidiuretic hormone), which reduces urine output and causes more water to remain in the blood. In cold weather at altitude, because the blood flow to the extremities is reduced, extra fluid accumulates outside blood vessels, causing swelling, or **oedema**. This may become apparent in swelling of the feet and legs, as well as the face, especially around the eyes. More seriously, fluid may accumulate in the lungs (**pulmonary oedema**), causing breathlessness and frothing at the mouth. An accumulation of fluid in the brain (**cerebral oedema**) causes it to swell and push against the cranium. Severe headaches may be followed by unconsciousness, and even death, if a return to low altitude does not occur soon enough.

8.7.2 Oxygen transport in the fetus

Even within a single species, haemoglobin can exist in more than one form at different stages of life. Consider the situation in mammals, when the fetus is being carried by the mother. At the placenta, the two blood systems are very close together, although the blood within them does not actually mix. If the mother and fetus had the same haemoglobin, then, under the same conditions, there would be no reason for the mother's haemoglobin to give up its oxygen to the fetal haemoglobin. In practice, fetal mammals produce their red blood cells in the liver, and the haemoglobin in these cells has two β polypeptide chains that differ from those of the adult. This gives fetal haemoglobin a higher affinity for oxygen than maternal haemoglobin (Fig 8.15). Oxygen therefore more readily transfers from the mother to the fetus at the placenta because fetal haemoglobin (HbF) associates with oxygen at partial pressures of oxygen that cause maternal haemoglobin to dissociate from it. After birth, red blood cell production moves from the liver to the marrow of bones such as the ribs. These produce the adult haemoglobin (HbA), which has a lower affinity for oxygen than fetal haemoglobin. Six months after birth almost all haemoglobin is of the adult type (HbA).

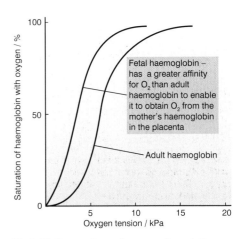

Fig 8.15 *Comparison of oxygen dissociation curves of fetal and adult haemoglobin*

SUMMARY TEST 8.7

The amount of oxygen at high altitude is **(1)**% of the total atmosphere. Its partial pressure is lower at high altitude, being only about the fraction **(2)** at 8500m compared with its partial pressure at sea-level. Mountaineers therefore need to **(3)** slowly to the conditions if they are to avoid a drop in blood oxygen levels, known as **(4)**. To maintain a high oxygen level, after a few weeks the kidneys produce the hormone **(5)**, which results in an increase in **(6)**. Haemoglobin in a fetus has a **(7)** affinity for oxygen and it is produced in the **(8)** rather than the **(9)**, as it is in an adult.

The carbon dioxide produced by respiratory tissues must be carried back to the gaseous exchange surface for removal from the body because its accumulation is harmful. This carbon dioxide is, however, essential to the efficient release of oxygen from **haemoglobin** to the tissues. This is the **Bohr effect**, named after the person who discovered it in 1904, Christian Bohr.

8.8.1 The Bohr effect

Haemoglobin has a reduced affinity for oxygen in the presence of carbon dioxide. The greater the concentration of carbon dioxide, the more readily it releases its oxygen. This is the Bohr effect, and explains the differing behaviour of haemoglobin in different regions of the body.

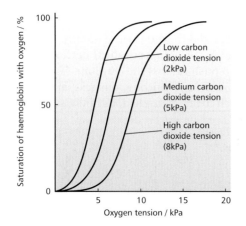

Fig 8.16 *The Bohr effect*

- At the gaseous exchange surface (e.g. lungs), the level of carbon dioxide is low because it diffuses across the exchange surface and is expelled from the organism. Haemoglobin's affinity for oxygen is increased, which, coupled with the high concentration of oxygen in the lungs, means that oxygen is readily absorbed by haemoglobin. The reduced carbon dioxide level has shifted the oxygen dissociation curve to the left (Fig 8.16).
- In the respiratory tissues (e.g. muscles), the level of carbon dioxide is high in the blood because of its production during respiration. Haemoglobin's affinity for oxygen is reduced, which, coupled with the low concentration of oxygen in the muscles, means that oxygen is readily released from the haemoglobin to the muscle cells. The increased carbon dioxide level has shifted the oxygen dissociation curve to the right (Fig 8.16). This is especially important during exercise because the more carbon dioxide that is produced, the more readily oxygen is supplied from the haemoglobin to meet extra energy demands due to exercise.

The Bohr effect is a consequence of the acidic nature of dissolved carbon dioxide: it forms hydrogen ions and hydrogencarbonate ions. It is the hydrogen ions that lower the pH and the affinity of haemoglobin for oxygen. Low pH caused by other chemicals, e.g. lactic acid, therefore also reduces haemoglobin's affinity for oxygen in the same way.

8.8.2 Transport of carbon dioxide

Carbon dioxide is carried from the tissues to the gaseous exchange surface in three ways:

- **In solution in the plasma** – although carbon dioxide is more soluble in the aqueous plasma than oxygen, only 5% of the total carbon dioxide is carried in this way.
- **In combination with haemoglobin** – carbon dioxide can combine with amino groups in the protein part of the haemoglobin molecule:

haemoglobin   carbon dioxide   carbamino-haemoglobin   hydrogen ions

About 10% of the total carbon dioxide is carried in this way.

- **As hydrogencarbonate ions** – the remaining 85% of the total carbon dioxide is transported in this form. The carbon dioxide combines with water to form carbonic acid, which then dissociates (splits) into hydrogen ions (H^+) and hydrogencarbonate ions (HCO_3^-). The reaction is catalysed by the enzyme carbonic anhydrase and is summarised as:

$$H_2O \ + \ CO_2 \xrightarrow[\text{anhydrase}]{\text{carbonic}} H_2CO_3 \longrightarrow H^+ \ + \ HCO_3^-$$

| water | carbon dioxide | carbonic acid | hydrogen ion | hydrogencarbonate ion |

This reaction takes place in red blood cells. The hydrogen ions produced combine with haemoglobin to form **haemoglobinic acid** and so cause it to release its oxygen, which diffuses out of the cell into the nearby respiratory tissue. In this way, haemoglobin acts as a buffer, helping to keep the pH of the blood around 7.4.

The loss of the negatively charged hydrogencarbonate ions would upset the ionic balance of the red blood cells if it were not for the movement of chloride ions from the plasma into red blood cells. This is known as the **chloride shift** (Fig 8.17).

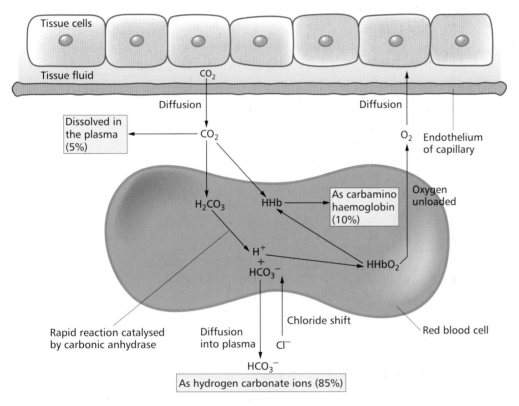

Fig 8.17 *Transport of carbon dioxide*

SUMMARY TEST 8.8

Carbon dioxide produced by tissues during the process of **(1)** is transported to the gaseous exchange surface in three ways. Firstly, it combines with haemoglobin to form **(2)** and **(3)** ions; this accounts for about **(4)**% of the total carbon dioxide carried. Secondly, around **(5)**% is transported in solution in the **(6)**. The remaining **(7)**% is carried in the form of **(8)**, which are formed from the dissociation of **(9)** by the enzyme **(10)**. One product of this reaction is hydrogen ions, which then combine with haemoglobin to form **(11)**, which acts as a **(12)** by helping to keep the blood pH neutral. The ionic balance of the red blood cells could be upset by this reaction if it were not maintained through a process called the **(13)**. The affinity of haemoglobin for oxygen is reduced in the presence of carbon dioxide. This change is known as the **(14)**.

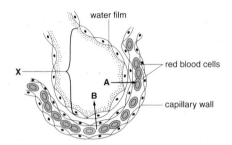

1 The following figure shows the structures involved in gaseous exchange in mammals.

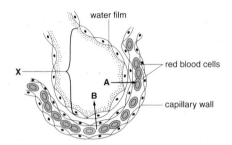

water film
red blood cells
X
A
B
capillary wall

a (i) Name the structure **X**. *(1 mark)*
  (ii) Name the process by which gases cross the gas exchange surface. *(1 mark)*
  (iii) **A** and **B**, on the figure, show the directions taken by gases crossing the gas exchange surface. Name the gases moving in directions **A** and **B**. *(2 marks)*

b Structure **X**, shown in the above figure, is very delicate. The surface tension of the water film shown in the diagram would cause **X** to collapse if it were not for a substance in the film that lowers the surface tension. This substance is secreted in the fetal lungs late in pregnancy. Premature babies sometimes display breathing difficulties, a condition known as Respiratory Distress Syndrome. Without treatment, premature babies may become exhausted.

Suggest the cause of this condition and suggest why premature babies may become exhausted. *(3 marks)*
(Total 6 marks)
OCR 2803/01 Jun 2003, B (T), No.5

2 The table compares the red blood cell count of a group of people when they were living at sea level and after they had spent several weeks at an altitude of 5000m.

altitude / m	number of red blood cells / 10^{12} dm^{-3}
0	4.90
5000	6.10

a Calculate the % increase in red blood cells after spending several weeks at high altitude. Show your working. *(2 marks)*

A company advertises a programme to athletes of living and training at altitude to improve their performance.

b Explain why the performance of an athlete at altitude would be expected to improve as a result of such training. *(3 marks)*

The figure shows the effect of different partial pressures of carbon dioxide on the dissociation curve for haemoglobin.

c With reference to the figure;
  (i) name this effect; *(1 mark)*
  (ii) calculate the difference in % oxygen saturation between the two partial pressures of carbon dioxide at a partial pressure of oxygen of 5 kPa; *(1 mark)*
  (iii) outline how this effect ensures more efficient delivery of oxygen to the tissues when exercising. *(3 marks)*

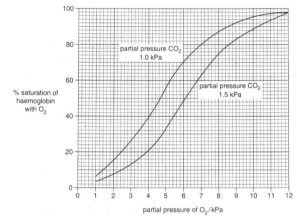

(Total 10 marks)
OCR 2803/1 Jun 2001, B (T), No.5

3 Figure 1 shows red blood cells (erythrocytes) in cross section and surface view.

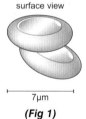

cross section
surface view
7μm
(Fig 1)

a Explain **three** ways in which the structure of a red blood cell is adapted to its function. *(3 marks)*

Figure 2 shows the dissociation curves for fetal and maternal haemoglobin in humans. The shape of the curves is described as sigmoid (S-shaped).

b Explain the advantage, in terms of oxygen supply to the tissues, of the fact that the **maternal curve** is sigmoid. *(3 marks)*

c State the difference in percentage saturation of haemoglobin with oxygen between maternal and fetal blood at an oxygen partial pressure of 4 kPa. *(1 mark)*

d Explain why it is essential for the survival of the fetus that the fetal curve is to the left of the maternal curve. *(3 marks)*

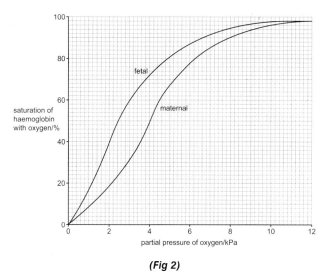

(Fig 2)

e Describe the structure of arteries and explain how their structure is related to their function.
(In this question, 1 mark is available for the quality of written communication.)

(6 marks)
QWC (1 mark)
(Total 17 marks)
OCR 2803/01 May 2002, B (T), No.2

4 Insert the most appropriate word or words in the paragraph below on mammalian gas exchange.

The surface area of the lungs is greatly increased by large numbers of structures called … which are the sites of gas exchange. The process of exchange is entirely by … and this is made more effective by the fact that the walls of these structures consist of … . Their surfaces are moist and once oxygen has entered the capillaries it passes into the … via the plasma and is rapidly transported away from the site of exchange, thus maintaining the … needed for efficient exchange.

(Total 5 marks)
OCR 2803/1 Jan 2002, B (T), No.3

5 Figure 1 shows the structure of a capillary.

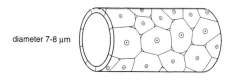

diameter 7-8 µm

(Fig 1)

a Complete the table to state **three** features of capillaries which allow them to function effectively and explain how each helps with exchange between the blood and tissue fluid.

feature	explanation of the role in exchange

(6 marks)

Figure 2 shows some of the pressures involved in exchange between a capillary and the tissue fluid around it. **A** and **B** are points at opposite ends of the capillary.

Effective blood pressure =
(SP + HP of blood) − (SP + HP of tissue fluid).

HP = hydrostatic pressure produced by the pumping of the heart.

SP = solute potential; an osmotic effect due to solutes dissolved in the blood or tissue fluid.

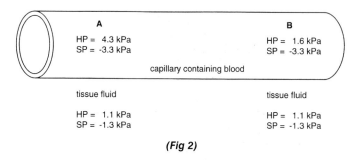

(Fig 2)

b The effective blood pressure at **A** is 1.2 kPa.

(i) Calculate, showing your working, the effective blood pressure at **B**. *(2 marks)*

(ii) Mark by means of an arrow on the diagram the direction the fluid will move between the blood and the tissue fluid at **A**. *(1 mark)*

Some of the tissue fluid does not return to the capillaries, but enters another set of vessels.

c Name the fluid in these vessels and describe its composition. *(4 marks)*

(Total 13 marks)
OCR 2803/1 Jan 2002, B (T), No.5

6 Multicellular animals have transport systems.

a Explain why multicellular animals need transport systems. *(3 marks)*

b Complete the table below by placing a tick (✓) or a cross (✗) in the boxes.

feature	red blood cell	lymphocyte	phagocyte
possesses a nucleus			
produces antibodies			
possesses endoplasmic reticulum			
contains haemoglobin			

(4 marks)
(Total 7 marks)
OCR 2803/1 Jun 2001, B (T), No.3

9

9.1 Structure of the heart

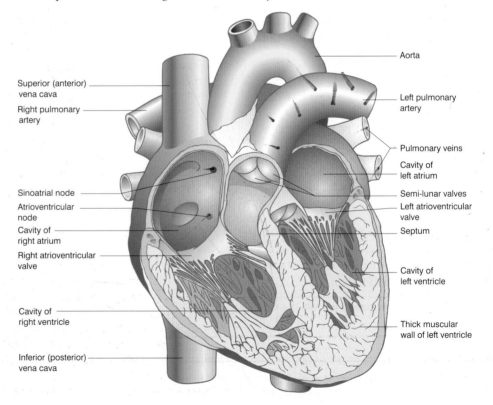

The heart is a muscular organ which, in humans, is between 250g and 350g in mass and pumps 13 000dm³ of blood each day – enough to fill a small road tanker. It operates continuously and tirelessly throughout the life of an organism – a period of up to 100 years. Lying in the thoracic cavity behind the sternum (breast-bone) and between the two lungs, the heart is made up of a unique type of muscle called **cardiac muscle**.

9.1.1 Structure of cardiac muscle

Between the outer layer of the heart wall (**epicardium**) and the inner layer (**endocardium**) is the **myocardium**. While the outer and inner layers are made up of epithelial cells and connective tissue, the myocardium comprises a specialised tissue – **cardiac muscle**. Cardiac muscle is capable of rhythmic contraction and relaxation over a long period without fatigue. Cardiac muscle appears striped under a light microscope and is made up of the proteins **actin** and **myosin**. The molecular mechanism of contraction is also the same as that of striated muscle (body muscle) and both have an abundance of mitochondria to supply **ATP**. Cardiac muscle is made up of short fibres. Where these meet end to end, there are distinct dark bands called **intercalated discs**. Because these discs resist the passage of electrical signals across them, they have points within them, called **gap junctions**, where the outer membranes of adjacent fibres fuse. These permit an almost unimpeded flow of electrical signals through them, thereby allowing a wave of contraction to spread along the muscle – an important feature in the beating of the heart. Each cardiac muscle fibre is made up of **myofibrils** up to 0.08mm in length and around 15µm in diameter.

The mammalian heart

Nucleus

Intercalated discs (junctions of muscle cells)

Striations

Connective tissue and blood capillaries between muscle fibres

Cardiac muscle fibre

LM of a longitudinal section through cardiac muscle (×300)

Aorta

Superior (anterior) vena cava

Right pulmonary artery

Left pulmonary artery

Pulmonary veins

Cavity of left atrium

Sinoatrial node

Atrioventricular node

Cavity of right atrium

Right atrioventricular valve

Semi-lunar valves

Left atrioventricular valve

Septum

Cavity of left ventricle

Cavity of right ventricle

Thick muscular wall of left ventricle

Inferior (posterior) vena cava

Fig 9.1 *Section through the human heart (VS)*

9.1.2 Structure of the human heart

The heart is covered by a tough membrane called the **pericardium**, which encloses the **pericardial fluid**. This surrounds the heart and lubricates its movement relative to the pericardium. The human heart is really two separate pumps lying side by side. The left-hand pump deals with oxygenated blood from the lungs, while the right-hand one deals with deoxygenated blood from the body. Each pump has two chambers.

- **The atrium** is thin-walled and elastic and distends as it collects blood. It only has to pump blood the short distance to the ventricle and therefore there is only a little muscle present.
- **The ventricle** has a much thicker muscular wall as it has to pump blood some distance, either to the lungs or the rest of the body.

As the right ventricle pumps blood to the lungs, a distance of only a few centimetres, it has a thinner muscular wall than the left ventricle. The left ventricle, in contrast, has a thick muscular wall, enabling it to create enough pressure to pump blood to the extremities of the body, a distance of about 1.5m. Although the two sides of the heart are separate pumps and there is no mixing of the blood in each after birth, they nevertheless pump in time with each other: both atria contract together, followed by both ventricles contracting together.

Between the atrium and ventricle are valves that prevent the backflow of blood into the atria when the ventricles contract. There are two sets of valves.

- **Left atrioventricular (bicuspid) valves** are formed of two cup-shaped flaps on the left side of the heart.
- **Right atrioventricular (tricuspid) valves** are formed of three cup-shaped flaps on the right side of the heart.

To prevent these valves turning inside out under pressure, they are attached to special pillars of muscle on the heart wall by fibres called the **tendinous cords (chordae tendinae)**. Each of the four chambers of the heart is served by large blood vessels that carry blood into or away from the heart. The ventricle walls are thicker than those of the atria because they pump blood away from the heart, and the ventricles are therefore always connected to arteries. The atrial walls are thinner because they receive blood and are therefore connected to veins (remember A / V: Atria / Veins and Arteries / Ventricles). Vessels connecting the heart to the lungs are called **pulmonary** vessels. The vessels connected to the four chambers are therefore as follows.

- **The aorta** is connected to the left ventricle and carries oxygenated blood to all parts of the body except the lungs.
- **The vena cava** is connected to the right atrium and brings deoxygenated blood back from the tissues of the body.
- **The pulmonary artery** is connected to the right ventricle and carries deoxygenated blood to the lungs, where its oxygen is replenished and its carbon dioxide is removed. Unusually for an artery, it carries deoxygenated blood.
- **The pulmonary vein** is connected to the left atrium and brings oxygenated blood back from the lungs. Unusually for a vein, it carries oxygenated blood.

The structure of the heart and its associated blood vessels is shown in figures 9.1 and 9.2.

9.1.3 Supplying the cardiac muscle with oxygen

Although oxygenated blood passes through the left side of the heart in vast quantities, the heart does not use this oxygen to meet its own great respiratory needs. Instead, the heart muscle is supplied by its own blood vessels, called the **coronary arteries,** which branch off the aorta shortly after it leaves the heart. Blockage of these arteries, e.g. by a blood clot, leads to **myocardial infarction,** or **heart attack**, because an area of the heart muscle is deprived of oxygen and so dies, leaving scar tissue.

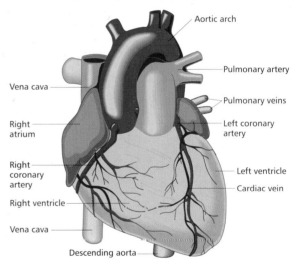

Fig 9.2 *External appearance of the human heart showing the blood supply to the heart muscle*

SUMMARY TEST 9.1

The mammalian heart is made up of (1) muscle and is covered by a tough layer called the (2). It is made up of four chambers: a pair of thin-walled elastic ones called (3) and a pair of thick muscular ones called (4). Between the chambers on the left side of the heart are the (5) valves, while those on the right side are called (6) valves. These valves are prevented from turning inside out by (7). Blood from the lungs passes to the heart by the (8) and into the chamber called (9); it leaves the heart via the vessel called the (10). Glucose absorbed from the intestines will first enter the (11) chamber of the heart. If this glucose is destined for the heart muscle itself, the next blood vessel that it passes along will be the (12). Oxygen is essential to heart muscle and is supplied via the (13) that branch off the aorta. A blockage of these vessels can lead to a heart attack, which is also known as a (14).

The cardiac cycle

1.
Blood enters atria and ventricles from pulmonary veins and venae cavae

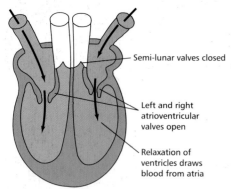

Semi-lunar valves closed

Left and right atrioventricular valves open

Relaxation of ventricles draws blood from atria

Diastole
Atria are relaxed and fill with blood. Ventricles are also relaxed.

2.

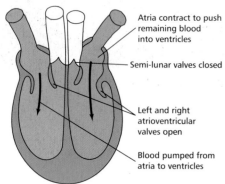

Atria contract to push remaining blood into ventricles

Semi-lunar valves closed

Left and right atrioventricular valves open

Blood pumped from atria to ventricles

Atrial systole
Atria contract, pushing blood into the ventricles. Ventricles remain relaxed.

3.
Blood pumped into pulmonary arteries and the aorta

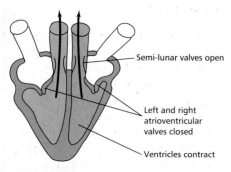

Semi-lunar valves open

Left and right atrioventricular valves closed

Ventricles contract

Ventricular systole
Atria relax. Ventricles contract, pushing blood away from heart through pulmonary arteries and the aorta.

Fig 9.3 *The cardiac cycle*

The heart undergoes a sequence of events that is repeated in humans around 70 times each minute when at rest. This is known as the **cardiac cycle**. There are two basic components to the beating of the heart – contraction, or **systole**, and relaxation, or **diastole**. Systole occurs separately in the ventricles and the atria and is therefore described in two phases but, for some of the time, diastole takes place simultaneously in all chambers of the heart and is therefore treated as a single phase in the account below, which is illustrated in figure 9.3.

9.2.1 Diastole

Blood returns to the atria of the heart through the pulmonary vein (from the lungs) and the vena cava (from the body). As the atria fill, the pressure in them rises, pushing open the atrioventricular valves and allowing the blood to pass into the ventricles. The **cardiac muscle** of both the atria and ventricles is relaxed at this stage. The relaxation of the ventricle's cardiac muscle reduces the pressure within the ventricle. This causes the pressure to be lower than that in the aorta and the pulmonary artery, and so the semi-lunar valves in the aorta and the pulmonary artery close, accompanied by the characteristic 'dub' sound of the heart beat.

9.2.2 Atrial systole

The cardiac muscle of the atria contracts, forcing the remaining blood that they contain (around 20% of the total in the heart) into the ventricles. The blood has only to be pushed a very short distance and therefore the muscle walls of the atria are very thin. During this stage, the cardiac muscle of the ventricle walls remains relaxed (ventricular diastole).

9.2.3 Ventricular systole

After a short delay, the ventricles contract simultaneously. This increases the blood pressure within them, forcing shut the atrioventricular valves and preventing backflow of blood into the atria. The 'lub' sound of these valves slamming shut is a characteristic of the heart beat. With the atrioventricular valves closed, the pressure rises further, forcing open the semi-lunar valves and pushing blood into the pulmonary artery and aorta. The walls of the ventricles are much thicker than those of the atria as they have to pump the blood much further. The wall of the left ventricle has to pump blood to the extremities of the body and so is much thicker than that of the right ventricle, which only has to pump blood as far as the adjacent lungs.

9.2.4 Valves in the control of blood flow

It is important to keep blood flowing in the right direction through the heart and around the body. This is achieved mainly by the pressure created by the heart muscle. Blood, as with all liquids and gases, will always move from a region of higher pressure to one of lower pressure. There are, however, situations within the circulatory system when pressure differences would result in blood flowing in the opposite direction from that which is desirable. In these circumstances valves are used to prevent any unwanted backflow of blood. Valves in the cardiovascular system are designed so that they open whenever the difference in blood pressure either side of them favours the movement of blood in the desired direction. When pressure differences are reversed, i.e. when blood would tend to flow in the opposite direction to that which is desirable, the valves are designed to close. Examples of such valves include:

- **The atrioventricular valves** between the left atrium and ventricle (bicuspid valves) and the right atrium and ventricle (tricuspid valves). These prevent backflow of blood when contraction of the ventricles means that ventricular pressure exceeds atrial pressure. Closure of these valves ensures that, when the ventricles contract, blood within them moves to the aorta and pulmonary arteries rather than back to the atria.
- **The semi-lunar valves** in the aorta and pulmonary arteries. These prevent backflow of blood into the ventricles when the recoil action of the elastic walls of these vessels (section 8.2.1) creates a greater pressure in the vessels than in the ventricles.
- **Pocket (semi-lunar) valves** in veins, which occur throughout the venous system. These ensure that when the veins are squeezed, e.g. when skeletal muscles contract, blood flows back to the heart rather than away from it.

The design of all these valves is basically the same. They are made up of a number of flaps of tough, but flexible, fibrous tissue, which are cusp-shaped, i.e. like deep saucers or bowls. When pressure is greater on the convex side of these cusps, rather than on the concave side, they move apart to let blood pass between the cusps. However, when pressure is greater on the concave side than on the convex side, blood collects within the 'bowl' of the cusps, pushing them together to form a tight fit that prevents the passage of blood (Fig 9.4). So great are the pressures created within the ventricles of the heart that the atrioventricular valves are at risk of becoming inverted. To prevent this, the valves have string-like tendons called the chordae tendinae (heart strings) that are attached to pillars of muscle in the ventricle wall (see unit 9.1, Fig 9.1). These tendons contract with the rest of the ventricle and so help to pull the valves closed.

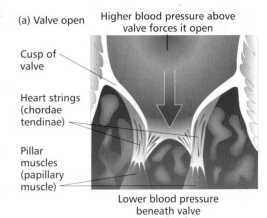

(a) Valve open — Higher blood pressure above valve forces it open

Cusp of valve

Heart strings (chordae tendinae)

Pillar muscles (papillary muscle)

Lower blood pressure beneath valve

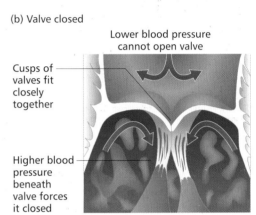

(b) Valve closed

Lower blood pressure cannot open valve

Cusps of valves fit closely together

Higher blood pressure beneath valve forces it closed

Fig 9.4 *Action of valves*

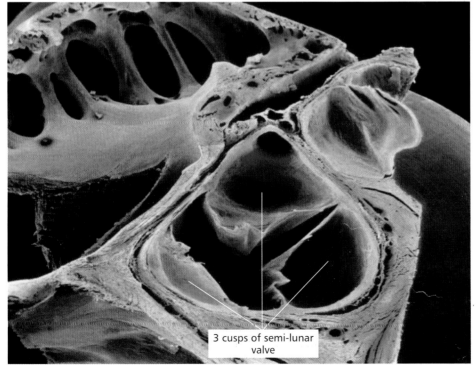

3 cusps of semi-lunar valve

False-colour SEM of the semi-lunar valve of the aorta

SUMMARY TEST 9.2

In humans the cardiac cycle repeats itself about **(1)** times each minute when the heart is at rest. The phase of the cycle when the atria and ventricles are relaxed is called **(2)**. When the atria contract during the phase called **(3)**, the remaining blood in them is pushed past the **(4)** valves into the chambers called **(5)**. Contraction of these chambers forces open the **(6)** valves and pushes blood into the **(7)**, which then goes to the lungs, and the **(8)**, which supplies blood to the rest of the body.

Control of the cardiac cycle

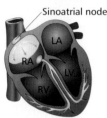

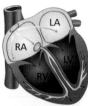

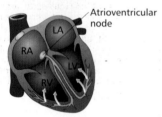

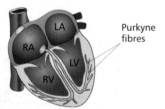

1. Wave of excitation spreads out from the sinoatrial node

2. Wave spreads across both atria causing them to contract and reaches the atrioventricular node

3. Atrioventricular node conveys wave of excitation between the ventricles to the Purkyne fibres

4. Wave of excitation is released by the Purkyne fibres and ventricles contract

Fig 9.5 *Control of the cardiac cycle*

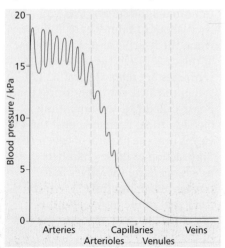

Fig 9.6 *Blood pressure in arteries, capillaries and veins*

For the heart to function efficiently, there must be careful control of the sequence of events that takes place during the **cardiac cycle**.

9.3.1 How is the cardiac cycle controlled?

Cardiac muscle is **myogenic**, i.e. its contraction is initiated from within the muscle itself, rather than by nervous impulses from outside (**neurogenic**), as is the case with other muscle. Within the wall of the right atrium of the heart is a distinct group of cells known as the **sinoatrial node (SAN)**. It is from here that the initial stimulus for contraction originates. The sinoatrial node has a basic rhythm of stimulation that determines the beat of the heart. For this reason it is often referred to as the **pacemaker**. The sequence of events that controls the cardiac cycle is as follows.

- A wave of excitation spreads out from the sinoatrial node across both atria, causing them to contract.
- A layer of non-conductive tissue (atrioventricular septum) prevents the wave crossing to the ventricles.
- The wave of excitation is allowed to pass through a second group of cells called the **atrioventricular node (AVN)**, which lies between the atria.
- The atrioventricular node, after a short delay, conveys a wave of excitation between the ventricles along a series of specialised muscle fibres called the **bundle of His**.
- The bundle of His conducts the wave through the atrioventricular septum to the base of the ventricles, where the bundle branches into smaller fibres known as **Purkyne tissue**.
- The wave of excitation is released from the Purkyne tissue, causing the ventricles to contract quickly at the same time, from the apex of the heart upwards.

These events are summarised in figure 9.5.

EXTENSION – ELECTROCARDIOGRAM

During the cardiac cycle, the heart undergoes a series of electrical current changes related to the waves of excitation created by the sinoatrial node and the heart's response to these. If picked up by a cathode ray oscilloscope, these changes can produce a trace known as an **electrocardiogram**. An example, related to the stages of the cardiac cycle, is shown as part of figure 9.7. Doctors can use this trace to provide a picture of the heart's electrical activity and hence its health. The electrocardiograms (ECG) below illustrate the difference between a normal ECG, one produced during a heart attack, and one in a person suffering **fibrillation** of the heart. In fibrillation, the cardiac muscle contracts in a disorganised way, causing different sections to contract and relax independently in an irregular manner. This effectively paralyses the heart and causes death if not treated immediately.

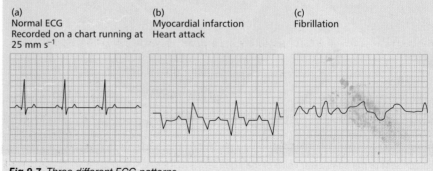

(a) Normal ECG
Recorded on a chart running at 25 mm s⁻¹

(b) Myocardial infarction
Heart attack

(c) Fibrillation

Fig 9.7 *Three different ECG patterns*

9.3.2 Pressure and volume changes of the heart

Mammals have a closed circulatory sytem, i.e. blood is confined to vessels and this allows the pressure within them to be maintained and regulated. Fig 9.6 shows the pressure within blood vessels while figure 9.8 illustrates the pressure and volume changes that take place in the heart during a typical cardiac cycle. Two facts will help you to understand this rather complex graph.

- Pressure and volume within a closed container are inversely related (Boyle's Law). When pressure increases, volume decreases, and vice versa.
- Blood, like all fluids, moves from a region where its pressure is greater to one where it is less, i.e. it moves down a pressure gradient.

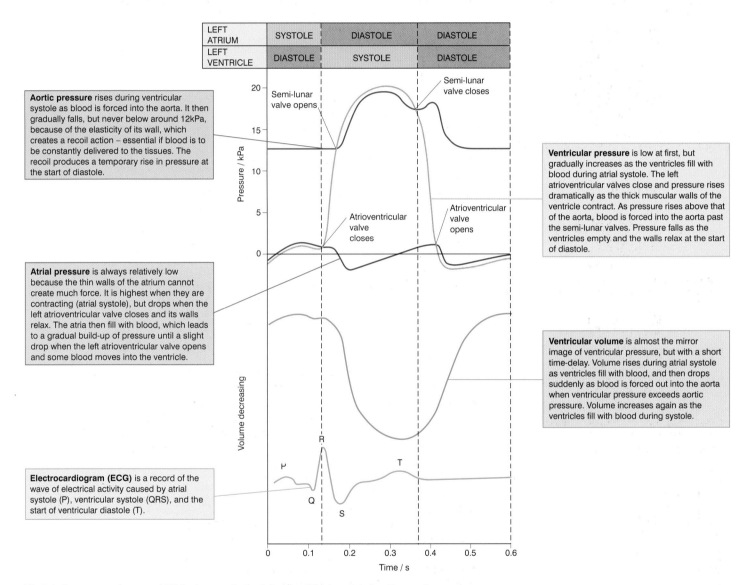

Aortic pressure rises during ventricular systole as blood is forced into the aorta. It then gradually falls, but never below around 12kPa, because of the elasticity of its wall, which creates a recoil action – essential if blood is to be constantly delivered to the tissues. The recoil produces a temporary rise in pressure at the start of diastole.

Atrial pressure is always relatively low because the thin walls of the atrium cannot create much force. It is highest when they are contracting (atrial systole), but drops when the left atrioventricular valve closes and its walls relax. The atria then fill with blood, which leads to a gradual build-up of pressure until a slight drop when the left atrioventricular valve opens and some blood moves into the ventricle.

Electrocardiogram (ECG) is a record of the wave of electrical activity caused by atrial systole (P), ventricular systole (QRS), and the start of ventricular diastole (T).

Ventricular pressure is low at first, but gradually increases as the ventricles fill with blood during atrial systole. The left atrioventricular valves close and pressure rises dramatically as the thick muscular walls of the ventricle contract. As pressure rises above that of the aorta, blood is forced into the aorta past the semi-lunar valves. Pressure falls as the ventricles empty and the walls relax at the start of diastole.

Ventricular volume is almost the mirror image of ventricular pressure, but with a short time-delay. Volume rises during atrial systole as ventricles fill with blood, and then drops suddenly as blood is forced out into the aorta when ventricular pressure exceeds aortic pressure. Volume increases again as the ventricles fill with blood during systole.

Fig 9.8 *Pressure, volume and ECG changes in the left side of the heart during the cardiac cycle*

SUMMARY TEST 9.3

The mammalian heart beat is initiated from within the heart muscle itself, which is therefore termed (1). The pacemaker of the heart is the (2), which lies in the wall of the chamber called the (3). A wave of excitation causes both (4) to contract. The wave is picked up by another group of specialised cells, called the (5), which in turn pass it down to the apex of the ventricles and out into the muscle via small branches of specialist muscle called (6).

1 a Figure 1 is a simplified plan of the mammalian circulatory system. The system is described as a double circulation.

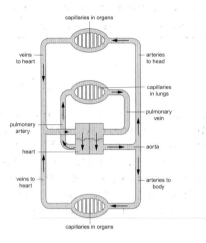

(Fig 1)

(i) Use the information in figure 1 to state what is meant by the term *double circulation*. *(2 marks)*

(ii) Suggest an advantage of the double circulation shown in figure 1. *(1 mark)*

b Figure 2 shows some detail of the external and internal structure of the mammalian heart and associated blood vessels.

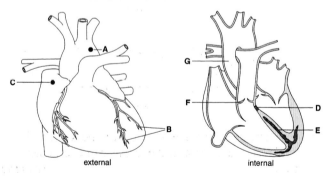

(Fig 2)

The table lists some possible functions of parts of the heart.

possible function	letter
contracts to force blood into the ventricle	P
stops impulses crossing the septum between the atria and the ventricles	Q
acts as a pacemaker	R
carries blood away from the heart to the body	S
conducts impulses to the apex of the heart	T
relays the impulses to the septum between the ventricles	U

Complete the table below by
- identifying each of the features listed, using an appropriate letter from figure 2
- matching a function to the feature, using the appropriate letter from the table

The first one has been done for you.

feature	letter on Fig 2	letter in table
aorta	G	S
sino-atrial node		
atrio-ventricular node		
Purkyne (Purkinje) fibres		

(6 marks)
(Total 9 marks)
OCR 2803/01 Jan 2003, B (T), No.3

2 a Figure 1 shows two blood vessels, **X**, and **Y**, in transverse section.

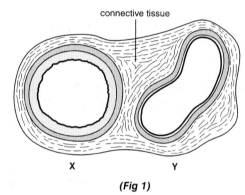

(Fig 1)

(i) State which of the blood vessels, **X** or **Y**, is a vein. *(1 mark)*

(ii) Give **two** reasons for your choice. *(2 marks)*

b Figure 2 gives information about blood pressure in various parts of the mammalian blood system.

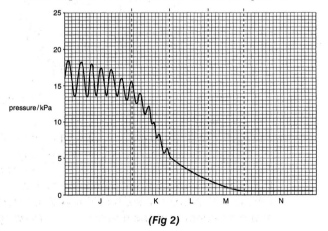

(Fig 2)

(i) Calculate the drop in systolic pressure in section **J** of figure 2. Show your working. *(2 marks)*

(ii) State which section, **J** to **N**, of figure 2 shows the pressure in the veins. *(1 mark)*

c Explain briefly how blood in the veins is returned to the heart. *(3 marks)*

(Total 9 marks)

OCR 2803/1 Jan 2003, B (T), No.5

3 a During the cardiac cycle the pressure in the different chambers of the heart varies in a regular pattern.

State in which one of the four chambers of the heart
(i) the highest pressure is generated; *(1 mark)*
(ii) the greatest change in pressure occurs. *(1 mark)*

b The following figure shows the changes in blood pressure during one cardiac cycle.

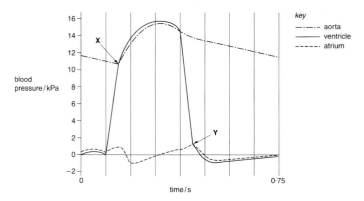

(i) Use the information in the figure to calculate the heart rate in beats per minute. Show your working. *(2 marks)*

(ii) Explain what is happening at **X** and **Y** in the above figure. *(6 marks)*

(Total 10 marks)

OCR 2803/1 Jun 2003, B (T), No.2

4 The figure shows a simplified diagram of the human heart and associated blood vessels in vertical section.

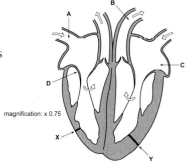

a Name structures **A** to **D**. *(4 marks)*

b (i) The actual width of the wall at **X** is 4 mm. Calculate the actual width of the wall at point **Y** using the scale provided. Show your working. *(2 marks)*

(ii) Explain the significance of this difference in thickness between the walls at **X** and **Y** in terms of the functioning of the heart. *(3 marks)*

c Complete the table below by adding T (true) or F (false) against the statements about the cardiac cycle.

	T or F
cardiac muscle is myogenic	
contraction occurs in the diastolic phase	
the left and right ventricles contract at the same time	
when the left ventricle contracts, the semilunar valve in the aorta shuts	
the semilunar valves have tendons to prevent inversion	

(5 marks)

(Total 14 marks)

OCR 2803/1 Jan 2002, B (T), No.4

5 An ECG (electrocardiogram) trace shows the electrical activity within a person's beating heart. The figure shows a normal trace, **A**, a trace after the administration of the drug digitalis, **B**, and a trace from someone whose heart has entered a state known as fibrillation, **C**. **P** represents activity in the atrial wall, **R** contraction of the ventricle and **T** recovery of the ventricle walls.

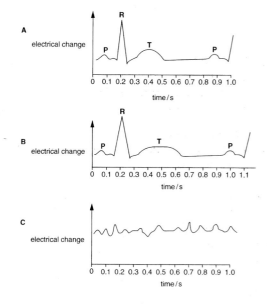

a With reference to trace **A**, calculate the length of a single cardiac cycle and the number of cycles per minute. *(2 marks)*

b With reference to trace **B**, state **two** effects of digitalis on heart activity. *(2 marks)*

c With reference to trace **C**,
(i) describe the effect of fibrillation on heart activity; *(1 mark)*
(ii) state **one** likely effect on the patient. *(1 mark)*

d Describe the events that occur in the cardiac cycle.
(In this question 1 mark is available for the quality of written communication.)

(9 marks)

(Total 15 marks)

OCR 2803/1 Jan 2001, B (T), No.4

Transport in multicellular plants

Distribution of vascular tissues in dicotyledonous plants

Side veins
Blade of leaf (lamina)
Stem
Main vein
Leaf stalk (petiole)
Node of stem (point where leaf is attached to stem)

Fig 10.1 *Leaf of a dicotyledonous plant, e.g. elm, showing arrangement of veins*

A flowering plant can be thought of as having two main functional areas: the leaves, which manufacture sugars by photosynthesis at one end, and the roots, which absorb water and minerals at the opposite end. Each relies on the other – the leaves needing water and minerals to photosynthesise, and the roots requiring sugar to respire and keep alive. No less important, therefore, are the communication channels between the two, namely the **vascular tissue**, of which there are two types:

* **xylem** – which carries water from the roots, up the plant to the aerial parts
* **phloem** – which carries sugars produced by leaves to other parts of the plant.

The two tissues occur together throughout the plant, sometimes with associated tissues, such as **sclerenchyma**, to form discrete areas, known as **vascular bundles**.

10.1.1 Distribution of vascular tissues in a leaf

The vascular tissues in a **dicotyledonous** leaf form a network of tiny vascular bundles throughout the blade, or **lamina**, of the leaf. These tiny bundles fuse to give a series of **side veins** that run parallel with one another. These side veins then merge into a central **main vein**, in much the same way that tributaries merge into a river. The main vein runs along the centre of the leaf, increasing in diameter towards the petiole, or leaf stalk. Within each vein, or vascular bundle, there is an area of xylem towards the upper surface of the leaf and an area of phloem towards the lower surface. This arrangement is illustrated in the section through the leaf shown in figure 10.1 (a). More detail of the cellular arrangement within a leaf is given in section 10.6.1.

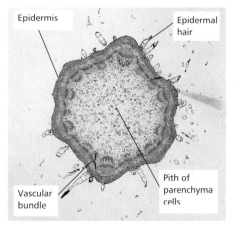

Epidermis
Epidermal hair
Pith of parenchyma cells
Vascular bundle

Young stem of the dicotyledon Helianthus *(TS) (×3)*

10.1.2 Distribution of vascular tissue in a stem

The xylem and phloem in a dicotyledonous stem form vascular bundles that are arranged towards the outside of the stem. The reason for this is that the vascular bundles, along with associated sclerenchyma, not only transport material but also provide support in **herbaceous** stems. The main forces acting on stems are lateral ones caused by the action of wind on them. Such forces are best resisted by an outer cylinder of supporting tissue. Hence the vascular bundles form a discontinuous ring towards the edge of them (Fig 10.1 (b)). Being discontinuous, this ring of supporting tissues allows the stem to be flexible and to bend in the wind. Within the vascular bundles, the xylem is to the inside of the stem and the phloem towards the outside. Between the two is a thin layer of dividing cells called **cambium**, which gives rise to both xylem and phloem.

10.1.3 Distribution of vascular tissue in a root

The vascular tissue in the root of a dicotyledonous plant is situated centrally rather than towards the outer edge, as in a stem. This is because roots are subject to pulling forces in a vertical direction, rather than in a lateral direction, as experienced by stems. Vertical forces are better resisted by a central column of supporting tissues, such as xylem, rather than an outer cylinder of tissue. The

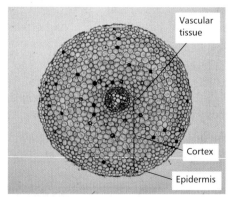

Vascular tissue
Cortex
Epidermis

Root of the dicotyledon Ranunculus *(buttercup) (TS) (×20)*

xylem is typically arranged in a single star-shaped block of tissue at the centre of the root, with the phloem situated in separate groups between each of the points of the star-shaped xylem. Around both is the **pericycle** and **endodermis**, more details of which are given in section 10.5.5.

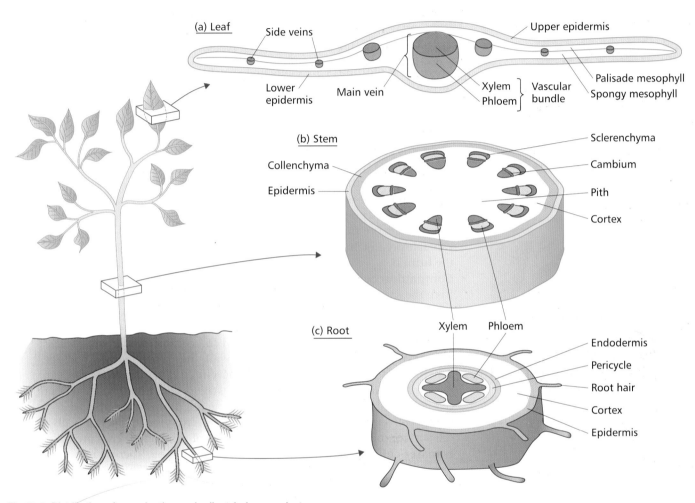

Fig 10.1 *Distribution of vascular tissues in dicotyledonous plants*

SUMMARY TEST 10.1

The vascular tissue in a leaf forms a network of bundles that finally merge to form one central **(1)**, which runs into the stalk of the leaf that is also called the **(2)**. The vascular tissue found nearest to the upper surface of the leaf is **(3)**. In stems, the vascular tissue nearest the outside of the stem is **(4)**. Inside this tissue is a layer of dividing tissue called **(5)** and outside it there is often a region of supporting tissue called **(6)**. The vascular tissue in stems forms a cylinder towards the outer edge as this arrangement best resists the **(7)** forces that stems are subjected to. In roots, the vascular tissue is central in order to resist the **(8)** forces experienced by roots.

10.2 Structure and function of phloem

Phloem is the principal food-conducting tissue in vascular plants. Details of this process are given in unit 10.3. It carries organic material, such as sugars and amino acids, from leaves and storage organs to other parts of the plant.

10.2.1 Structure of phloem

Phloem is composed of a number of cell types:

- **Sieve tube elements** are elongated structures that are joined end to end to form long tubes. The cells are living and retain a thin layer of cytoplasm within their plasma membrane, which lies against the cellulose cell wall. Within the cytoplasm are mitochondria and a modified form of endoplasmic reticulum. However, unlike most cells, there is no nucleus or Golgi apparatus and there are no ribosomes. These structures are broken down in order to make the sieve tubes more hollow and so reduce resistance to the flow of liquid within them. The end walls of the sieve tubes are perforated by large pores, 2–6μm in diameter. These perforated end walls are called **sieve plates**. The pores are lined with a carbohydrate called **callose**. This callose seals the sieve plate if it is damaged and so prevents the loss of valuable organic material. The central space within the sieve tube is called the **lumen**.
- **Companion cells** are always associated with sieve elements and both come from the same cell division. As the sieve tube elements lack structures like a nucleus, ribosomes and Golgi apparatus, they are unable to carry out many of the metabolic processes essential for their survival. The companion cells are the site of these processes. With their full complement of organelles, dense cytoplasm and thin cellulose cell wall, they perform the metabolic activities for both themselves and the sieve tube elements. Materials can easily pass through the many **plasmodesmata** that link the two types of cell. So close is this relationship, that if one cell dies, so does the other. At the tips of veins in the leaf, companion cells have very folded cell walls and cell surface membranes. These special types of companion cells are called **transfer cells** and are thought to **actively transport** sucrose into the sieve tube elements.

Two other cell types are associated with phloem: phloem **parenchyma** and phloem fibres. Both provide support to the phloem.

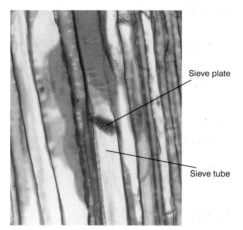

Phloem as seen under a light microscope (×400 approx.)

Labels: Sieve plate, Sieve tube

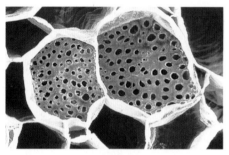

False-colour SEM of sieve plates

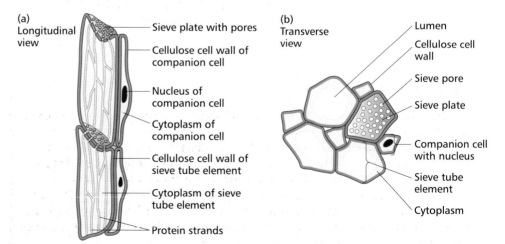

(a) Longitudinal view — Sieve plate with pores; Cellulose cell wall of companion cell; Nucleus of companion cell; Cytoplasm of companion cell; Cellulose cell wall of sieve tube element; Cytoplasm of sieve tube element; Protein strands

(b) Transverse view — Lumen; Cellulose cell wall; Sieve pore; Sieve plate; Companion cell with nucleus; Sieve tube element; Cytoplasm

Fig 10.2 *Phloem as seen under a light microscope*

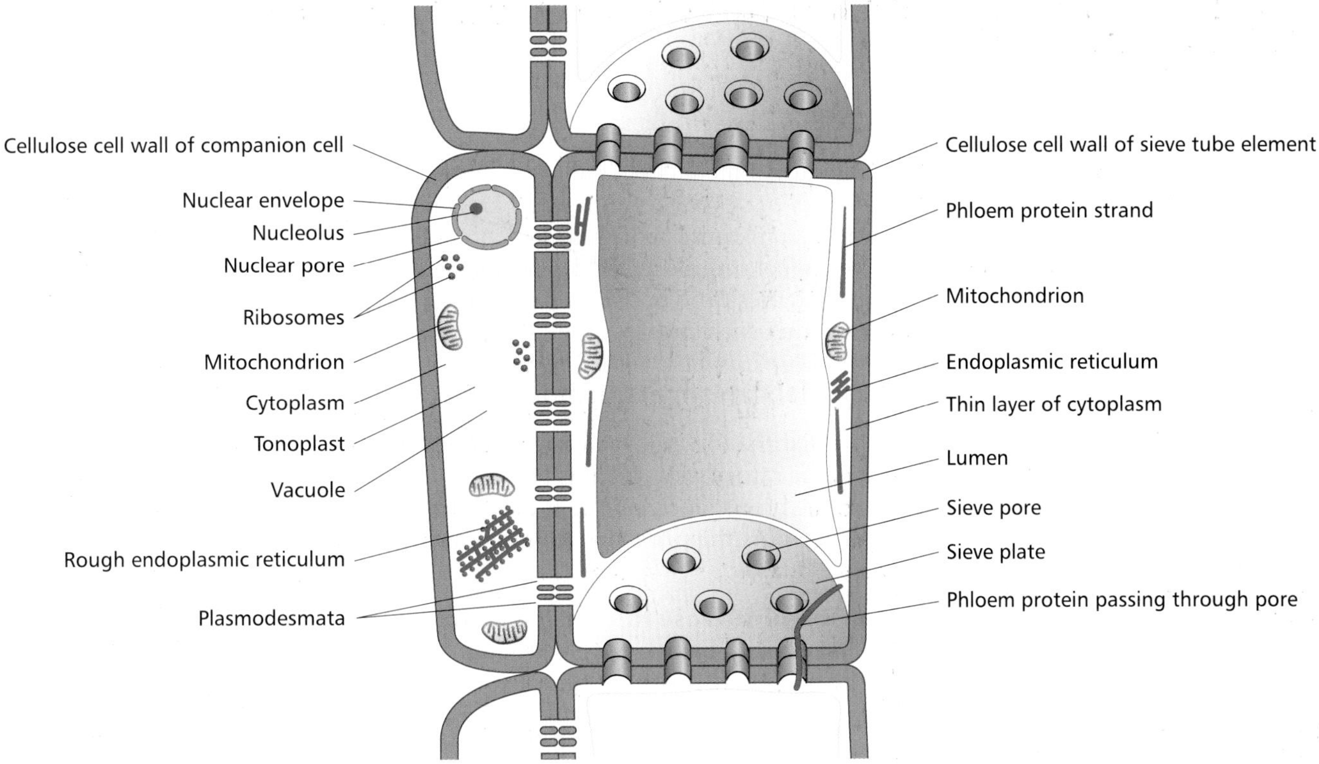

Cellulose cell wall of companion cell
Nuclear envelope
Nucleolus
Nuclear pore
Ribosomes
Mitochondrion
Cytoplasm
Tonoplast
Vacuole
Rough endoplasmic reticulum
Plasmodesmata
Cellulose cell wall of sieve tube element
Phloem protein strand
Mitochondrion
Endoplasmic reticulum
Thin layer of cytoplasm
Lumen
Sieve pore
Sieve plate
Phloem protein passing through pore

Fig 10.3 *Sieve tube element and companion cell structure as revealed by an electron microscope*

10.2.2 How phloem's structure is related to its function

The structure of sieve tubes has evolved to suit their function of transporting organic materials in solution.

- The elements are elongated and arranged end to end to form a continuous column.
- The nucleus and many of the organelles are located in the companion cells, leaving the lumen of the sieve tube elements more open and lessening obstruction to the flow of liquid.
- The end walls are perforated, so give less resistance to liquid flow.
- The companion cells have many mitochondria to release the **ATP** needed for translocation of organic materials (unit 10.3).
- The walls contain cellulose microfibrils that run around the cells, giving strength and preventing the tubes bursting under pressure.

SUMMARY TEST 10.2

Phloem carries organic materials such as **(1)** and **(2)** from the leaves and **(3)** regions to other parts of the plant. It is made up of two main cell types. The sieve tube elements form long, vertical tubes. Each element has perforated end walls called **(4)**. The pores within them have strands of cytoplasm made up of **(5)** running through them. Sieve tube elements lack important structures found in other cells. These include **(6)**, **(7)** and **(8)**. Associated with sieve tube elements are cells that carry out the metabolic activities of sieve tube elements. These cells are called **(9)** and transfer material to and from sieve tube elements via **(10)** in the cellulose cell walls between them.

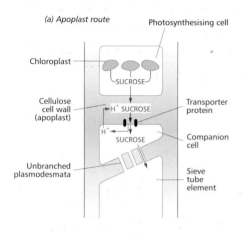

(a) Apoplast route

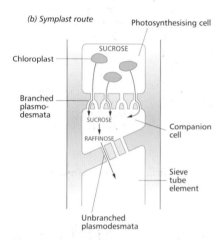

(b) Symplast route

Fig 10.4 *Loading of phloem with sucrose*

Having produced sugars during photosynthesis, the plant needs to transport them from the sites of production, known as **sources**, to the places where they will be used directly or stored for future use – known as **sinks**. As sinks can be anywhere in a plant – sometimes above and sometimes below the source – it follows that the translocation of organic molecules in phloem can be in either direction. Other organic materials to be transported include amino acids. The phloem also transports inorganic **ions** such as potassium, chloride, phosphate and magnesium ions.

10.3.1 Mechanism of translocation

It is accepted that organic materials are transported in the phloem (section 10.3.2) and that the rate of movement is too fast to be explained by diffusion. What is in doubt is the precise mechanism of how translocation is achieved. Current thinking favours the **pressure flow hypothesis**, a theory which can be divided into three phases:

- **Transfer of sucrose into sieve elements from photosynthesising tissue**. This takes place by two possible routes:
  - **The apoplast route** (section 10.5.2) occurs in those plants that have few, if any, cytoplasmic connections between the sieve element–companion cell complex (SE–CC complex) and the photosynthesising cell. In this case, hydrogen ions are actively pumped from companion cells into the apoplast (spaces within the cell walls) using ATP. These hydrogen ions then flow down a concentration gradient through a transporter protein, which also transports the sucrose at the same time (Fig 10.4 (a)).
  - **The symplast route** (section 10.5.3) occurs in those plants that have branched **plasmodesmata** between the photosynthesising cells and the sieve element–companion cell complex. The sucrose moves freely along these plasmodesmata into the companion cells, which then convert it into larger sugars, such as raffinose, which cannot flow back, but easily move into the sieve tube. The plasmodesmata therefore act as a type of valve. As the conversion into larger sugars needs energy, the overall process is an active one (Fig 10.4 (b)).
- **Mass flow of sucrose through sieve tube elements.** The sucrose produced by photosynthesising cells (source) causes them to have a lower (more negative) **water potential**. Water therefore moves into them by osmosis from the xylem, which has a very much higher (less negative) water potential. At the sink, sucrose is either used up during respiration or converted to starch for storage. These cells therefore have a low sucrose content, giving them a higher (less negative) water potential. As a result of water entering cells at the source and leaving at the sink, there is a hydrostatic pressure created which causes a mass flow of the sucrose solution along the phloem. This is a passive process. A model of this theory is shown in figure 10.5.
- **Transfer of sucrose from the sieve tube elements into storage or other sink cells**. The sucrose is actively transported by companion cells, out of the sieve tubes and into the sink cells.

Some of the evidence for and against the pressure flow theory is listed in table 10.1.

10.3.2 Evidence that translocation of organic molecules occurs in phloem

- When phloem is cut, a solution of organic molecules is exuded.
- Plants provided with radioactive carbon dioxide can be shown to have radioactively labelled carbon in phloem after a short time.

Table 10.1 *Evidence for and against the pressure flow theory*

Evidence supporting the pressure flow theory	Evidence questioning the pressure flow theory
• There is a pressure within sieve tubes, as shown by sap being released when they are cut. • The concentration of sucrose is higher in leaves (source) than in roots (sink). • Downward flow in the phloem occurs in daylight, but ceases when leaves are shaded, or at night. • Increases in sucrose levels in the leaf are followed by similar increases in sucrose levels in the phloem a little later. • Metabolic poisons and/or lack of oxygen inhibit translocation of sucrose in the phloem. • Companion cells possess many mitochondria and readily produce ATP.	• The function of the sieve plates is unclear, as they would seem to hinder mass flow (it has been suggested that they may have a structural function, helping to prevent the tubes from bursting under pressure). • Not all solutes move at the same speed – they should do so if movement is by pressure flow. • Sucrose is delivered at more or less the same rate to all regions, rather than going more quickly to the ones with the lowest sucrose levels, which the pressure flow theory would suggest.

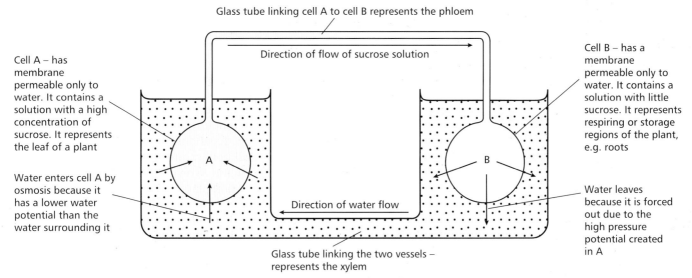

Provided sucrose is continually produced in A (leaf) and continually removed at B (e.g. root), the mass flow of sucrose from A to B continues

Fig 10.5 *Model illustrating the movement of sucrose solute by pressure flow in phloem*

• Aphids which have penetrated the phloem with their needle-like mouthparts can be used to extract the contents of the sieve tubes. These contents show diurnal variations in the sucrose content of leaves that are mirrored a little later by identical changes in the sucrose content of the phloem (Fig 10.6).
• The removal of a ring of phloem from around the whole circumference of a stem leads to the accumulation of sugars above the ring and their disappearance from below it.

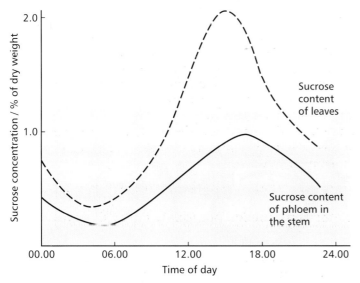

Fig 10.6 *Diurnal variation in sucrose content of leaves and phloem*

SUMMARY TEST 10.3

Transport of sucrose in plants occurs in the tissue called (1), from places where it is produced, known as (2), to places where it is used up or stored, called (3). One theory of how it is translocated is called the (4) theory. Initially the sucrose is transferred into (5) elements by either the (6) route or the (7) route. The sucrose is produced by (8) cells that therefore have a (9) water potential due to this sucrose. Water therefore moves into them from the nearby (10) tissue that has a (11) water potential. The opposite occurs in those cells (sinks) using up sucrose, and water therefore leaves them by the process of (12). Water entering at the sources and leaving at the sinks creates a (13) pressure that causes the mass flow of sucrose solution along the phloem.

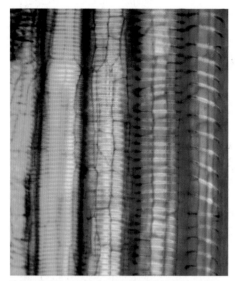

Xylem in LS as seen under a light microscope

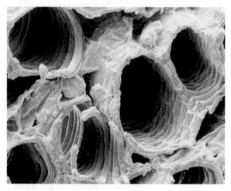

Leaf of tobacco showing xylem vessels (SEM) (×500 approx.)

Xylem is the principal water-conducting tissue in vascular plants. Details of this process are given in units 10.5 and 10.6.

10.4.1 Structure of xylem

Xylem (Figs 10.8 and 10.9) performs the functions of both supporting the plant and transporting water and minerals within it. **Parenchyma** cells and **sclerenchyma fibres** in the xylem all contribute to support, whereas the **vessels** and **tracheids** have both support and transport roles.

- **Xylem parenchyma** is composed of unspecialised cells that act as packing tissue around the other components of the xylem. They are roughly spherical in shape, but when they are turgid they press upon and flatten each other in places. In this way they provide support.
- **Xylem fibres** are elongated sclerenchyma cells with walls that are thickened with **lignin**; these features suit them to their role of support.
- **Vessels** (Fig 10.9(b)) vary in structure, depending on the type and amount of thickening of their cell walls, but are all hollow and elongated. As they mature their walls become impregnated with lignin, which causes them to die. The end walls break down to form a perforation plate which allows the cells to form a continuous tube. (The word 'element' is sometimes used rather than 'cell' because a cell is a living structure, whereas mature xylem vessels are dead). Sometimes the lignin forms rings (annular thickening) around the vessel; in other cases it forms a spiral or a network (reticulate thickening), see figure 10.7. This arrangement is better than a continuous thickening, because it allows elongation of the vessels as the plant grows. Pits in the lignified walls allow for lateral (sideways) movement of water. In angiosperms (flowering plants), vessels are the structures through which the vast majority of water is transported.
- **Tracheids** (Fig 10.9(a)) have a similar structure to vessels, except they are longer and thinner, and have tapering ends. They, too, are thickened with lignin and therefore die when mature. As with vessels, the end walls break down, and their side walls possess bordered pits which allow lateral movement of water between adjacent cells. Tracheids are found in all plants and are the main conducting tissue in ferns and conifers.

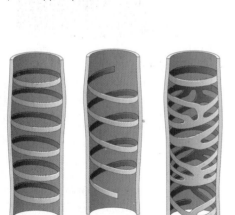

Annular   Spiral   Reticulate

Fig 10.7 *Types of thickening in xylem vessels*

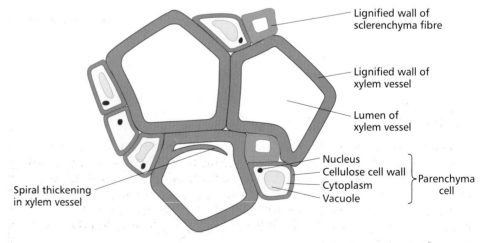

Lignified wall of sclerenchyma fibre

Lignified wall of xylem vessel

Lumen of xylem vessel

Nucleus
Cellulose cell wall
Cytoplasm
Vacuole
} Parenchyma cell

Spiral thickening in xylem vessel

Fig 10.8 *Xylem in TS as seen under a light microscope*

10.4.2 How xylem's structure is related to its functions

The vessels and tracheids of which xylem is mostly made up are structurally suited to the transport of water in a number of ways.

- The cells are long and arranged end to end to form a continuous column.
- The cell contents die when mature, which means that:
  - there is no cytoplasm or nucleus to impede water flow
  - the end walls can break down, so that there is no barrier to water flow between adjacent cells.
- Cell walls are thickened with lignin, which
  - makes them more rigid and therefore less likely to collapse under the tension created by the transpiration pull (section 10.6.2)
  - increases the **adhesion** of water molecules, enabling them to rise by capillarity (section 10.6.2).
- Annular, reticulate and spiral thickening allow xylem vessels to elongate during growth, and make them more flexible, so that branches can bend in the wind.
- There are pits throughout the cells, to allow lateral movement of water.
- The narrow **lumen** of the cells increases the height to which water rises by capillarity.

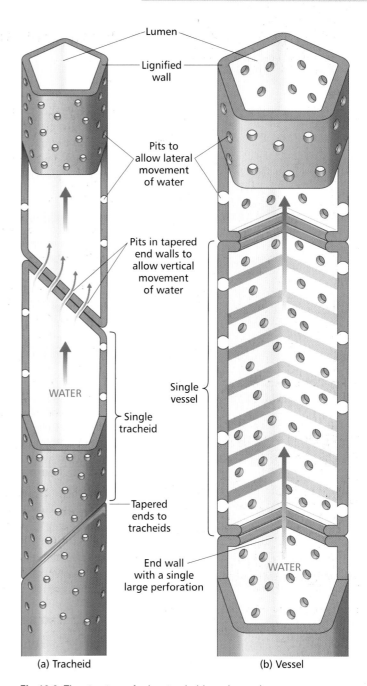

(a) Tracheid   (b) Vessel

Fig 10.9 *The structure of xylem tracheids and vessels*

SUMMARY TEST 10.4

Xylem transports water within the plant in two series of elongated structures called vessels and **(1)**. The walls of both are thickened with a substance called **(2)**, which forms three different patterns of thickening, known as **(3)**, **(4)** and **(5)**. Water moves through the central cavity, called the **(6)**, of the xylem elements but can move sideways between adjacent elements through structures called **(7)**. Also found within the xylem are living cells, called **(8)**, and supporting tissue, called **(9)**.

In section 4.6.3 we saw how root hairs are adapted to their role of absorbing water and mineral **ions**. These root hairs take up water, which then passes across the root cortex into the water-conducting tissue of the plant, the xylem. This, in turn, carries it up the stem to the aerial parts of the plant. The general structure of roots is described in section 10.1.3, while the structure of stems is covered in section 10.1.2.

10.5.1 Uptake of water by root hairs

Root hairs arise from **epidermal** cells a little way behind the tips of young roots. These hairs grow into the spaces around soil particles. In damp conditions they are surrounded by a soil solution which contains small quantities of mineral ions, but which is mostly water and therefore has a very high **water potential** – only slightly less than zero. By contrast, the root hairs, and other cells of the root, have sugars, amino acids and inorganic ions dissolved within them. These cells therefore have a much lower water potential. As a result, water moves by **osmosis** from the soil solution into the root hair cells down this water potential gradient.

Having been absorbed into the root hair cell, water continues its journey across the root cortex in three ways:

* the apoplast pathway
* the symplast pathway
* the vacuolar pathway.

10.5.2 The apoplast pathway

As water is drawn into endodermal cells, it pulls more water along behind it, due to the cohesive properties of the water molecules. This creates a tension that draws water along the cell walls of the cells of the root cortex (cortical cells), in exactly the same way as water moves through the xylem in the stem (section 10.6.2). The mesh-like structure of the cellulose cell walls of the cortical cells has many water-filled spaces and so there is little or no resistance to this pull of water along the cell walls.

10.5.3 The symplast pathway

This takes place across the cytoplasm of the cortical cells as a result of osmosis. The water passes through the cell walls along tiny openings called **plasmodesmata**. Each plasmodesma (singular) is filled with a thin strand of cytoplasm. There is therefore, in effect, a continuous column of cytoplasm extending from the root hair cell to the xylem at the centre of the root. Water moves along this column as follows:

* Water entering the root hair cell by osmosis (section 10.5.1) makes its water potential higher.
* The root hair cell now has a higher water potential than the adjacent cortical cell.
* Water therefore moves from the root hair cell to the cortical cell by osmosis, down the water potential gradient.
* This first cortical cell now has a higher water potential than its neighbour to the inside.
* Water therefore moves into this neighbouring cortical cell by osmosis along the water potential gradient.
* This second cortical cell now has a higher water potential than its neighbour to the inside, and so water moves from the second to the third cell by osmosis along the water potential gradient.

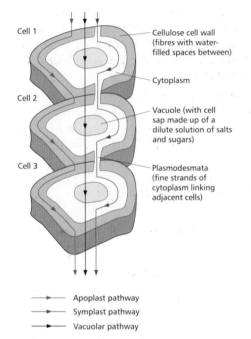

Cell 1

Cellulose cell wall (fibres with water-filled spaces between)

Cytoplasm

Cell 2

Vacuole (with cell sap made up of a dilute solution of salts and sugars)

Cell 3

Plasmodesmata (fine strands of cytoplasm linking adjacent cells)

Apoplast pathway
Symplast pathway
Vacuolar pathway

Fig 10.10 *Routes of water across cells such as those in the root cortex*

- At the same time, this loss of water from the first cortical cell lowers its water potential, causing more water to enter it by osmosis from the root hair cell.
- In this way, a water potential gradient is set up across all the cells of the cortex, which carries water along the cytoplasm from the root hair cell to the endodermis.

10.5.4 The vacuolar pathway

This occurs as a result of the same water potential gradient that arises from the process described for the symplast pathway (section 10.5.3). In this case, however, water moves through the vacuoles of the cortical cells as well, rather than being confined to the cytoplasm alone. These three pathways are illustrated in figure 10.10.

10.5.5 Passage of water into the xylem

Water reaching the endodermis by the apoplast pathway finds its further progress along the cell wall prevented by the waterproof band of **suberin** that makes up the Casparian strip in endodermal cells. At this point, water is forced into the living protoplast of the cell, where it joins water that has reached there by the symplast and vacuolar pathways (Fig 10.11). **Active transport** of salts is the most likely mechanism by which water now gets into the xylem. By pumping nitrates into the xylem, a much lower

(more negative) water potential is created and this causes water to enter the xylem from the endodermal cells by osmosis. As this active process requires energy, it can only occur within living tissue. It therefore takes place along protein carriers in the cell membrane and so this explains why the water from the apoplast pathway is forced into the living protoplast of the cell, by the suberin band of the endodermal cells.

The active transport of mineral salts into the xylem creates a lower water potential in the xylem and the subsequent osmotic movement of water into it. This creates a force that helps to move water up the plant. This force is called **root pressure**, and it can be measured at around 150kPa. While its contribution to water movement up a large tree is minimal, root pressure can make a significant contribution to water movement in small, **herbaceous** plants. Evidence for the existence of root pressure due to the active pumping of ions into the xylem includes:

- The pressure increases with a rise in temperature and decreases at lower temperatures.
- Metabolic inhibitors, such as cyanide, prevent most **ATP** production by respiration and therefore cause root pressure to cease.
- A decrease in the availability of oxygen or respiratory substrates cause a reduction in root pressure.

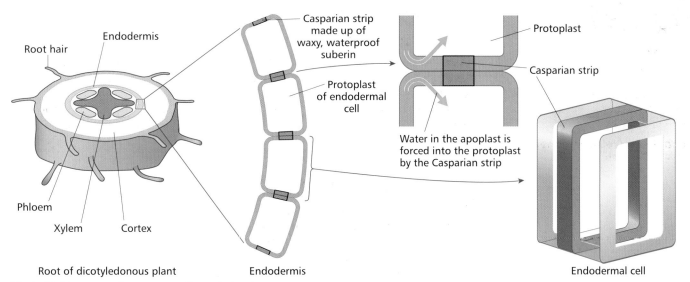

Fig 10.11 *Movement of water across the endodermis*

SUMMARY TEST 10.5

Water is absorbed from the soil solution by tiny extensions of root epidermal cells called **(1)**. This water is taken up because these cells have a **(2)** water potential than the soil solution that surrounds them. Water now passes into the cortical cells of the root in three ways: through the cytoplasm and vacuoles, a route called the **(3)** pathway; through the water-filled spaces of the cellulose cell wall, a route called the **(4)** pathway; and through the cytoplasm alone, a route called the **(5)** pathway, in which water crosses the cell walls through strands of cytoplasm called

(6). Around the vascular tissue at the centre of the root is a one-cell-thick ring of cells, called the **(7)**, the cell walls of which are impregnated with a band called the **(8)**, which is made up of the waxy, waterproof substance called **(9)**. Water from all pathways is now forced through the living portion, or **(10)**, of the endodermal cells. Water enters the xylem as a result of the active pumping of **(11)** into the xylem, which creates a **(12)** water potential that draws water into the xylem by the process of **(13)**. The movement of this water helps push water up the plant – a force known as **(14)**.

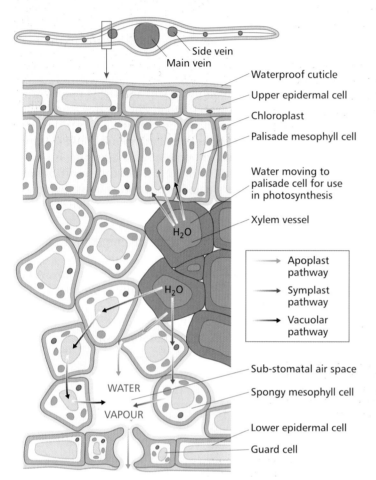

Fig 10.12 *Movement of water across a leaf*

Labels in figure:
- Side vein
- Main vein
- Waterproof cuticle
- Upper epidermal cell
- Chloroplast
- Palisade mesophyll cell
- Water moving to palisade cell for use in photosynthesis
- Xylem vessel
- H_2O
- H_2O
- Apoplast pathway
- Symplast pathway
- Vacuolar pathway
- Sub-stomatal air space
- Spongy mesophyll cell
- Lower epidermal cell
- Guard cell
- WATER VAPOUR

The main force that pulls water up the stem of a plant is the evaporation of water from leaves – a process called **transpiration** (unit 10.7). It is therefore logical to begin this unit from the point where water molecules evaporate from leaves, through the tiny openings, called **stomata**, on the surface of a leaf.

10.6.1 Movement of water across the leaf

The humidity of the atmosphere is usually less than that of the sub-stomatal air-space and so, provided that the stomata are open, water diffuses out of the air-spaces into the surrounding air. Water lost from the air-spaces is replaced by water evaporating from the cell walls of the surrounding **spongy mesophyll** cells. This water is replaced by water reaching the spongy mesophyll cells from the xylem by either the apoplast, symplast or vacuolar pathways (unit 10.5). In the cases of the symplast and vacuolar pathways, the water movement occurs because, once the spongy mesophyll cells have lost water to the sub-stomatal air-space, they have a lower (more negative) **water potential** and so water enters by **osmosis** from the neighbouring cells. The loss of water from these adjacent cells causes them to have a lower (more negative) water potential and so they, in turn, take in water from their neighbours by osmosis. In this way, a water potential gradient is established that pulls water from the xylem, across the leaf mesophyll, and finally out into the atmosphere. These events are summarised in figure 10.12.

10.6.2 Movement of water up the stem in the xylem

A number of factors are jointly responsible for the movement of water up the xylem from the roots to the leaves – cohesion, tension, capillarity, **adhesion** and root pressure. The main mechanism by which water moves up the xylem is known as the **cohesion-tension theory**. It operates as follows:

- Water evaporates from leaves as a result of transpiration (unit 10.7).
- Water molecules form **hydrogen bonds** between one another and hence tend to stick together – this is known as **cohesion.**
- Water forms a continuous, unbroken path across the mesophyll cells and down the xylem.
- As water evaporates from mesophyll cells in the leaf into the sub-stomatal air space, more molecules of water are drawn up behind it as a result of this cohesion.
- Water is hence pulled up the xylem as a result of transpiration. This is called the **transpiration pull**.
- The transpiration pull puts the xylem under **tension**, i.e. there is a negative pressure within the xylem – hence the name cohesion-tension theory.

Such is the force of the transpiration pull that it can easily raise water up the 100m or more of the tallest trees. One piece of evidence for the cohesion-tension theory is the change that occurs in the diameter of trees according to the rate of transpiration. During the day, when transpiration is at its greatest, there is more tension (more negative pressure) in the xylem. This causes the trunk to shrink in diameter. At night, when transpiration is at its lowest, there is less tension in the xylem and so the diameter of the trunk increases. Other evidence includes the fact that, if a **xylem vessel** is broken and air enters it, it can no longer draw up water. The broken vessel does not exude water, as would be the case if it were under pressure, but rather draws in air, which is consistent with it being under tension.

Capillarity is the result of **adhesion** between water molecules and the walls of the xylem vessels. This is sufficient in theory to raise water to a height of about 3m. However, adhesion also causes some frictional drag on the upward movement of water and hence its overall influence is very small. **Root pressure** (section 10.5.5) makes some contribution to the movement of water up the xylem, especially in small **herbaceous** plants.

False-colour SEM of spongy mesophyll

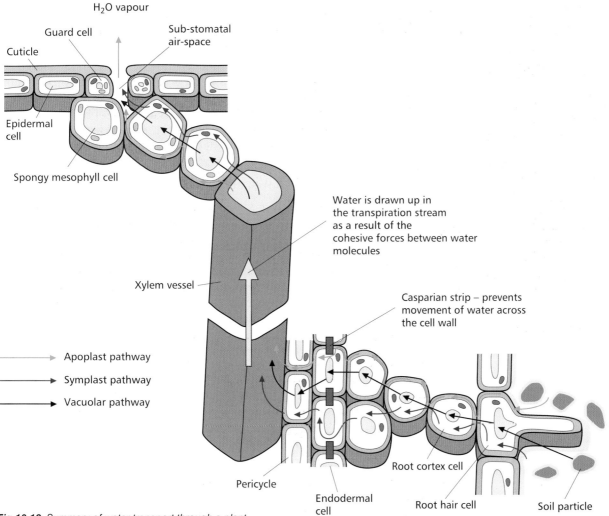

Fig 10.13 *Summary of water transport through a plant*

SUMMARY TEST 10.6

Water evaporates from the air-spaces in the plant in a process called **(1)**. This evaporation takes place mainly through pores called **(2)** in the epidermis of the leaf, each of which is surrounded by a pair of **(3)**. Water evaporates into the air-spaces from spongy mesophyll cells. As a result these cells have a **(4)** water potential and so draw water by **(5)** from neighbouring cells. In this way, a water potential gradient is set up that draws water from the xylem. Water is pulled up the xylem because water molecules stick together – a phenomenon called **(6)**. Other forces helping to move water up the stem include **(7)**, which is the result of adhesion between water molecules and the walls of the xylem vessels, and **(8)**, which is the result of the active movement of water into the xylem in the root.

Transpiration is the evaporation of water vapour from plants. It takes place at three sites:

- **Stomata** occur in leaves and **herbaceous** stems, and account for 90% of water loss.
- **Cuticle** is a waxy external layer on plant surfaces which serves to limit water loss through cell walls, although up to 10% of water nevertheless escapes by this route.
- **Lenticels** are areas of loosely packed cells on the surface of woody stems through which gas exchange, and therefore water loss, take place.

10.7.1 Role of transpiration

Although transpiration is universal in flowering plants, it is the unavoidable result of plants having leaves adapted for photosynthesis. Leaves have a large surface area to absorb light, and **stomata** to allow adequate inward diffusion of carbon dioxide. Both features result in an immense loss of water – up to 700dm^3 a day in a large tree. Transpiration is not essential as a means of bringing water to the leaves – osmotic processes could achieve this. It does, however help, although less than 1% of water moved in the transpiration stream is used by a plant. What then, if any, are the benefits of transpiration?

- It contributes to the supply of water to aerial parts.
- On hot days, it cools the plant.
- It speeds the process of mineral absorption and transport.

10.7.2 Measurement of water uptake under different conditions

It is virtually impossible to measure transpiration as it is extremely difficult to condense and collect all the water vapour that leaves all the parts of a plant. What we can easily measure, however, is the amount of water that is taken up in a given time by a part of the plant such as a leafy shoot. As around 99% of the water taken up by a plant is lost during transpiration, this measure is almost the same as the rate at which transpiration is occurring. We can then measure water uptake by the same shoot under different conditions, e.g. various humidities, wind speeds or temperatures. In this way we get a reasonably accurate measure of the effects of these conditions on the rate of transpiration.

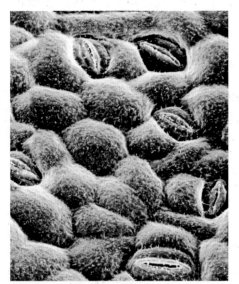

Surface view of stomata on a rose leaf (SEM) (×500)

Under natural conditions, the transpiration rate of both species increases in the middle of the day because the higher temperature speeds up the outward diffusion of water and the increased light intensity causes stomata to open. The species living in a dry environment is better adapted to conserve water.

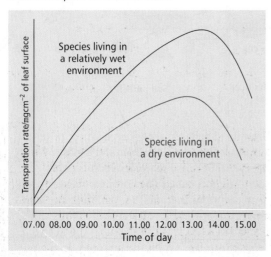

Fig 10.14 *Transpiration rates of two species of plants*

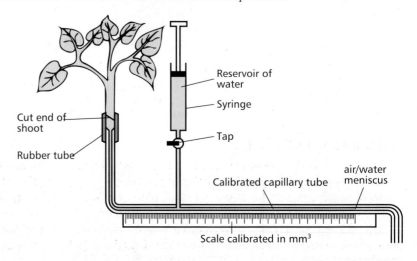

Fig 10.15 *A potometer – an instrument that measures water uptake which is more or less equivalent to water loss*

The rate of water loss in a plant can be measured using a **potometer** (Fig 10.15). The experiment is carried out in the following stages:

- A leafy shoot is cut under water to prevent air entering the xylem. Care is taken not to get water on the leaves.
- The potometer is filled completely with water, making sure there are no air bubbles.
- Using a rubber tube the leafy shoot is fitted to the potometer under water.
- The potometer is removed from under the water and all joints are sealed with petroleum jelly.
- The distance moved by the air / water meniscus in a given time is measured, and from this the volume of water lost can be calculated ($\pi r^2 l$).
- The volume of water lost against the time in minutes can be plotted on a graph.
- Once the air / water meniscus nears the junction of the reservoir tube and the capillary tube, the tap on the reservoir is opened and the syringe is pushed down until the meniscus is pushed back to the start of the calibration on the capillary tube. Measurements then continue as before.

The experiment can be repeated to compare the rates of water loss under different conditions, e.g. at different temperatures, humidities, light intensity, or differences in water loss between different species under the same conditions (Fig 10.14). To obtain reliable results, we need to take the following precautions when carrying out the experiment.

- The leafy shoot is cut under water to prevent air-bubbles being drawn into the xylem (which is under tension) and blocking the flow of water through the shoot.
- The end of the leafy shoot should be cut at the start to prevent any dissolved air from the water collecting at the end of the shoot.
- All joints must be watertight to prevent air breaking the continuous column of water. Tight-fitting rubber tubing and / or petroleum jelly is therefore used at all joints.
- When making comparisons, e.g. the rate of transpiration at different temperatures, all other factors, e.g. humidity, wind speed etc. must be kept constant.

10.7.3 Factors affecting transpiration

Both internal and external factors influence the rate of transpiration (table 10.2). Plants living in conditions that lead to high transpiration rates often have adaptations which allow them to reduce water loss. Such plants are called xerophytes; their features are considered in unit 10.8.2.

Table 10.2 Summary of factors affecting transpiration

Type	Factor	How factor affects transpiration	Increase in transpiration caused by	Decrease in transpiration caused by
External	Light	Stomata open in the light and close in the dark	Higher light intensity	Lower light intensity
	Humidity	Affects the diffusion gradient between the air-spaces in the leaf and in the atmosphere	Lower humidity	Higher humidity
	Temperature	Alters the kinetic energy of the water molecules and the relative humidity of the air	Higher temperatures	Lower temperatures
	Wind speed	Changes the diffusion gradient by altering the rate at which moist air is removed from around the leaf	Higher speeds	Lower speeds
	Water availability	Influences water potential gradient between soil and leaf	Wetter soils	Drier soils
Internal	Leaf area	Some water is lost over the whole surface of the leaf	Larger leaf area	Smaller leaf area
	Cuticle	Forms a waterproofing layer over the leaf surface	Thinner cuticle	Thicker cuticle
	Number of stomata	Most water is lost by evaporation through stomata	More stomata	Fewer stomata
	Distribution of stomata	Upper surface is more exposed to environmental factors that increase the rate of transpiration.	Greater proportion of stomata on upper surface of leaf.	Greater proportion of stomata on lower surface of leaf

SUMMARY TEST 10.7

Transpiration is the **(1)** of water from plants and around 90% of this occurs through **(2)**, with the remainder occurring either through the waxy **(3)** on plant surfaces, or loosely packed cells called **(4)** found on the surfaces of woody stems. The amount of water taken up by a plant can be measured using a piece of apparatus called a **(5)**. To produce reliable results it is essential that all seals are made **(6)**. Various external and internal conditions affect the rate of transpiration. In each of the following cases state whether the rate of transpiration is increased or decreased: reduced temperature **(7)**, higher humidity **(8)**, thicker cuticle **(9)**, more stomata **(10)**, lower light intensity **(11)**, greater wind speed **(12)**.

Xerophytes (xero = 'dry', phyte = 'plant') are plants that are adapted to living in areas where their water losses due to **transpiration** may exceed their water uptake. Without these adaptations these plants would become desiccated and die.

10.8.1 Xerophytic plants

Xerophytes are typically thought of as desert plants, showing a whole range of adaptations to cope with hot, dry conditions. However, similar adaptations may also be seen in plants found in sand dunes or other dry, windy places in temperate climates where rainfall is high and temperature relatively low. These adaptations are essential because the rainfall quickly drains away through the sand and out of the reach of the roots, making it difficult for these plants to obtain water. At the same time, coastal areas where sand dunes typically occur have salty soils and this lowers the **water potential** of the soil solution, reducing the water potential gradient between the soil solution and the root hair cells. Water uptake by **osmosis** is therefore very slow. Plants living on salt marshes may have their roots drenched in water but its saltiness means that this water is hard to obtain and so only xerophytic plants survive in these conditions. In addition, coastal regions are exposed to greater wind speeds, which increase transpiration rates. Plants living in cold regions often have difficulty obtaining water because it is frozen in the soil for much of the year. These plants also show xerophytic modifications to enable them to survive. There are therefore many habitats where plants show structural **(xeromorphic)** and physiological modifications designed to increase water uptake, store water and reduce transpiration. These modifications are called xerophytic features and some examples are listed in table 10.3.

10.8.2 Xerophytic adaptations of leaves that reduce transpiration

One way of surviving in habitats with an unfavourable water balance is to reduce the rate at which water can be lost through transpiration. As the vast majority of transpiration occurs through the leaves, it is these organs that show most modifications. Examples include:

- **Having a thick cuticle.** Although the waxy cuticle on leaves forms a waterproofing barrier, up to 10% of transpiration can still occur by this route. The thicker the cuticle, the less water can escape by this means. Many evergreen plants, such as *Ilex* (holly), have thick cuticles to reduce water loss, especially during the winter when water is difficult to absorb.
- **Rolling up of leaves.** Most leaves have their **stomata** largely, or entirely, confined to the lower epidermis. The rolling of leaves in a way that protects this lower epidermis from the outside helps to trap a region of still air within the rolled leaf. This region becomes saturated with water vapour and so there is no water potential gradient between the sub-stomatal air space and the outside, and so transpiration is considerable reduced. Plants such as *Ammophila* (marram grass) roll their leaves when transpiration rates are high, e.g. in hot or windy conditions.
- **Having hairy leaves.** A thick layer of hair on leaves, especially on the lower epidermis, traps moist air next to the leaf surface. The water potential gradient between the inside and the outside of the leaves is reduced and therefore less water is lost by transpiration. *Calluna* (ling) is a plant with this modification.
- **Having stomata in pits or grooves.** These again trap moist air next to the leaf and reduce the water potential gradient. Examples of plants using this mechanism include *Ilex* (holly) and *Pinus* (pine).

Conifers, such as this cedar, have needle-like leaves to reduce water loss

Holly has leaves with a thick waxy cuticle that reduce water loss

Succulents, such as this cactus, store water in their swollen stem and have sharp, needle-like leaves, both to reduce water loss and protect themselves from being eaten by animals in search of a juicy meal

Table 10.3 *Xerophytic adaptations of plants*

Xeromorphic feature	Mechanism	Examples
Long root system	• absorption of water from deep underground	Cacti
	• absorption of water from near surface after rain	*Acacia*
Reduction in transpiration rate	• reduction in leaf area	Needles of pine and spruce
		Spines of cacti and gorse
	• increasing humidity around stomata	Rolled leaves of marram grass
		Sunken stomata of holly and pine
	• thick cuticle	Evergreens such as pine and holly
Storage of water	• retention of water in stem for use in droughts	Prickly pear and many cacti
Resistance to wilting	• having smaller cells, which makes the proportion of cell wall material greater	Many xerophytes
	• more lignified material in leaf, enabling photosynthesis to continue	*Hakea*

- **Reducing the surface area to volume ratio of leaves.** We saw in section 4.6.1 that the smaller the surface area to volume ratio, the slower the rate of diffusion. By having leaves that are small and roughly circular in cross-section, as in pine needles, rather than ones that are broad and flat, the rate of water loss can be considerably reduced. This reduction in surface area must always be balanced against the need for a sufficient area for photosynthesis to meet the needs of the plant.
- **Closing stomata when transpiration rates are very high.** Plants such as cacti can close stomata during the hottest parts of the day. Some plants, called C_4 plants, use a modified form of photosynthesis that makes more efficient use of carbon dioxide, and so this closure of stomata does not unduly affect rates of photosynthesis. Other plants produce abscisic acid in response to the stress of dehydration and this causes the stomata to close.
- **Having a very low (more negative) water potential in leaf cells.** The accumulation of salts in cells reduces the water potential gradient between leaf cells and the air spaces around them, and so less water is lost from them.

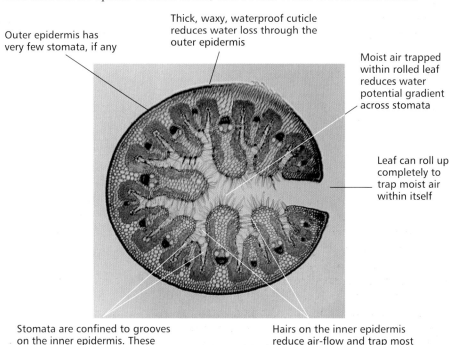

Outer epidermis has very few stomata, if any

Thick, waxy, waterproof cuticle reduces water loss through the outer epidermis

Moist air trapped within rolled leaf reduces water potential gradient across stomata

Leaf can roll up completely to trap moist air within itself

Stomata are confined to grooves on the inner epidermis. These grooves trap moist air

Hairs on the inner epidermis reduce air-flow and trap most air next to the leaf surface

Xerophytic modifications of the leaf of Ammophila *(marram grass)*

SUMMARY TEST 10.8

Xerophytes are plants adapted to living in areas where transpiration rates may exceed water uptake. In addition to hot and cold deserts, xerophytes can be found in temperate habitats such as **(1)** or **(2)**. The leaves of xerophytic plants show adaptations to reduce water loss. These include having a thick, waxy **(3)** or closing stomata when transpiration rates are high. Accumulating salts in leaf cells makes their water potential **(4)** and makes it harder for water to be lost from them. Increasing humidity around stomata is an effective way of reducing transpiration and three methods by which leaves achieve this are **(5)**, **(6)** and **(7)**.

1 Figure 1 is a vertical section through part of the leaf of a dicotyledon.

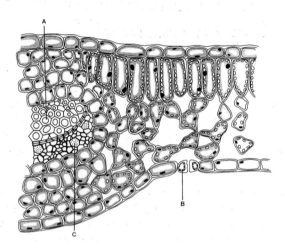

(Fig 1)

a Identify **A**, **B**, and **C**. *(3 marks)*

b Plants typically lose large quantities of water each day by transpiration.

Explain why the loss of large quantities of water by transpiration is inevitable. *(2 marks)*

c Figure 2 shows the rate of transpiration of the **same plant** on two consecutive mornings, day 1 and day 2.

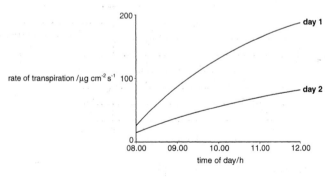

(Fig 2)

(i) Suggest **two** environmental factors that could account for the difference between day 1 and day 2 and explain how they could have caused the difference. *(4 marks)*

(ii) Describe an experimental technique that could have been used to obtain the data in figure 2.
(In this question, 1 mark is available for the quality of written communication.)

(6 marks)
QWC (1 mark)
(Total 16 marks)
OCR 2803/01 Jan 2003, B (T), No.2

2 Complete the following passage using the most appropriate word or words.

Some plants, such as cacti, inhabit dry areas. These plants of dry areas are known as … . Reduction of water loss by the process of … can be achieved by employing a variety of adaptations. In some species the leaves are needle-like, which reduces the … … ratio, whilst in others the epidermis is covered by a thick layer of … . In order to conserve the greatest amount of water, may species shut their stomata during the … .

(5 marks)
(Total 5 marks)
OCR 2803/01 Jun 2003, B (T), No.3

3 The figure shows the outline structure of some cells from the phloem of a dicotyledonous plant.

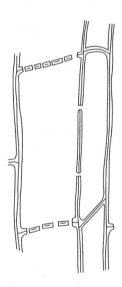

a (i) Label the following structures on the diagram using label lines:

sieve tube, sieve plate, sieve pore, companion cell, plasmodesma.
(5 marks)

(ii) Complete the drawing by adding **cytoplasm** and a **nucleus** or **nuclei** where appropriate. Label these components. *(2 marks)*

Movement in phloem occurs between **sources** and **sinks**.

b Use examples of regions in a plant to explain the meaning of these terms. *(2 marks)*

The table shows some of the typical components of phloem sap.

c State the form in which carbohydrate is translocated in the phloem. *(1 mark)*

component	concentration/mg cm^{-3}
sucrose	80–160
protein	1.45–2.20
amino acids	5.20
phosphate ions	0.35–0.55
potassium ions	2.30–4.40
ammonia	0.03
ATP	0.24–0.36
auxin	10.5×10^{-6}

Translocation is considered to be an active process.

d State,
  (i) what is meant by an *active process*; *(1 mark)*
  (ii) two pieces of evidence supporting the idea that translocation in the phloem is active. *(2 marks)*
  (Total 13 marks)
  OCR 2803/1 May 2002, B (T), No.1

4 The figure shows the pathways taken by water across the cells in the root of a dicotyledonous plant.

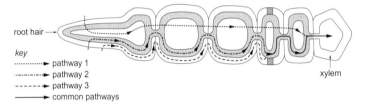

a (i) Label the endodermis on the diagram. *(1 mark)*
  (ii) Using the information in the figure, describe the pathways taken by water across the root. *(3 marks)*
  (Total 4 marks)
  OCR 2803/01 May 2002, B (T), No.3

5 The figure shows a section through a typical dicotyledonous stem.

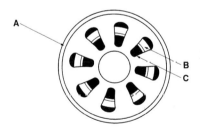

a With reference to the figure,
  (i) name the tissues **A** to **C**; *(3 marks)*
  (ii) describe how the distribution of **B** and **C** would differ in a dicotyledonous root. *(2 marks)*

b Describe the pathways and explain the mechanisms by which water is absorbed from the soil and transported across the root and up to the leaves.
  (In this question, 1 mark is available for the quality of written communication.) *(10 marks)*
  (Total 15 marks)
  OCR 2803/01 Jan 2002, B (T), No.2

6 a Explain what is meant by the term *transpiration*.
  (2 marks)

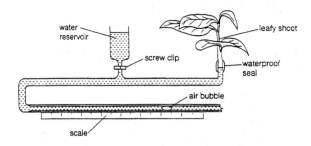

The figure shows a potometer which can be used to measure the rate of water uptake in a leafy shoot. The rate of movement of a bubble through the potometer is measured as water is taken up by the shoot. This may be used as a measure of the rate of transpiration by the shoot.

b (i) List **three** practical precautions which should be taken when using a potometer. *(3 marks)*
  (ii) Suggest an assumption that has to be made if the potometer is used to measure the rate of transpiration. *(1 mark)*

A lack of water in herbaceous plants results in wilting. Plant cells are no longer turgid and mechanical strength is lost. The graph shows the results of an investigation to compare the rates of transpiration and water absorption in a plant during a hot day in summer. There was adequate soil moisture available to the plant throughout the investigation, which began at midnight.

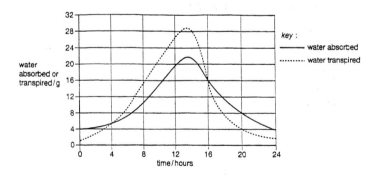

c With reference to the graph,
  (i) describe how the rate of transpiration varied during the investigation, and compare this with the rate of water absorption.
  (In this question, 1 mark is available for the quality of written communication.) *(8 marks)*
  (ii) suggest, in terms of transpiration and absorption, when wilting would be most likely to occur in this investigation. *(1 mark)*
  (iii) explain, in terms of water potential, when plant cells will no longer be turgid and wilting may occur. *(2 marks)*
  (Total 17 marks)
  OCR 2803/1, 2000 Specimen paper, B (T), No.3

PART THREE

Human health and disease

Introduction to health and disease

Types of disease – 1

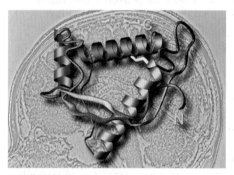

Computer-generated model of a prion molecule, the mutated form of a normal cell protein that causes Creutzfeldt-Jacob Disease (CJD). CJD illustrates the problem of classifying disease because it has mental and physical symtoms, is infectious and degenerative and, in some cases, can be inherited

11.1.1 Definition of health

It is not easy to define health. Some might say that it is the absence of disease. However, would it be correct to describe a person whose bodily systems are functioning normally but who feels depressed as healthy? What if they are just 'unhappy'? Can an alcoholic or drug addict whose habits have not yet caused any physical harm be called healthy? The World Health Organisation (WHO) defines health as: **'a state of complete physical, mental and social well-being and not merely the absence of disease or infirmity'**. In other words, to be healthy, a person should not only have all body organs working efficiently, but should also feel well. By the WHO definition, health includes:

- **physical well-being** – the body is able to function to its maximum potential and individuals are physically fit through taking exercise, having a balanced diet and taking appropriate rest and sleep
- **mental well-being** – individuals are free of mental illness and enjoy a high level of personal contentment as well as feeling good about themselves
- **social well-being** – individuals have their basic social needs met, e.g. proper housing and sanitation, and are properly integrated into the society to which they belong
- **absence of disease** – there are no symptoms of any disease that may prevent the efficient functioning of the body systems.

Health is clearly a complex concept. In addition to the physical, mental and social aspects dealt with in the WHO definition, there are also emotional, spiritual and sexual dimensions. Often the 'feel good' factor that is important to good health comes from having control over one's life. This usually involves having the means to

Table 11.1 Main groups of disease and examples of each

Category	Disease						
	Scurvy	Measles	Lung cancer	Coronary heart disease	Alzheimer's disease	CJD	Cystic fibrosis
Physical	✔	✔	✔	✔	✔	✔	✔
Mental					✔	✔	
Social			✔	✔			
Infectious		✔				✔	
Non-infectious	✔		✔	✔	✔		✔
Degenerative				✔	✔	✔	
Inherited						✔	✔
Self-inflicted			✔	✔			
Deficiency	✔						

exercise this control and may therefore include factors such as economic security, i.e. having employment or some other means of income. Overall, there are three main determinants of health – genetics, lifestyle and environment. We shall explore in the following units the often complex way in which these interact in determining our state of health.

11.1.2 Definition of disease

If it is hard to say what is meant by 'health', to define 'disease' is even more difficult. Disease is not so much a single entity, but rather a description of certain symptoms, either physical or mental, or both. Disease suggests a malfunction of body or mind which leads away from good health. Like health, it has mental, physical and social aspects. Some diseases, like malaria, have a single cause, but others, like heart disease, have a number of causes and are said to be **multifactorial**. Diseases may be divided into two forms depending upon their duration:

- **acute** – a sudden onset but short-duration condition from which the patient recovers quickly
- **chronic** – an on-going condition which may recur over a number of years.

These terms do not describe the severity of a disease. A chronic case of indigestion may be hardly noticeable, while an acute attack of hepatitis may be fatal. There are many ways of classifying diseases, but whichever is used, the groups tend to overlap, so that any one disease may be put in a number of different categories. The nine commonly recognised groups of disease, and how examples of each overlap, are given in table 11.1.

11.1.3 Physical diseases

Almost all diseases have some physical component. In other words, they involve permanent or temporary damage to some part of the body. Examples include pulmonary tuberculosis, which damages the lungs, and arthritis, which damages the joints. The only diseases that have no physical component are those psychological disorders that do not involve any physical damage to the brain.

11.1.4 Mental diseases

Mental diseases cover a broad range of disorders that cause psychological, personality or behavioural symptoms. What constitutes a mental illness may depend upon the 'normal' behaviour of a particular society. What is considered abnormal or deviant in one group might be perfectly acceptable in another. There is no clear distinction between physical and mental disease because each type often displays symptoms of the other. One example of a mental disease is schizophrenia, which affects around half a million people in the UK. People with schizophrenia display severe distortion and disorder of thought, leading to delusions, hallucinations, bizarre behaviour and social withdrawal. The causes of schizophrenia are not fully understood but may be linked to the brain neurotransmitter, dopamine. A mental disease with a known cause is new variant Creutzfeldt–Jacob disease (CJD), the human form of BSE (bovine spongiform encephalopathy) in cattle. This results from infection by protein molecules called prions, which cause spaces to appear in the brain, giving it a spongy appearance. As a result the patient experiences dementia and loss of coordination. As prions are proteins and not microorganisms, the body's immune system does not respond to their presence. CJD illustrates the overlapping nature of disease classification because it has mental and physical symptoms, is infectious as well as degenerative and, in 15% of cases, is inherited. Schizophrenia and CJD are examples of severe mental disorders known as **psychoses**. Less severe disorders, called **neuroses**, include compulsive or obsessive behaviour and phobias such as agoraphobia – a fear of open spaces.

11.1.5 Social diseases

Some diseases result from the social environment in which individuals live or the behaviour which they exhibit. They can include almost all infectious diseases because poor sanitation and over-crowding increase the risk of infections spreading in a population. Pulmonary tuberculosis can be spread in this way. Sexually transmitted diseases, such as acquired immune deficiency syndrome (AIDS), which are the result of sexual behaviour, are often referred to as social diseases. Social conditions may add to the risk of contracting certain diseases. Cardiovascular disease is more prevalent amongst the poorer groups of the developed world, while deficiency diseases, such as kwashiorkor, arise as the result of a poor diet amongst the poor of the developing world. Some occupations carry a direct risk of certain diseases, e.g. mesothelioma (a form of lung cancer) in asbestos workers.

SUMMARY TEST 11.1

Disease is not merely the absence of health, but also the complete (1) well-being, (2) well-being and (3) well-being. Some diseases, like heart disease, have many causes and so are said to be (4). Regardless of its severity, a disease may either be short-lived, in which case it is said to be (5), or be on-going over many years, in which case it is called (6). Physical diseases cause temporary or permanent damage to some part of the body. For example, the disease called (7) can damage joints. Mental diseases cause psychological, personality or behavioural symptoms. Where symptoms are severe, as in the case of schizophrenia, the diseases are called (8). More minor symptoms include phobias, such as fear of open spaces, a condition known as (9). Proteins called prions are responsible for a human form of BSE, called (10). Social diseases result from the environment in which people live. Examples include the sexually transmitted disease called (11) and the deficiency disease known as (12) that results in a swollen abdomen.

11.2.1 Infectious diseases

The human body makes an ideal habitat for many microorganisms. It provides a warm environment of constant temperature, a near-neutral pH, a ready supply of food and water, a constant supply of oxygen and mechanisms for removing wastes. Not surprisingly therefore, our bodies are naturally colonised by a wide variety of microorganisms. Many live more or less permanently in or on our bodies, causing us no harm; others, however, cause disease. Any organism that lives on or in a host organism and gains an advantage while causing harm to the host is called a **parasite**. Parasites include worms such as *Wucheria*, which causes elephantiasis, and insects such as fleas. Parasitic disease-causing microorganisms, such as viruses, bacteria, fungi and protoctists, are called **pathogens**.

Diseases that can be spread from person to person, or from animals to people, are called **infectious** or **communicable diseases**. The parasites may be transmitted in a variety of ways, e.g. through water, food, sexual contact or other social interactions. Examples of infectious diseases include influenza, measles, cholera, tuberculosis, malaria and AIDS. The last four are covered in detail in chapter 12.

11.2.2 Non-infectious diseases

This is an umbrella category which includes all those diseases that are not caused by pathogens. Also called **non-communicable diseases**, they include examples that are found in most of the other groups, namely mental, social, degenerative, inherited, self-inflicted and deficiency. Some diseases, e.g. sickle-cell anaemia, have a single cause while others, e.g. stroke, are multifactorial.

11.2.3 Degenerative diseases

Degenerative diseases are the result of the gradual breakdown in the functioning of tissues or organs as the result of deterioration. This deterioration may be the result of ageing, as in the case of senile dementia, or it may occur much earlier in life, e.g. multiple sclerosis. Those associated with old age commonly include problems with poor circulation, reduced mobility and memory loss. They may also occur when the immune system (unit 16.2) starts to attack the body's own cells. One form of senile dementia is **Alzheimer's disease**, where there is a series of degenerative changes in the brain leading to memory loss and confusion. Another example is **rheumatoid arthritis**, a common disease of joints. This begins with inflammation of the synovial membrane of a joint. This thickens and becomes filled with white blood cells that attack and erode away the cartilage at the ends of the bones in the joint. Examples like this, where the body's own immune system is misdirected against its own tissues rather than foreign ones, are known as **autoimmune diseases**.

11.2.4 Inherited diseases

Inherited diseases are those that are caused by **genes** and can be passed from parents to their children. In the UK, the most common genetic disease is cystic fibrosis, with one person in 2000 suffering its debilitating effects. It results from an individual inheriting two recessive **alleles** for the condition. This leads to an abnormality in the channel proteins that transport chloride **ions** across cell membranes (section 4.2.3). As a result, chloride ions are not transported out of the epithelial cells in places like the lungs, pancreas and testes. As water

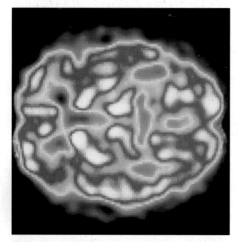

Colour-coded scan of normal brain

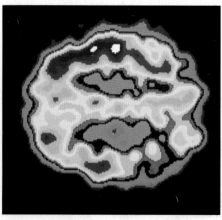

Colour-coded scan of brain of patient with Alzheimer's disease

normally follows the chloride ions out of the cell due to **osmosis**, the consequence is that the epithelial membranes are dry and the mucus they produce is stickier and more viscous than normal. This results in mucus congestion in the lungs, reduced gaseous exchange and blocked pancreatic and sperm ducts. A potential cure is to mask the defective genes by adding normal ones cloned from healthy individuals in a technique called **gene therapy**. In this way the recipient's cells produce the normal / functional version of the protein as opposed to the non-functioning one.

Other inherited conditions include sickle-cell anaemia, haemophilia and Huntington's disease – the latter being the result of a dominant, rather than a recessive, allele.

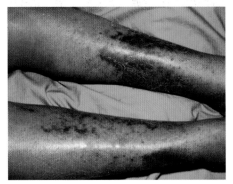
Bleeding of the shins is one symptom of scurvy

11.2.5 Self-inflicted diseases

There are a number of disorders that result from an individual's own actions and behaviour. In some cases the harmful consequences are known at the outset – few people who begin smoking can be unaware of the increased risk of lung cancer and emphysema that can result. In other cases, the damage may only become apparent later – exercise, normally beneficial, can lead to **osteoarthritis** if it is excessive or inappropriate.

Self-inflicted diseases cover a variety of conditions, ranging from attempted suicide, linked to poor mental health, to the misuse of drugs such as alcohol, nicotine or heroin. Obesity and heart disease may result from a high intake of fatty food, and skin cancer may be the consequence of excessive sunbathing. **Anorexia nervosa** has underlying psychological causes that result in an obsessive desire to be thin which leads individuals to starve themselves.

11.2.6 Deficiency diseases

Deficiency diseases are caused by the shortage of some essential nutrient in the diet. Perhaps the most dramatic is **kwashiorkor**, which results from a deficiency of protein in children. Another example is vitamin D deficiency. This is rare in developed countries because vitamin D is added to margarine and other spreads. Vitamin D is also manufactured naturally through the action of sunlight on the skin. Problems may arise for dark-skinned people in temperate countries, or for those who, for whatever reason, do not expose their skin to the sun. In these cases, calcium absorption is poor, leading to softer bones and a bowing of the legs in the young – a condition known as **rickets**, or **osteomalacia** (soft bones that are easily fractured) in adults.

GENETIC OR INHERITED DISEASES
The terms 'genetic' and 'inherited' are often used synonymously when referring to disease. Strictly speaking, however, not all genetic diseases are inherited. The form of Down's syndrome where a child has three copies of chromosome 21, for example, is a genetic disorder but it is not inherited. All **cancers** could be considered as genetic diseases because they involve uncontrolled cell division (unit 6.4) and yet very few forms of cancer are inherited. Many, possibly all, diseases could be said to have a genetic component because our genes partly determine such factors as our susceptibility to infectious disease and our personality, which in turn affect our lifestyle.

SUMMARY TEST 11.2

Any organism that lives in or on a host organism, causing harm while gaining an advantage itself, is called a (1). Disease-causing microorganisms are known as (2). Where these diseases are spread from individual to individual, the disease is said to be (3). Some diseases are the result of deterioration due to old age and these are termed (4) diseases. One example is a form of senile dementia, called (5), and another example which affects joints is called (6). Genetic disorders are examples of (7) diseases. An example of this is a condition causing mucus congestion in the lungs and blocked pancreatic ducts, which is called (8). It is the result of two (9) alleles being acquired from the parents. Diseases that result from one's own abuse of the body are known as (10) diseases, of which the obsessive desire to be thin, known as (11), is an example. The lack of some essential nutrient in the diet may lead to a (12) disease, such as (13), which results from a lack of vitamin D in children, or (14) where protein is the missing nutrient.

The incidence of diseases alters in time and space. In our efforts to combat disease, it is therefore helpful to collect statistics on the patterns and distribution of disease.

11.3.1 Epidemiology

Epidemiology is the study of the spread of disease and the factors that affect it. It involves an analysis of the incidence and patterns of disease with a view to finding various means of prevention and control. Originally it was applied to infectious diseases but modern epidemiology is equally concerned with environmental and lifestyle factors in the incidence of diseases such as cancer and cardiovascular diseases.

Epidemiologists try to identify the cause of any new disease – for example AIDS, which first appeared in the 1980s, and severe acute respiratory syndrome (SARS), which appeared in 2003. The collection of statistics over many years may help to establish a link between a certain factor and disease, e.g. smoking and lung cancer, mesothelioma and asbestos. Such links are then followed up with scientific experiments to confirm whether there is a biological basis for such a link and that it is more than a statistical coincidence.

11.3.2 Terminology

In order to refer clearly and unambiguously to health statistical data, a number of terms need to be understood.

- **Incidence** – this is the number of new cases of a disease in a population in a given time, for example in one month or one year. It is expressed as follows:

$$\text{Incidence} = \frac{\text{number of new cases of a given disease in a certain period}}{\text{number of individuals in a population}}$$

- **Prevalence** – this is the number of people in a population suffering from a particular disease at a particular time, regardless of when the disease was first recognised.
- **Mortality rate** – this is the number of people who have died of a particular disease in a given time, e.g. in one month or one year. This is usually expressed as:

$$\text{Mortality rate} = \frac{\text{number of deaths as a result of a given disease}}{\text{number of individuals in a population}}$$

It can, however, be calculated using only those individuals who are known to have the disease, in which case it is expressed as:

$$\text{Mortality rate} = \frac{\text{number of deaths as a result of a given disease}}{\text{number of individiuals in the population suffering from the same disease}}$$

- **Morbidity rate** – this is the number of people who are suffering from a particular disease in a given time. It is normally expressed as the number of people suffering from the disease divided by the number of individuals in the population.
- **Endemic** – infectious diseases are referred to as endemic if the disease is always present in the population. Tuberculosis is endemic in many countries, and most people carry the bacteria that cause it, even though they do not show any symptoms of the disease.

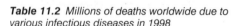

Table 11.2 *Millions of deaths worldwide due to various infectious diseases in 1998*

Infectious disease	Deaths / millions
Acute respiratory infections (e.g. pneumonia and influenza)	3.5
HIV/AIDS	2.3
Diarrhoeal diseases (e.g. cholera, typhoid and dysentery)	2.2
Tuberculosis	1.5
Malaria	1.1
Measles	0.9

Health statistics helped to establish the link between smoking and lung cancer

- **Epidemic** – refers to any disease that spreads suddenly to a large number of people over a widespread area. Influenza epidemics arise periodically, affecting large numbers of people, often across different parts of the globe. This, however, is distinct from the regular, expected rise in influenza cases during the winter in many countries, including the UK.
- **Pandemic** – this term is used to describe worldwide epidemics, such as acquired immune deficiency syndrome (AIDS). Severe occurrences of influenza may affect much of the globe and so are referred to as a pandemic.

11.3.3 Using health statistics

Statistics on the health of populations, rather than of individuals, are collected and expressed in a form that allows fair comparisons to be made, e.g. death rate per thousand of the population per year. The information gives a valuable insight into the health of a population or a country. Comparisons can be made between different geographical areas, ethnic groups, ages or occupations. This enables governments to set priorities for spending, targeting particular diseases or areas, or setting up screening programmes. The World Health Organisation (WHO) operates similar procedures on a global scale, identifying areas with specific problems and coordinating international action. The collection of accurate data not only enables suitable action to be taken, but also enables the effectiveness of various programmes to be monitored. Such data inform us that no more than six deadly infectious diseases (table 11.2) caused half of all premature deaths in 1998.

11.3.4 Analysing health statistics

Mild food poisoning can be caused by various bacteria, including the bacterium *Campylobacter*. Humans usually become infected with this bacterium by eating contaminated meat, especially chicken, or by drinking contaminated milk. The incidence of food poisoning caused by *Campylobacter* in England and Wales over a typical 3-year period is given in figure 11.1. What does this graph tell us and what are the explanations for the fluctuations in the cases of food poisoning?

Firstly the graph shows an annual variation, with the highest incidence in the summer months and the lowest in the winter months. Clearly, it is warmer in the summer and therefore:

- bacteria can multiply more rapidly and build up in chicken and milk that is not kept refrigerated
- more barbecues take place in the summer months. Meat may not be as thoroughly cooked on barbecues as it would be in a thermostatically controlled oven, and so bacteria such as *Campylobacter* may not be killed during cooking.

The collection of health statistics for diseases such as food poisoning can therefore provide patterns of disease

incidence that cause questions to be asked, such as those above. The answers to these questions allow health professionals to provide guidance, e.g. keep milk refrigerated and cook meat thoroughly on barbecues, which can lead to a reduction in the incidence of disease.

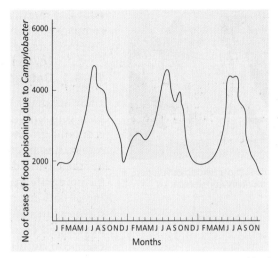

Fig 11.1 *Health statistics for food poisoning caused by* Campylobacter

11.3.5 Presenting and comparing health statistics

When making comparisons between the numbers of deaths due to a disease in one **population** and another, it can be misleading simply to state the number of people affected or who die. This is because it takes no account of the total numbers in each population. Population A may have twice as many people affected by a disease as population B, but if population A is 10 times larger than population B, then the proportion of individuals affected is actually much lower. For this reason, health statistics are usually expressed as the number of deaths, or people affected, per 100 000 of the population.

SUMMARY TEST 11.3

The study of the spread and distribution of disease is called (**1**). The number of people in a population suffering from a disease at any one time is called its (**2**), while the number dying from it is called its (**3**). Where a disease is always present in a population, it is said to be (**4**), and an example of such a disease is (**5**). Diseases such as (**6**), which occur all over the world, are said to be (**7**). To make health statistics easy to compare, the number of people dying or affected by a disease is usually expressed per (**8**) of the population.

Global patterns of disease distribution

Health-care resources have helped developed countries reduce mortality from infectious diseases

Health statistics tell us that the types of disease, their **prevalence** and their mortality rates vary from one part of the world to another. Before we can make comparisons between the health of one country and another, we must find both a way of categorising countries and a means of measuring health.

11.4.1 Categorising countries

There are many ways of dividing the countries of the world into groups – by their geographical position, by their climate, according to race or religion, on economic grounds etc. Each has a valid basis and each, to a greater or lesser extent, has an effect on health. One method is to divide the world into developed and developing countries. This takes account of a number of factors, including life expectancy, educational attainment and a measure of wealth (gross national product, or GNP). Countries that are ranked highly on most of the measures are called **developed countries** (or **more economically developed countries (MEDC)**), and include industrialised nations such as European countries, Russia, North America, Japan and Australia. Countries that are ranked low down on most of the measures are referred to as **developing countries** (or **less economically developed countries (LEDC)**), and include most countries of Africa, Asia and Central and South America. Such a method of dividing up the world is inevitably somewhat crude and inaccurate. For example, some parts of many large countries, e.g. China, are far more industrialised and highly developed than other parts. The division does, however, give us a general basis on which to compare the standard of health in different countries.

11.4.2 Measurement of health

As we saw in section 11.1.1, it is not easy to define what exactly we mean by health – measuring it is therefore at least as difficult. We can give descriptive accounts of health, but these do not provide the sort of precise statistical data needed to make comparisons between countries. Instead we have to look to a series of measures that we can quantify; these include:

- **the mortality rate** (section 11.3.2) of the whole **population** from different diseases
- **the prevalence of diseases** (section 11.3.2) within a population
- **the infant mortality rate from different diseases**. This is calculated as the number of babies that die during the first year of life per 1000 live births in the population. Babies are more vulnerable to disease as they have not acquired **immunity** to many **pathogens**. Infant mortality therefore gives a measure of the level of infectious disease within a population
- **life expectancy** is the number of years that a newborn can, on average, expect to live.

11.4.3 Diseases in developing countries

Figure 11.2 shows the main causes of death in developing countries. You will notice that 45% of deaths are the result of infectious diseases. The high mortality rate for these diseases has many causes, including:

- **poor sanitation** leading to the spread of water-borne diseases like cholera, diarrhoea, dysentery and typhoid
- **unsafe water** that often contains the pathogens of the above diseases
- **densely populated cities and over-crowded accommodation** increasing

the likelihood of air-borne diseases, such as tuberculosis and influenza being transmitted from person to person

- **widespread poverty** leading to poor diet and a lack of doctors and health facilities to treat disease. Malnourished individuals are less able to fight infections.
- **climate** – many developing countries are situated in warmer areas of the world where pathogens, and the insects that can spread them (e.g. mosquitoes carrying malaria), can reproduce rapidly and build up large populations.

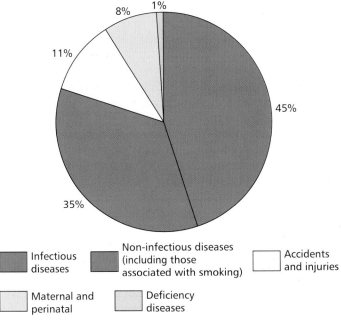

Fig 11.2 Main causes of death in low-income countries (South-East Asia and Africa) in 1998 (source: World Health Organisation Report 1999)

In developing countries there are also higher incidences of deficiency diseases and accidents, and many more deaths at birth or in the early years of life. It is poverty more than any other factor that accounts for these differences compared to developed countries. In 1999, the World Health Organisation (WHO) reported that about one-third of the world's population (1.3 billion people) had an income of less than one euro (approximately 70p) a day. This results in malnutrition leading to starvation of the very young and deficiency diseases in those who survive the first few years of life. It also means that poor countries lack the medical facilities to treat diseases and the resources to prevent disease through health education programmes. Similarly the lack of education accounts for the larger number of

accidents, although the use of old and unsafe equipment (another consequence of poverty) is also a major contributor. Of concern is the rise in self-inflicted diseases in developing countries, notably those associated with smoking, considered by the WHO to be reaching **epidemic** proportions.

11.4.4 Diseases in developed countries

In developed countries, both the **incidence** of infectious disease and the mortality associated with it have been reduced. Living conditions are substantially better than in developing countries, with improved hygiene, sanitation and nutrition. There are successful **vaccination** programmes, and **antibiotics** are readily available to cure bacterial infections. As can be seen in figure 11.3, only about 2% of deaths result from infectious diseases. However, the relative affluence of developed countries brings an increase in deaths from cardiovascular diseases, **cancer** and road accidents, albeit that cardiovascular disease is more prevalent amongst the poorer sections of developed countries. Also, most people live longer than in developing countries and so the degenerative diseases associated with old age are often seen.

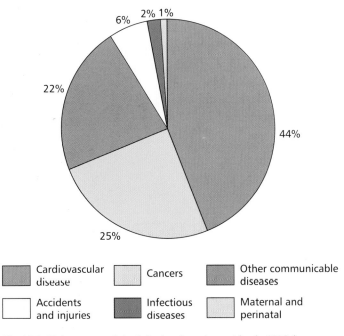

Fig 11.3 Main causes of death in developed countries in 1998 (source World Health Organisation Report 1999

SUMMARY TEST 11.4

The world can be divided into developing and developed countries based upon three main measures: **(1)**, **(2)** and **(3)**. Measuring the health of people within these countries is difficult and we have to use quantitative measures, four of which are: **(4)**, **(5)**, **(6)** and **(7)**. The major cause of death in developing countries is **(8)**. The reasons for this include poor sanitation and unsafe water, which lead to the spread of water-borne diseases such as **(9)** and **(10)**.

Overcrowding increases the chances of contracting an air-borne disease such as **(11)**. Overall, it is poverty that increases the likelihood of an early death, as it leads to malnutrition, poor education and health facilities and the use of unsafe equipment and machinery. The latter results in the high death rate from **(12)**. In developed countries, the main cause of death, by far, is **(13)**, with **(14)** as the second largest cause.

11.5 Health and the human genome project

The human genome project is arguably the most exciting piece of scientific research in recent times. The information it provides will help mankind to tackle problems of pollution, disease and ageing, and provide immense benefit to health and the environment. It also presents mankind with social, economic and ethical problems as to how the information revealed might be used.

11.5.1 What is the human genome project?

The human genome project is the largest piece of international collaboration in biology that has ever taken place. A **genom**e is all the genetic material in an organism, i.e. all its **genes**. The aim is to sequence accurately every one of the 3.2 billion bases which make up the DNA of a single human being (unit 5.2). The project, which began in 1990, was so vast that it was agreed that specific laboratories in more than 50 countries would each take a different portion of a human **chromosome** and simultaneously sequence its DNA. Even then, it was expected to take 15 years at least but, in the event, the first draft of the human DNA base sequence was published on 26 June 2000. Certain details still need to be finalised, but the complete picture should be available shortly. So vast is the amount of information that it is never likely to be published on paper. To do so would take nearly 500 000 pages – about 2000 books the size of this one. Instead, the information will be stored electronically and made available over the internet. One discovery has been that there are only around 30 000 genes in the human genome, and some recent estimates suggest it may be as few as 25 000 – even a nematode worm has 15 000 genes. What the human genome project has revealed is just how many genes we share with other organisms – all but 300 of our 25 000–30 000 genes are found in the genome of a mouse and some 230 of our genes are found in our most distant of relatives, the bacteria. Some important applications of the sequencing of the human genome are in the field of medicine, examples of which are discussed in the rest of this unit.

11.5.2 The prevention and treatment of human genetic diseases by gene therapy

Many human genetic diseases, such as haemophilia, sickle-cell anaemia, thalassaemia and **cystic fibrosis**, are the result of a gene not expressing itself, or being missing altogether. There are around 4000 such diseases, affecting 1–2% of the population. Some of the diseases carried on the X chromosome are illustrated in figure 11.4. The human genome project means that we know the base sequence of the normal genes. We can therefore make copies of these normal genes and supplement the defective genes with normal ones. This treatment is called **gene therapy** and is currently being used in clinical trials to treat genetic disorders such as cystic fibrosis, haemophilia, **rheumatoid arthritis** and AIDS. In the case of cystic fibrosis the process is as follows.

- A normal and therefore dominant cystic fibrosis transmembrane-conductance regulator (CFTR) gene is identified from the sequencing of the human genome.
- The CFTR gene is inserted into a bacterial **plasmid** and re-introduced into a bacterium for **cloning**.
- Cloned copies of plasmids with the normal CFTR gene are either introduced into a harmless virus or wrapped in lipid molecules to form liposomes in order to allow them to easily pass across cell surface membranes.
- The viruses or liposomes are introduced into the lungs of the cystic fibrosis sufferer, where the dominant normal CFTR gene they contain over-rides the detrimental effects of the recessive mutant gene. The effect, however, is localised

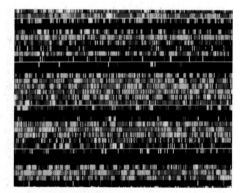

Human DNA sequences

– being limited to the area treated, namely the epithelial lining of the lungs. As this lining is constantly sloughed off over time the treatment has to be repeated.

11.5.3 The treatment of diseases that are influenced by heredity

Diseases like **cancer**, heart disease and Alzheimer's disease are multifactorial, i.e. they have a number of causes. In many cases, individuals have a genetic make-up that means they are susceptible to the disease. Knowing the sequence of the human genome will allow these individuals to be identified and the genetic component of their disease to be treated in the same way, e.g. by gene therapy, as other genetic diseases. While this will not act as a cure, it will reduce the likelihood of an individual getting a particular disease.

11.5.4 Genetic testing

Knowledge of the human genome allows reliable tests to be developed which can identify whether an individual is a carrier of a genetic disorder, e.g. thalassaemia. One key element in this process will be a new form of one-stop genetic test, called a **DNA chip**. A DNA chip (biochip) contains hundreds of normal human genes constructed by copying precisely the sequences identified by the human genome project. If a small amount of a person's DNA, e.g. a drop of blood, is flushed over the DNA chip, at the points where the genes in the fluid match those on the chip, they will stick to one another. A computer can be used to identify these matches and also any differences – called SNPs (single **nucleotide** polymorphisms). In this way, a record can be made each time an individual nucleotide differs from the reference sequence on the DNA chip. Some differences will be harmless, e.g. the production of red hair, but others will cause disease, e.g. sickle-cell anaemia, or thalassaemia. The individuals concerned can be offered treatment (if it is available) and be made aware of the probability of their offspring being affected and so decide whether or not to have children. DNA chip technology is likely to become the main diagnostic tool of medicine over the coming decades. There are plans to create global data bases of all SNPs that are detected so that doctors will easily be able to screen patients for those genes that may cause genetic diseases.

11.5.5 Improved diagnosis of disease

Knowing the DNA sequence of all human genes will make it possible to test individuals using a DNA chip (section 11.5.4) for those genes that make them susceptible to other diseases, e.g. cancer, Alzheimer's disease and heart disease. These individuals can then take additional precautions to reduce the level of other risk factors. For example, people with a genetic susceptibility to heart disease can reduce their salt intake, take regular exercise, refrain from smoking etc., and so offset the additional risk of heart disease that they have inherited. In time, gene therapy may be available to eliminate the genetic risk altogether.

11.5.6 Designing more effective drugs

With our knowledge of the nucleotide base sequence for each human gene, we can determine the amino acid sequence of the polypeptide it codes for. In turn we can find the three-dimensional structure of every protein coded for by DNA. Many of these proteins will be enzymes involved in metabolic processes. We can therefore design drugs that will precisely match or fit these enzymes. In this way the drugs can supplement or inhibit the action of enzymes, and so treat diseases by controlling the metabolic processes that cause the disease or give rise to unwanted symptoms of the disease. Individuals respond differently to the same drug. Some drugs are ineffective on one person but highly effective on another; some people get side effects, some do not. With improved knowledge on how genes and drugs interact, we can produce 'designer' medicines that are individually tailored to a person's own genetic profile and therefore are highly effective and produce no side effects.

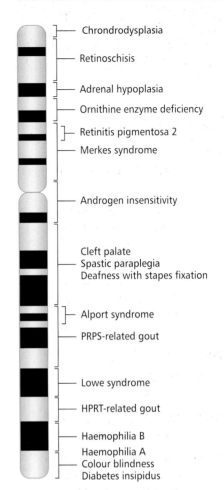

Fig 11.4 Some diseases that have been mapped on the human X chromosome. At least 60 diseases have been mapped to segments of the X chromosome. Many of these disorders have forms on other chromosomes and are also influenced by genes on other chromosomes.

Labels on chromosome:
- Chrondrodysplasia
- Retinoschisis
- Adrenal hypoplasia
- Ornithine enzyme deficiency
- Retinitis pigmentosa 2
- Merkes syndrome
- Androgen insensitivity
- Cleft palate
- Spastic paraplegia
- Deafness with stapes fixation
- Alport syndrome
- PRPS-related gout
- Lowe syndrome
- HPRT-related gout
- Haemophilia B
- Haemophilia A
- Colour blindness
- Diabetes insipidus

SUMMARY TEST 11.5

The human genome project is a massive international effort to sequence all the nucleotide bases, which number around (**1**), on the 23 pairs of chromosomes in each human cell. The knowledge gained can be used in the treatment of genetic disorders such as (**2**) and (**3**), in a process known as (**4**). It can also help in the treatment of multifactorial diseases like (**5**) and (**6**). Other benefits include improved disease diagnosis as well as (**7**) and (**8**). There are, however, moral, ethical and cultural implications that come with this new and immense body of knowledge.

1 Diphtheria is an acute infectious disease that affects the trachea and bronchi and sometimes the skin. The bacterium that causes diphtheria releases a toxin that affects nervous tissues and the heart. The disease is spread by droplet infection.

The World Health Organisation (WHO) collects statistics on infectious diseases, such as diphtheria. The figure shows the number of reported cases of diphtheria worldwide between 1980 and 2000.

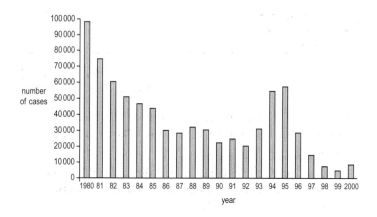

a Describe the trends in the number of reported cases of diphtheria between 1980 and 2000.

Credit will be given if you use figures to illustrate your answer. *(3 marks)*

b Explain why the WHO collects statistics for major infectious diseases, such as diphtheria. *(3 marks)*

c In 1940, the number of reported cases of diphtheria in England and Wales was 42 281, with 2480 deaths. A vaccine was introduced to give protection against diphtheria in that year. The number of reported cases in England and Wales is now very small indeed.

Explain how vaccination has reduced the number of cases of infectious diseases, such as diphtheria.

(3 marks)

d Antibiotics, such as penicillin, are chemical substances used in the treatment of infectious diseases.

Explain why the widespread use of antibiotics may be considered to be undesirable. *(3 marks)*

(Total 12 marks)

OCR 2802 Jan 2003, B (HHD), No.6

2 The first draft of the human genome was announced on 26th June 2000. The first comprehensive overview and analysis of the human genome was published in the scientific journal *Nature* on 15th February 2001.

a Define the term *genome*. *(1 mark)*

The following statements have been made recently about the Human Genome Project.

> Genomics, the science of acquiring genome knowledge and determining what genes do, will dramatically influence healthcare as further research should lead to improvements in diagnosis, prevention and treatment.
>
> **BBC News Online**

> I am concerned that there are some who will want to use this new knowledge as a basis for discrimination.
>
> Dr Craig Venter,
> **Celera Genomics**

> Now we can look down the list of genes and see which one is causing the problem.
>
> Professor John Burn,
> **Newcastle University**

> The human genome will exist on the world's computers for as long as we exist.
>
> Dr John Sulston,
> **Sanger Centre, Cambridge**

> We need an international agreement that genetic information needs to be obtained by consent.
>
> Dr Arthur Caplan,
> Bioethics Dept.,
> **Pennsylvania University**

b Outline the implications of the Human Genome Project for human health.

Credit will be given for answers that deal with a variety of issues rather than concentrating on just one. You may refer to the statements above in support of any points that you make.

(In this question, 1 mark is awarded for the quality of written communication.)

(7 marks)
QWC *(1 mark)*
(Total 9 marks)

OCR 2802 May 2002, B (HHD), No.4

3 a Explain the meaning of the following terms as they apply to infectious diseases.
endemic   *epidemic*   *(2 marks)*

The table shows the diseases which cause death in developing and developed countries.

developing countries		developed countries	
disease	percentage deaths	disease	percentage deaths
diarrhoea	42	heart diseases	32
respiratory infections: e.g. tuberculosis (TB)	25	cancers	23
		strokes	12
malnutrition	10	bronchitis	6
malaria	7	pneumonia	5
measles	15	others	22
others	11		

b With reference to the table,

(i) explain why infectious diseases are leading causes of death in developing countries; *(4 marks)*

(ii) explain why degenerative diseases are leading causes of death in developed countries. *(4 marks)*
(Total 10 marks)
OCR 2802, 2000 Specimen paper, B (HHD), No.1

4 a Complete the table below to show which of the three statements about disease transmission apply to cholera, tuberculosis (TB) and HIV/AIDS. Put a tick (✓) to show if the statement applies.

statement	cholera	tuberculosis (TB)	HIV/AIDS
causative organism is a bacterium			
transmission is via drinking water			
sexually transmitted			

(3 marks)

b Suggest **two** reasons why antibiotics are **not** suitable for treating all infectious diseases. *(2 marks)*

Tuberculosis is a disease considered to be endemic throughout the world.

c State the meaning of the term *endemic*. *(1 mark)*

d Explain the advantage of expressing numbers of deaths as *per 100 000 of the population*. *(2 marks)*

The following table shows the estimated number of deaths in 1998 for five diseases in developing countries in South-East Asia and developed countries in Europe.

e With reference to this table, describe and explain the differences between standards of health in developing countries and developed countries.
(In this question, 1 mark is awarded for the quality of written communication.) *(8 marks)*

disease	number of deaths per 100 000	
	developing countries in South-East Asia	developed countries in Europe
tuberculosis	51.90	1.78
HIV/AIDS	23.66	3.05
diarrhoeal diseases, e.g. cholera	36.39	1.01
lung cancer	17.10	53.00
chronic obstructive pulmonary disease e.g. chronic bronchitis and emphysema	11.93	35.67

(Total 16 marks)
OCR 2802 Jun 2001, B (HHD), No.3

5 Diseases are classified into categories. It is sometimes appropriate to classify a disease into more than one category.

a Complete the table below by indicating with a tick (✓) the category or categories in which each disease is classified.

disease	categories of disease		
	infectious	deficiency	degenerative
cholera			
night blindness			
lung cancer			
stroke			

(4 marks)

b (i) State **one** example of an inherited disease. *(1 mark)*

(ii) Explain why the disease you name is classified as an inherited disease. *(3 marks)*

Some diseases are classified as self-inflicted diseases. Smoking is considered by the World Health Organisation to be a self-inflicted disease of epidemic proportions.

c State what is meant by a *self-inflicted disease*. *(1 mark)*

d Suggest why the World Health Organisation considers smoking to be a disease of epidemic proportions.
(1 mark)
(Total 10 marks)
OCR 2802 Jan 2001, B (HHD), No.1

12 *Diet*

12.1 Balanced diet

In order to remain healthy, we need to eat and drink adequate amounts of nutrients:

- to provide energy
- for the growth and repair of cells
- for the proper functioning of all our vital organs.

12.1.1 Food and nutrients

A **food** is any substance taken into the body that, once processed, can supply:

- material from which the body can release energy
- material for growth, repair and reproduction
- substances to regulate energy release, growth, repair and reproduction.

Foods are a mixture of substances known as **nutrients**. There are five main groups of nutrients:

- **carbohydrates** to provide energy
- **lipids (fats)** to provide energy and essential fatty acids
- **proteins** which are mainly used to provide **essential amino acids** for the growth and replacement of tissues, but may also provide energy
- **vitamins** to regulate metabolic processes
- **mineral (inorganic) ions** that may be used for growth and repair, and to regulate metabolic processes.

Some of these nutrients are required in relatively large amounts and are called **macronutrients**. Macronutrients are carbohydrates, lipids and proteins. Other nutrients are needed in much smaller quantities and are called **micronutrients**; these are vitamins and mineral salts.

Macronutrients are often broken down during digestion. Carbohydrates such as polysaccharides can be **hydrolysed** into disaccharides and monosaccharides (units 2.2 and 2.3). Lipids can be hydrolysed into fatty acids and glycerol (unit 2.5). Proteins can be hydrolysed into amino acids (unit 2.6). Some products of these breakdowns can be converted into other forms by the cell, or built up into different macronutrients. The monosaccharide glucose, for example, can be converted to lipids, and vice versa. Micronutrients may be added to macronutrients to give different molecules. For example, the replacement of one fatty acid in a lipid molecule by the mineral ion phosphate gives rise to a **phospholipid**, which is an essential component of cell surface membranes (unit 4.1). In this way, many of the body's nutritional needs can be obtained from certain essential nutrients that must be provided in the diet. These essential nutrients are:

- essential fatty acids
- essential amino acids
- most vitamins
- mineral ions.

The functions of these essential nutrients are described in unit 12.4. Two other dietary components are necessary to remain healthy, although they are not considered as nutrients because they are not usually digested by the body. These are water and dietary fibre.

A balanced diet requires a range of foods that provide carbohydrates, fats, protein, minerals, vitamins, water and dietary fibre

A vegetarian diet can provide all the essentials of a balanced diet

A **balanced diet** is one that provides all the necessary components (nutrients, water and dietary fibre) needed to maintain good health. These components must be provided in the correct amounts, neither too little nor too much, for a given individual, and therefore vary with age, gender, activity and particular circumstances (e.g. pregnancy). These variations are discussed in unit 12.3.

12.1.2 Water

We saw in section 2.9.7 why water is essential to living organisms. A typical human needs to consume around 2.5dm³ of water a day in their diet, to replace that lost in expired air, urine, faeces and sweat, as well as through evaporation from the skin. The daily water balance for a typical human is shown in table 12.1.

Table 12.1 Daily water balance in a typical human

Volume of water gained / dm³ day⁻¹		Volume of water lost / dm³ day⁻¹	
Diet	2.3	Urine	1.5
Metabolism (e.g. from respiration)	0.2	Expired air	0.4
		Evaporation from skin	0.35
		Faeces	0.15
		Sweat	0.1
Total	2.5	Total	2.5

12.1.3 Dietary fibre

Dietary fibre, or **roughage**, comes from the cell walls of plants and includes a number of **non-starch polysaccharides (NSPs)**, including cellulose, hemicellulose, pectins and gums. It is not digested by enzymes in the gut, although bacteria in the large intestine may break it down to fatty acids, carbon dioxide, methane and hydrogen. Some of the fatty acids may be absorbed and used by the body but their contribution is minimal. The main functions of dietary fibre are:

* to speed up the passage of food through the alimentary canal
* to produce larger, softer faeces
* to reduce the risk of bowel diseases such as constipation, diverticular disease and cancer of the colon
* to reduce blood **cholesterol** levels.

12.1.4 Energy expenditure and basal metabolic rate

The amount of energy used by the body varies at different ages, different times of day and with various activities. However, even when we are at rest, a certain amount is needed for ventilation (breathing), pumping the heart, maintaining body temperature and various chemical reactions. These processes, needed to keep us alive, are referred to as **basal metabolism**. The energy used up in maintaining them is called the **basal metabolic rate (BMR)**. The BMR accounts for about three-quarters of an individual's energy needs.

BMR can be assessed either directly, by determining the amount of heat lost, or indirectly, by measuring oxygen consumption in a given time. Measurements are made under standard conditions, when the individual is lying down in a warm environment and after fasting for at least 12 hours. Results are expressed as kJ of heat energy per kg of body mass.

The main variations in BMR are as follows.

* Men usually have a higher BMR than women, partly as a result of the lower proportions of fatty tissue they have; fatty, or adipose, tissue is less metabolically active than other body tissues.
* BMR increases with an increase in body mass because there is a greater mass of actively respiring tissue.
* Infants and small children have a proportionately higher BMR for their size due to their rapid growth and development. They also have proportionately greater heat losses from the surface of the body (since smaller children have a greater ratio of surface area to body mass) and so the BMR must increase in order to maintain body temperature.
* Older adults have a lower BMR than younger adults since the amount of muscle tends to decrease with age.
* BMR increases during pregnancy and remains high during lactation. This is because the infant's energy needs must be supplied by the mother.
* BMR increases due to increased production of the hormone thyroxine, which raises the metabolic rate.
* BMR increases by about 13% for every 1°C rise during fevers associated with illness.
* BMR may fall by as much as 10% in very hot climates.

SUMMARY TEST 12.1

A balanced diet provides all the components needed to maintain good health. Adequate amounts of nutrients are needed to provide **(1)**, to allow **(2)** and **(3)** of cells and for the proper functioning of vital organs. There are five main nutrients in a balanced diet. Three are needed in large quantities and are therefore called **(4)**. These three nutrients are **(5)**, **(6)** and **(7)**. The remaining two are needed in tiny quantities and are known as **(8)**. These two nutrients are **(9)** and vitamins. The role of vitamins is to regulate **(10)**. Water is also essential in the diet and most of this is consumed as food and drink, but a little is obtained from **(11)**. Most water is lost from the body as **(12)**. Also essential in the diet is **(13)**, which comes from **(14)** of plants, which contain non-starch polysaccharides such as **(15)**.

Dietary reference values

It is clear from the number of newspaper, radio and television items that food and diet are playing an increasingly important role in our lives. From celebrity chefs to advertisements for crisps, those of us in developed countries are surrounded by information and advice on what we should eat and drink. How then can we make sense of this vast amount of information, especially when it is often conflicting? Even from the same source, the advice seems to change from one day to the next. We must firstly appreciate that, as with all science, current information on diet is constantly reviewed. New technologies and data allow us to challenge established ideas and to modify them in the light of new information. Health statistics accumulated over decades have allowed us to establish connections between particular diets and foods and many diseases, such as cancer and heart disease. Since the early 1940s, therefore, governments have commissioned studies to evaluate the effect our diet has on health. This led to the setting up of the Committee on Medical Aspects of Food Policy (COMA) which, in 1991, published a far-reaching and influential report that first introduced **dietary reference values (DRVs)**.

Dietary reference values allow cooks, caterers and individuals to provide an appropriate menu for those they serve

12.2.1 Dietary reference values (DRVs)

Dietary reference values (DRVs) were introduced to replace the recommended dietary allowances (RDAs) of food, energy and nutrients which had been in use in the UK since 1950. The problem with RDAs was that they were often wrongly used to assess whether an individual's diet was adequate, when, in fact, the needs of one individual often differ greatly from those of another. A DRV provides enough of any nutrient for 95% of the population, whereas the RDA, being an average, was too little for 50% of the population and too much for the remaining 50%. The DVAs do not attempt to give recommended daily intakes of particular nutrients for individuals. Instead, they refer to groups of people. Dietary reference value is a general term. In practice, it includes three more specific values.

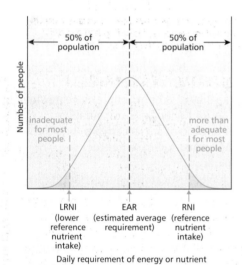

Fig 12.1 *Graph illustrating three dietary reference values*

- **Estimated average requirement (EAR)** – this is the estimated average quantity of energy, protein, vitamin or mineral that individuals in a population need to maintain good health. As an average, this means that 50% of the population will need more than this value and 50% will need less (Fig 12.1).
- **Reference nutrient intake (RNI)** – this is the quantity of protein, vitamin or mineral that is sufficient for almost everybody in the population, including those with the greatest dietary need for a particular nutrient. Any individual consuming this quantity is highly unlikely ever to be deficient in that nutrient.
- **Lower reference nutrient intake (LRNI)** – this is the quantity of protein, vitamin or mineral that is sufficient only for those individuals with the lowest needs for a particular nutrient. Most people require more than this value and, if they constantly take in this quantity, they will almost certainly suffer symptoms of its deficiency.

Provided there are enough data, it is possible to produce a graph showing the requirement for energy or any particular nutrient for a population. This graph is a normal distribution curve and is illustrated in figure 12.1. Where there are insufficient data, another group of DVDs, called **safe intakes**, is given. For example, low levels of vitamin A cause deficiency diseases (section 12.6.1), but a very high level of vitamin A can be toxic. The DRV is therefore set well below the danger level. In other words, a safe intake level is quoted.

Table 12.2 *Dietary reference values for fat*

Fatty acid	Population average intake as % energy
Saturated fatty acids	11
Monounsaturated fatty acids	13
Polyunsaturated fatty acids	6.5
Trans fatty acids	2.0
Total fatty acids	32.5
Equivalent amount of total fat	35

12.2.2 Energy and dietary reference values

Energy requirements vary much more widely between individuals than the requirement for any particular nutrient. Only the EAR is therefore given in tables of recommended energy intake. The recommended energy intakes according to age, activity and sex, are given in table 12.4 in unit 12.3. One crucial question, however, is how is the energy being obtained? It can be derived from four sources.

- **Alcohol** – as some people do not consume alcohol, the DRVs provide two sets of data: one that assumes no alcohol intake and a second set that assumes 5% of energy needs are provided by alcohol. For simplicity, the data given here exclude alcohol as a source of energy.
- **Carbohydrate** – it is recommended that 50% of our daily energy needs are derived from carbohydrate. Three forms of carbohydrate are recognised for this purpose:
  - **Starches** – polysaccharides such as those found in potatoes.
  - **Intrinsic sugars** – sugars found within the cell membranes of foods (e.g. apples). These include glucose, fructose and sucrose.
  - **Extrinsic sugars** – sugars not found within cell membranes at the point of consumption. These are mostly refined sugar, either used directly or in foods such as biscuits and jam. Extrinsic sugars also include the milk sugar, lactose.

It is recommended that starches, intrinsic sugars and lactose should make up 39% of total energy intake, while non-milk extrinsic sugars should make up 11%. Concerned about health problems such as obesity in the young and dental caries, the World Health Organisation recommended in 2003 that this figure should not be above 10%. The sugar industry, however, disputes the evidence on which this advice was founded.

- **Fats (lipids)** – it is recommended that 35% of total energy requirements should come from fats. The contribution that each form of fat should make to the overall energy requirement is shown in table 12.2. Because of evidence that saturated fatty acids contribute to heart disease through raising blood **cholesterol** levels, it is recommended that only 11% of total energy intake should come from this source.
- **Protein** – it is accepted that around 15% of energy will come from protein, although the main function of protein is to build up cells during growth and repair, rather than to provide a source of energy.

The percentage of daily energy intake that should be derived from each source (both including and excluding alcohol) is given in table 12.3.

12.2.3 Uses of dietary reference values (DRVs)

Although DRVs are not designed for individual use, it makes sense that, if you are eating a nutrient at around the EAR, it is unlikely to cause harm. If, however, you are eating more than the RNI for a particular nutrient, you may be having too much, as such a small percentage of the population has needs as high as this. DRVs are useful for helping:

- chefs and caterers to design appropriate menus for groups of people living in communities such as schools, old peoples' homes and prisons
- managers to plan food supplies for large groups of people
- dieticians and others to assess the dietary needs of individuals, e.g. the elderly, pregnant women
- individuals to calculate their own dietary requirements and so maintain, or improve, health
- food manufacturers to provide appropriate nutritional information on food labels.

Table 12.3 *Dietary reference values (DRVs) for fats, carbohydrates, protein and alcohol as a percentage of daily total energy intake*

Source	% Daily energy intake	
	Excluding alcohol	Including alcohol
Non-milk extrinsic sugars	11	10
Intrinsic sugars, milk sugars and starches	39	37
Fat	35	33
Protein	15	15
Alcohol	0	5
Total	100	100

This typical ready-made meal contains over 30 different ingredients, including half the daily reference nutrient intake of sodium

SUMMARY TEST 12.2

Dietary reference values (DRVs) provide recommended daily intakes of energy and nutrients for groups of people. It is a general term that includes the quantity of a nutrient that is sufficient only for individuals with lowest needs. This is called **(1)**. It also includes the quantity of a nutrient that is sufficient for almost everyone in a population. This is called **(2)**. Another value relates to the average quantity of a nutrient that individuals in a population need. This is called **(3)** and it relates not only to nutrients but also to **(4)** intake. Where data are insufficient to give average values, a DRV called **(5)** is used. Energy in the diet can be provided by carbohydrates, fats, protein or **(6)**. The daily percentage that should come from all carbohydrate sources is **(7)**. Sugars that are found within cell membranes at the point of consumption are called **(8)** sugars and, along with starches and the milk sugar, **(9)**, should make up **(10)**% of the daily energy intake. The contribution of saturated fatty acids to daily energy intake should not exceed **(11)**%.

Variations in dietary requirements

Table 12.4 *Estimated average requirements (EARs) for energy of males and females of various ages*

Age	EAR / MJ day^{-1}	
	Male	Female
0–3 months	2.28	2.16
4–6 months	2.89	2.69
7–9 months	3.44	3.20
10–12 months	3.85	3.61
1–3 years	5.15	4.86
4–6 years	7.16	6.46
7–10 years	8.24	7.28
11–14 years	9.27	7.72
15–18 years	11.51	8.83
19–50 years	10.60	8.10
51–59 years	10.60	8.00
60–64 years	9.93	7.99
65–74 years	9.71	7.96
75+ years	8.77	7.61

The advantage of dietary reference values (unit 12.2) is that they provide information on the different needs of individuals according to gender, age, level of activity and particular circumstances, e.g. pregnancy and lactation. In this unit we shall explore these variations and the scientific basis for them.

12.3.1 Gender variations

From birth, the energy requirements of males are higher than those of females (table 12.4). This is because, on average, males have a larger body mass than females of the same age. They therefore need additional energy to maintain the extra quantity of living tissue. Even for a male and female of the same mass, the male would normally require more energy. This is because females have a larger proportion of adipose (fat) tissue in their bodies, and this respires only very slowly and hence requires little energy for its maintenance.

In the case of proteins, vitamins and minerals, the dietary requirements of boys and girls are the same up to, and including, the age of 10 years. From age 11 years onwards, certain differences (table 12.5) arise for the following reasons.

- Males require more protein because their growth is greater than females during adolescence, and after that stage they have a greater mass of tissue to repair and replace.
- Males require more of the vitamin B complexes that are needed for respiration and metabolism, e.g. B_1, B_2 and B_3, because they have a greater mass of respiring tissue.
- Males require more calcium and phosphorus in adolescence because they develop larger bones than females during this stage of growth.

Table 12.5 *Dietary reference values (DRVs) for protein and selected vitamins and minerals*

Age	Protein / g day^{-1}	Vitamin B$_1$ (thiamine) / mg day^{-1}	Vitamin B$_2$ (riboflavin) / mg day^{-1}	Vitamin B$_3$ (niacin) / mg day^{-1}	Vitamin C / mg day^{-1}	Calcium / mmol day^{-1}	Phosphorus / mmol day^{-1}	Iron / mmol day^{-1}	Sodium / mmol day^{-1}
1–3 years	14.5	0.5	0.6	8	30	8.8	8.8	120	22
4–6 years	19.7	0.7	0.8	11	30	11.3	11.3	110	30
7–10 years	28.3	0.7	1.0	12	30	13.8	13.8	160	50
Males:									
11–14 years	42.1	0.9	1.2	15	35	25.0	25.0	200	70
15–18 years	55.2	1.1	1.3	18	40	25.0	25.0	200	70
19–50 years	55.5	1.0	1.3	17	40	17.5	17.5	160	70
50+ years	53.3	0.9	1.3	16	40	17.5	17.5	160	70
Females:									
11–14 years	41.2	0.7	1.1	12	35	20.0	20.0	260	70
15–18 years	45.0	0.8	1.1	14	40	20.0	20.0	260	70
19–50 years	45.0	0.8	1.1	13	40	17.5	17.5	260	70
50+ years	46.5	0.8	1.1	12	40	17.5	17.5	160	70
Pregnancy	+6	+0.1	+0.3	–	+10	–	–	–	–
Lactation:									
0–4 months	+11	+0.2	+0.5	+2	+30	+14.3	+14.3	–	–
4+ months	+8	+0.2	+0.5	+2	+30	+14.3	+14.3	–	–

- Females require more iron from puberty until the menopause because iron is part of the **haemoglobin** molecules found in red blood cells, and females lose blood at each **menstrual** period. Without sufficient iron, individuals suffer from anaemia, which makes them pale and tired. Around 50% of women in the Western world are estimated to be iron-deficient (not quite anaemic) due to a combination of menstruation and dieting.

12.3.2 Age variations

Tables 12.4 and 12.5 show some effects of age on dietary requirements. Variations with age that occur include the following.

- Energy requirements increase with age, up to and including adolescence, as growth is rapid during these years and / or physical activity is generally at a high level. These energy requirements remain almost constant up to the age of 60 years, after which they decrease as physical activity diminishes and body mass often decreases. Basal metabolic rate also decreases with age.
- Protein requirements increase with age, especially around puberty. This is because additional protein is needed for the rapid growth around adolescence and, thereafter, to repair and replace cells.
- Calcium and phosphorus requirements are greater in the first year of life as they are laid down in the bones of infants. The rate again increases in adolescence when the second growth spurt involves relatively rapid elongation of the bones. After adolescence, the requirements remain constant.
- Requirements for other minerals and vitamins increase as body mass increases up to the age of 20 years, but thereafter remains relatively constant.

12.3.3 Level of activity variations

The more physically active a person is, the greater their energy requirement. It follows that, compared to an office worker, a manual labourer requires a greater energy intake, especially of carbohydrate food.

12.3.4 Pregnancy variations

It is recommended that women planning to become pregnant should supplement their diet with folic acid tablets. Folic acid is known to protect against spina bifida – a condition resulting from a defect in the formation of the nerve cord during embryonic development. During pregnancy, the mother's diet must provide not only for the growth and development of her fetus, but also for the laying down of extra tissue in her own body. Pregnant women are therefore advised to increase their intake of:

- energy during the last 3 months of pregnancy, when fetal growth is at its greatest and the mother has this additional mass to carry around

- protein to supply the growth needs of the rapidly dividing cells of the fetus
- vitamins A, C and D, although too much vitamin A can be harmful to the fetus in the early stages of pregnancy. For this reason pregnant women are recommended to avoid liver, which is very rich in vitamin A.

12.3.5 Lactation variations

Mothers who are breast-feeding not only need to maintain increased levels of energy, protein and vitamins, for the same reasons as when they were pregnant, but also need:

- more calcium, phosphorus and magnesium, as these are needed by the newly born to develop their bones
- more zinc because it is present in a number of enzymes, including those involved in growth and metabolism.

12.3.6 Problems associated with measuring dietary reference values (DRVs)

DRVs, such as the estimated average requirement (EAR) for energy, are calculated for each age group using information on basal metabolic rate (BMR) (unit 12.1), the energy needed for growth and the extent of physical activity. Obtaining accurate figures for just one, e.g. energy, in just one age group, e.g. 11–14 years, is difficult because:

- measuring BMR accurately is not easy
- measuring the extent and duration of an individual's physical activity is difficult
- individuals vary, e.g. there are religious, racial and ethnic differences that influence diet
- growth rates vary as individuals undergo 'growth spurts' at different times
- it is difficult to obtain sufficient data from a large enough sample to make the findings statistically reliable.

SUMMARY TEST 12.3

Amounts of energy or nutrients needed in the diet vary. Males at puberty require more minerals, such as **(1)** and **(2)**, to aid the growth of **(3)**, while menstruating females need more **(4)**. During pregnancy, vitamins **(5)**, **(6)** and **(7)** are essential, while breast-feeding mothers are recommended to increase their levels of the mineral **(8)**, which is present in certain essential enzymes.

Functions of essential nutrients

12.4.1 The functions of carbohydrates

Carbohydrates comprise sugars and starches, and their functions are to provide:

- **energy** – the oxidation of glucose yields approximately $16\,kJ\,g^{-1}$
- **storage** – glucose in excess of immediate energy requirements can be converted to:
  - glycogen and stored in the liver and muscles
  - fat
- **dietary fibre** – to increase the bulk and reduce the transit time of food in the alimentary canal.

The **dietary reference values** (DRVs) for carbohydrates show that 10–11% of an individual's total energy needs should come from sugars such as sucrose. In 2003, the World Health Organisation, concerned about rising levels of obesity, **diabetes** and tooth decay, recommended that not more than 10% of energy intake should come from this source. The DRV for milk sugars and starch is 37–39% of total energy intake. One major problem in reducing dietary sugars is the quantity of sugar added to processed foods. These **hidden sugars** may go unnoticed and are difficult for individuals to exclude from the diet. The sources of certain carbohydrates are given in table 12.6.

Table 12.6 Sources of some carbohydrates

Carbohydrate	Sources
Glucose	Grapes, honey
Fructose	Honey, fruits
Sucrose	Sugar cane, sugar beet, and therefore jams, biscuits and chocolate
Starch	Potatoes, cereals, and therefore bread and pasta

12.4.2 The functions of lipids

Lipids are made up of fatty acids and glycerol (unit 2.5). Most of the fatty acids we require can be made within the body, but some cannot. These include **linoleic acid** and **linolenic acid**. They can only be made by plants and therefore have to be included as part of the diet in humans. They are therefore known as **essential fatty acids**. They are **polyunsaturated fatty acids (PUFAs)**, which means that they have more than one double bond in their chemical structure. The roles of these fatty acids include:

- **the provision of energy stores** – as part of a lipid these fatty acids yield $37\,kJ\,g^{-1}$ on oxidation and therefore form a more concentrated store of energy than carbohydrates
- **the provision of components of cell membranes** in the form of **phospholipids**
- **the synthesis of some hormones**
- **the metabolism of cholesterol**.

The lipid content of certain foods is given in table 12.7.

Table 12.7 The lipid content of foods

Food	% Lipid
Cooking oils	100
Butter	82
Peanuts	49
Cheddar cheese	34
Beef	24
Eggs	11
Cod	0.7

12.4.3 The functions of proteins

The structure of proteins is discussed in unit 2.7. Although proteins may be broken down to release energy ($17\,kJ\,g^{-1}$), their main function is as a source of amino acids for the synthesis of new proteins for growth and repair of cells. Some amino acids (**non-essential amino acids**) can be made by human cells, such as those of muscle, liver, kidney and brain tissue. However, eight amino acids (**essential amino acids**) cannot be made in the body and must be supplied in the diet. Table 12.8 lists the essential amino acids required by adults and children. Although histidine is made by adult humans, it has to be provided in the diet of young children to ensure they gain in body weight. Tyrosine is an amino acid that can be made from phenylalanine. It only becomes essential if phenylalanine is absent from the diet. Similarly, cysteine only becomes essential if there is not enough methionine in the diet. Most animal protein, such as meat,

Table 12.8 Essential amino acids required by adults and children

For adults and children	Additional requirement of small children
Isoleucine	Histidine
Leucine	
Lysine	
Methionine	
Phenylalanine	
Threonine	
Tryptophan	
Valine	

eggs and milk, contains all the essential amino acids in the proportions required by an adult. However, many vegetable proteins can have a low level of one or more essential amino acids. This does not pose a problem if a variety of vegetable products is eaten. For example, beans (lacking methionine) may be eaten with bread (lacking lysine).

12.4.4 The functions of vitamin A

Vitamins are organic substances required by the body in very small amounts. Although vitamins D and K can be made in the body, all others are required in the diet. Vitamin A is found in:

- milk, eggs, liver and fish-liver oils
- mango and papaya.

It can also be formed in the body from the related compound, **β-carotene**, found in vegetables such as carrots and spinach. High concentrations of vitamin A in the blood are toxic, but the liver is able to store enough to last 1–2 years. Vitamin A is required by the body to make:

- **rhodopsin**, used by the rod cells in the eye to detect low light intensities
- **retinoic acid**, used for the growth and maintenance of epithelial tissues.

The problems associated with a lack of vitamin A are considered in section 12.6.1.

12.4.5 The functions of vitamin D

Vitamin D (calciferol) is found in:

- eggs and oily fish
- meat and meat products
- margarine and low-fat spreads (to which vitamin D has to be added, by law).

It can also be made in the body by the action of sunlight on the skin, although little is made in this way by dark-skinned people in temperate climates. Vitamin D is required by the body:

- to stimulate epithelial cells in the intestine to absorb calcium
- to act on bone cells to regulate the deposition of calcium.

Vitamin D can be stored in muscles and fat. The problems associated with its deficiency are considered in section 12.6.2.

Oily fish like mackerel are good sources of vitamins A and D

EXTENSION – ROLES OF MINERAL IONS

Mineral ions may only be required in tiny amounts but their importance in the body far outweighs the quantity required. They play a wide variety of roles that affect all major functions in the body. Table 12.9 lists four of the most important mineral ions and the functions they perform.

Table 12.9 *Sources and functions of some mineral ions required by humans*

Mineral ion	Major food source	Function
Calcium (Ca^{2+})	Dairy foods, eggs	Constituent of bones and teeth, needed for blood clotting
Phosphate (PO_4^{3-})	Dairy foods, eggs, meat	Constituent of nucleic acids, ATP, phospholipids, bones and teeth
Iron (Fe^{2+}/Fe^{3+})	Liver, green vegetables	Constituent of haemoglobin and myoglobin
Sodium (Na^+)	Table salt, dairy foods, meat, eggs, vegetables	Needed for nerve and muscle action and maintenance of anion / cation balance

SUMMARY TEST 12.4

A balanced diet requires carbohydrates to provide (**1**) and these occur in a number of forms. Starch is found in foods such as (**2**) and (**3**). Honey contains the two monosaccharide sugars called (**4**) and (**5**), while cane sugar is an almost pure source of (**6**). Even more energy is available from the food type (**7**), which is made up of (**8**) and glycerol. Proteins, essential for (**9**) and (**10**), are broken down into amino acids, of which a total of (**11**) cannot be synthesised in the body and must be supplied in the diet. These are called (**12**) amino acids. Vitamins play many roles. Vitamin (**13**) is essential for the functioning of rod cells in the eye, while vitamin (**14**) is essential for calcium absorption by the gut and can be manufactured by the (**15**) of humans.

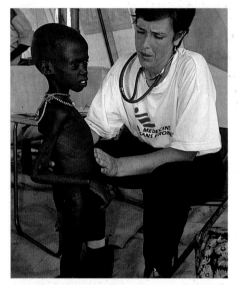

Child with marasmus; note the muscle-wasting on the arms.

Child with kwashiorkor; note the enlarged abdomen as a result of a swollen liver

Young woman with anorexia nervosa

Any imbalance of nutrients in the diet is referred to as **malnutrition**. The word means 'badly nourished', and may be due to too much of one or more things – **over-nutrition** – or to too little – **under-nutrition**. Under-nutrition is characterised by inadequate intake of protein, calories and micronutrients and, if it persists, leads to starvation. Globally there is enough food for the whole population of the world, and yet around 800 million people do not have access to sufficient food – this is 20% of the population of the developing world (section 11.4.1). Malnutrition contributes to more than half of the 10 million deaths of children under 5 years old each year in developing countries.

12.5.1 Protein-energy malnutrition

Protein-energy malnutrition, also known as protein-calorie malnutrition, is due to a reduction in all macronutrients (section 12.1.1). It results from having a protein and / or energy intake somewhere between an adequate quantity and starvation (no food intake). It can occur in individuals of any age, in any country, but is seen mainly in poorer countries of the developing world. The children grow slowly and have a high mortality rate. The two main symptoms of protein-energy malnutrition are:

- **Stunting** – results from chronic protein-energy malnutrition. It occurs in children from 2–5 years of age who are therefore in their early phase of rapid growth. The lack of protein in particular stunts growth and this cannot usually be reversed, even if a protein-rich diet is later provided. The highest prevalence occurs in Guatemala, Bangladesh and India, where around 65% of children suffer from stunting.
- **Wasting** – occurs when there is acute protein-energy malnutrition and is found in individuals over 5 years of age. It is characterised by rapid weight loss in those who had near normal weight and a failure to put on weight in those already below average weight. It is most prevalent in South East Asia but also occurs sporadically in many regions of the world when war or famine lead to temporary interruptions in food supply.

Table 12.10 shows estimates of the extent of stunting and wasting in children in three major regions of the world.

If prolonged and extreme, protein-energy malnutrition gives rise to two forms of disease, called **marasmus** and **kwashiorkor**.

Table 12.10 *Global estimates of the prevalence of stunted and wasted children in developing regions of the world*

Region	Stunting		Wasting	
	% of children	Number of children (millions)	% of children	Number of children (millions)
Africa	38	44	7	8
Asia	47	173	11	40
Latin America	22	12	2	1
All developing countries	43	230	9	50

12.5.2 Marasmus

Marasmus is the most common type of protein-energy malnutrition in most developing countries. It is seen mainly where breast-feeding is stopped early and the babies are weaned on a diet low in protein. Without breast-feeding, some **immunity** that is ordinarily acquired from the mother's milk is lost. Infections such as gastroenteritis often arise, especially where hygiene and sanitation are poor. These infections add to the problem of a poor diet because they mean that food is not properly absorbed by the alimentary canal. As energy intake is inadequate for the body's needs, it draws on its stores – firstly glycogen in the liver, then fat stored around the body and, finally, the protein in **striated muscle**, which is converted into amino acids and then glucose, ready for **oxidation** during respiration. Symptoms of marasmus include:

- **low weight** for an individual's age – normally 60% or less of the average weight
- **muscle wasting**, leading to visible ribs, facial bones and limb joints through the skin
- **almost no subcutaneous fat**, again making bones visible through the skin
- **thin, wrinkled skin**
- **sunken eye**s
- **pot-belly**
- **irritability and fretful behaviour**.

12.5.3 Kwashiorkor

Kwashiorkor is an African word meaning 'first child – second child'. It was used because people observed the disease in the first child when a second child was born. This was because there was frequently insufficient breast milk to feed both children, and so the first child was weaned, usually on a weak gruel of very poor nutritional quality. In kwashiorkor, the protein deficiency is more marked than the energy deficiency, leading to much reduced protein synthesis. One effect of this is that the liver cannot make enough blood plasma protein. The blood therefore has a higher **water potential** than normal and so does not re-absorb as much water from the tissues. This retention of water in the tissues causes them to swell – a condition called **oedema**, which is characteristic of kwashiorkor. The symptoms of kwashiorkor include:

- **low body weight** due to decreased muscle mass
- **oedema** due to low blood plasma protein (e.g. albumin) levels
- **'moon-face'** due to oedema
- **swollen abdomen** as a result of an enlarged liver
- **dry, brittle hair**, often reddish in colour
- **skin conditions**, such as dermatitis and changes in pigmentation
- **irritabilit**y coupled with lethargy or apathy.

In both marasmus and kwashiorkor, there is a marked increase in infections such as measles, diarrhoea, pneumonia and sepsis, because white blood cells, and the antibodies they produce, are based on protein. Without sufficient levels of these, the immune system is easily overwhelmed by even minor infections.

12.5.4 Treatment of protein-energy malnutrition

The immediate treatment for protein-energy malnutrition is to correct the salt and fluid balance by **oral rehydration therapy** and to treat infections with antibiotics. The next stage is **dietary therapy**, typically with milk-based formulas. It is important to provide energy foods initially and only to increase the proportion of protein gradually as the patient recovers. This is because it is essential to provide the energy needed to operate the body's systems as these are needed to permit protein to be metabolised. It can be likened to a car that has run out of fuel – you first need to add energy (petrol) to make the other systems work. Micronutrients, especially iron, must be provided. In the long run it is essential to improve the nutrition of lactating mothers, so that their breast milk is nutritious enough to support their offspring, to improve sanitation and, finally, to improve general education, especially on diet and hygiene.

12.5.5 Anorexia nervosa

Although developed countries rarely have cases of marasmus or kwashiorkor, one form of wasting disease, anorexia nervosa, is increasingly common. This so-called 'slimmer's disease' is most common in Western developed countries and affects girls more often than boys. Underlying psychological causes lead to an obsessive desire to be thin. The weight loss can become so severe that:

- muscles waste (including heart muscle)
- periods stop
- blood pressure falls
- hair becomes thin and sparse
- hands and feet are cold
- there is increased susceptibility to infection
- there are personality changes.

Anorexics often believe themselves to be fat, even when they are dangerously thin, and are slow to realise that they need help. Once this is recognised, prolonged psychotherapy is often needed to assist recovery.

SUMMARY TEST 12.5

The form of malnutrition in which too little of one or more substances is present is known as **(1)**. One form of this is called protein-energy malnutrition, two symptoms of which are **(2)** and **(3)**. Prolonged and extreme malnutrition leads to marasmus or kwashiorkor. Marasmus arises when breast-feeding stops early and the baby is weaned on a diet low in **(4)**. Without breast milk the baby is also more prone to **(5)**. As energy intake is insufficient, the baby uses its reserves, firstly in the form of **(6)** from the **(7)**, then fat, and finally from the breakdown of protein found in **(8)**. In kwashiorkor, one problem is that a lack of protein intake reduces the level of **(9)** in the plasma of the blood and ultimately leads to water retention in the tissues – a condition called **(10)**. The initial treatment of protein-energy malnutrition begins with correcting the **(11)** balance and then providing **(12)** foods. The amount of **(13)** in the diet should be steadily increased throughout the treatment. A particular form of protein-energy malnutrition found more often in developed than developing countries is **(14)**.

Vitamins are a group of essential organic compounds that are needed in small amounts for normal growth and metabolism. Lack of any vitamin in the diet results in a disorder called a deficiency disease (section 11.2.6). In this unit we shall look at deficiency diseases that arise from the lack of two vitamins – vitamin A and vitamin D.

12.6.1 Vitamin A deficiency

We saw in section 12.4.4 that vitamin A was found in milk, eggs, liver, fish liver oils, mango and papaya, as well as being manufactured by the body from the β-carotene found in vegetables such as carrots. There are therefore two types of vitamin A deficiency:

- **Primary vitamin A deficiency** results from insufficient vitamin A or β-carotene in the diet.
- **Secondary vitamin A deficiency** results from an inability to manufacture vitamin A from β-carotene, or inadequate absorption, transport or storage of it in the liver. This may be the consequence of disorders such as **cystic fibrosis**, pancreatic disease or cirrhosis of the liver.

Vitamin A deficiency is common in individuals suffering from marasmus (section 12.5.2) or kwashiorkor (section 12.5.3), both as the result of a lack of vitamin A in the diet and defective storage and transport. The effects of vitamin A deficiency are much more severe in young children. Vitamin A deficiency:

- is the leading cause of preventable blindness
- affects between 100 million and 140 million children worldwide
- causes between 250 000 and 500 000 children to become blind each year, of which half die within a year of losing their sight
- is involved in many of the 600 000 childbirth-related deaths of women that occur each year.

Vitamin A deficiency is much more common in developing countries because the animal products which contain it are expensive and the vegetables that contain β-carotene are not available. Diets in these countries are often based on cereals such as maize or rice, which are low in vitamin A and β-carotene. The use of genetically modified varieties of these crops that contain **genes** for vitamin A production could help prevent these deficiencies. The symptoms of vitamin A deficiency are:

- **Xerophthalmia** – a drying and scarring of the cornea of the eye. There is also inflammation that may lead to blindness.
- **Night blindness** – the reduced ability to see in dim light, which can affect both children and adults. Vitamin A (retinol) can be converted to retinal, a compound that combines with the visual pigment opsin to make rhodopsin in the rod cells of the retina. This pigment is bleached back to opsin when light enters the eye. In the absence of vitamin A, the reformation of rhodopsin is slow or non-existent, and so the rod cells, which respond to dim light, cannot operate.
- **Dry, rough skin** – because vitamin A is needed to make retinoic acid, which is needed to maintain epithelial tissues like the skin.
- **Increased susceptibility to infections** – because the lack of maintenance of epithelial tissues means that surfaces such as the intestines and the lungs are weakened and **pathogens** can more easily enter the body through them.

Table 12.11 shows the estimated number of people with night blindness in various regions of the world

Table 12.11 *Estimated numbers of people with night blindness in various regions of the world*

Region	Millions of people with night blindness
Africa	1.3
Americas	0.1
South and South East Asia	10.0
Eastern Mediterranean	1.0
Western Pacific and China	1.4

Eye showing symptoms of xerophthalmia caused by inadequate amounts of vitamin A in the diet

Child with xerophthalmia; one of up to half a million children who are blinded in this way each year

12.6.2 Vitamin D deficiency

In section 12.4.5 we learned that vitamin D is found in eggs, oily fish, meat and margarine, as well as being manufactured by the body in response to the action of sunlight on the skin. As a result, vitamin D deficiency is rare amongst people in developed countries. Problems may, however, arise for dark-skinned people in temperate climates, or people whose religion requires them to wear clothes that entirely cover the skin. Although there may not be enough winter sun in temperate regions to produce adequate vitamin D, enough is usually stored in muscles and fat during the summer to prevent any deficiency. We also saw in section 12.4.5 that the main role of vitamin D is the absorption of calcium from the intestines and its subsequent assimilation into bone tissue. As a consequence, vitamin D deficiency may be linked to poor calcium uptake and / or poor assimilation into bone. The two main disorders due to vitamin D deficiency are:

Child with rickets caused by a lack of vitamin D; note the bow-shaped legs typical of the disease

- **Rickets** – the result of vitamin D deficiency in children. Their bones become soft and grow irregularly, the joints become swollen and limbs and the chest may be distorted. Typically the legs are bow-shaped (see photo).
- **Osteomalacia** – the result of vitamin D deficiency in adults. This is the progressive softening of bones, making them tender and painful. Fractures of bone occur more easily. Muscle weakness as well as loss of appetite and weight can also arise. Women may suffer osteomalacia during pregnancy because the fetus makes demands on the mother for vitamin D. During the period of breast-feeding, there is an increased demand for both vitamin D and calcium because the mother's milk is rich in both to ensure proper bone development in the growing child. It is therefore important that the intake of both nutrients is increased during lactation (table 12.12) and that vitamin D is increased during pregnancy.

Table 12.12 Dietary reference values for vitamin D and calcium

Time of life	Vitamin D / μg day^{-1}	Calcium / mmol day^{-1}
Child 0–6 months	8.5	13.1
Child 6–12 months	7.0	13.1
Child 1–3 years	7.0	8.8
Child 4–6 years	*	11.3
Child 7–10 years	*	13.8
Adults	*	17.5
During lactation period	10	additional 14.3
During pregnancy	10	no addition

*No dietary requirement as sufficient is normallly manufactured by the action of sunlight on the skin

SUMMARY TEST 12.6

Foods rich in vitamin A include **(1)**, **(2)** and **(3)**. It can also be manufactured by the body from a substance called **(4)**, found in vegetables such as **(5)**. A lack of vitamin A causes deficiency diseases, such as a drying and scarring of the cornea – a condition called **(6)**, and night blindness, which is the result of cells called **(7)** in the retina of the eye being unable to reform a pigment called rhodopsin. The skin may also become dry and infections become more frequent because without vitamin A the maintenance of **(8)** tissues is poor. Vitamin A deficiency occurs mostly in the **(9)** regions of the world. Vitamin D is found in **(10)** and **(11)**, as well as being produced by the action of sunlight on the skin. Vitamin D is needed for the absorption of **(12)** and its assimilation into **(13)**. If it is not in adequate supply the deficiency disease of **(14)** occurs in children, and **(15)** in adults.

Table 12.13 Body mass index (BMI)

BMI	Category
Under 20	Under-weight
20–24.9	Acceptable
25–29.9	Over-weight
30–39.9	Obese
Over 40	Very obese

Obesity is on the increase in both developed and developing countries

Over-nutrition is as much a form of malnutrition as under-nutrition. If people regularly consume more energy than they use, they will put on mass and if they continue to do so, they will become obese.

12.7.1 Body mass index (BMI)

Body mass index (BMI) is a mathematical calculation used to find out whether an individual is under-weight or over-weight. It is calculated by dividing a person's body mass in kilograms by his/her height in metres squared.

$$\text{Body mass index (BMI)} = \frac{\text{body mass in kilograms}}{(\text{height in metres})^2}$$

Table 12.13 shows how this index can be used to determine whether a person is under-weight, over-weight or obese. The BMI should only be used for adults because younger people store fat as part of their normal growth and this distorts the figure produced. Even with adults the index can occasionally be misleading; for example, some adjustment is needed for very muscular individuals and for pregnant or lactating women.

12.7.2 How common is obesity?

Obesity is very much more common in the developed countries, especially those of North America, Europe and Australasia. Even among the poorer households in these countries there is a tendency to spend scarce income on high-calorie fat and carbohydrate foods, rather than on fruit and vegetables. There is worrying evidence that obesity is becoming increasingly common in the developing nations, especially those with emerging economies, e.g. China. It is estimated that 60% of men and 40% of women in the UK are over-weight or obese. In North America nearly 65% of the adult population, some 127 million people, is categorised as over-weight or obese, and the proportion is increasing. Some general trends about obesity include the following.

- Globally, women have higher rates of obesity than men, although men have a higher incidence of being over-weight.
- Obesity in children and adolescents is on the rise in both developed and developing countries.
- Obesity is uncommon in African and Asian developing countries; where it does occur, it is more prevalent in urban than rural areas.
- In the majority of countries obesity has increased by between 10% and 40% in the past 10 years.
- As a country's economic development increases, so does the BMI of its people.
- In some developing countries, obesity occurs alongside starvation.

12.7.3 Why is obesity on the increase?

Some of the possible reasons for the increasing prevalence of obesity in developed countries are:

- **lack of physical exercise** – entertainment is often sedentary, e.g. watching TV, playing computer games
- **sedentary occupations** – machines now carry out much physical labour that was previously performed manually
- **well-heated homes and work places** – mean that little energy is lost as heat and therefore less food needs to be consumed
- **better transportation** – people drive or take public transport rather than walk, and take lifts or escalators rather than climb stairs

SUMMARY TEST 12.7

Body mass index (BMI) is calculated by dividing a person's **(1)** by his/her **(2)** squared. A person is considered to be obese if he / she has a BMI above **(3)**, and under-weight if their BMI is below **(4)**. People who are obese are more likely to suffer from coronary heart disease due to increased **(5)** and to high levels of **(6)** in the blood that are associated with obesity. They are also prone to a form of diabetes called **(7)**. Two degenerative diseases that are associated with obesity are **(8)** and **(9)**. High blood cholesterol levels can increase the risk of CHD, as can a high intake of **(10)** and fatty acids – especially **(11)** ones.

- **increased wealth** – food is readily available and easily afforded
- **convenience / fast foods** – beefburgers and pizzas, for example, are relatively cheap, readily available, heavily marketed and very high in calories
- **increased stress** – leads to the use of convenience foods (they are quick to prepare in a busy world) and 'comfort eating'
- **increased alcohol consumption** – alcohol is nutritionally poor, but high in calories
- **urbanisation** – means food stores / burger bars etc. are close by and often open 24 hours a day. Fear of crime in cities can prevent people, especially women and the elderly, from walking outside, especially at night.

12.7.4 Health problems associated with obesity

People who are obese are at greater risk of developing one or more serious medical conditions, and of dying prematurely. There is scientific evidence that obesity is associated with more than 30 medical conditions and it is linked with at least 15 others. These conditions include:

- **Coronary heart disease** – may be caused by increased blood pressure and blood **cholesterol** levels that commonly occur with obesity.
- **Type II (mature onset) diabetes** – food intake, especially carbohydrate, in obese individuals can be greater than normal insulin production can cope with. As a result they are unable to control their blood sugar levels properly. As many as 90% of type II diabetes sufferers are over-weight.
- **Cancer** – breast cancer, especially in post-menopausal women, is more common if they are over-weight. Cancers of the oesophagus, colon, rectum and the kidney are also more common in the obese, especially men.
- **Osteoarthritis** – a degenerative disease that affects especially the hips, spine and knees of obese individuals because the extra body weight puts additional strain on the bones in these regions.
- **Rheumatoid arthritis** – another degenerative disease which is more likely to arise when people are obese because extra weight puts stress on joints.
- **Hypertension** – around three-quarters of all cases of high blood pressure occur in people who are obese.

Table 12.14 *Prevalence of medical conditions by body mass index (BMI) for men (source: NHANES III, 1988–1994)*

Condition	% Prevalence of medical conditions at different BMIs			
BMI	18.5–24.9	25–29.9	30–34.9	≥40
Type II diabetes	2.03	4.93	10.10	10.65
Coronary heart disease	8.84	9.60	16.01	13.97
High blood pressure	23.26	38.77	47.95	63.16
Osteoarthritis	5.22	8.51	9.94	17.19

Table 12.15 *Prevalence of medical conditions by body mass index (BMI) for women (source: NHANES III, 1988–1994)*

Condition	% Prevalence of medical conditions at different BMIs			
BMI	18.5–24.9	25–29.9	30–34.9	≥40
Type II diabetes	2.38	7.12	7.24	19.89
Coronary heart disease	6.87	11.13	12.56	19.22
High blood pressure	23.26	38.77	47.95	63.16
Osteoarthritis	5.22	8.51	9.94	17.19

12.7.5 Diet and coronary heart disease

Tables 12.14 and 12.15 show that there is an increased risk of **coronary heart disease** (CHD) in both men and women who are over-weight, with those who are obese being about twice as likely to suffer from CHD than those with an acceptable BMI. CHD arises when a fatty substance called plaque builds up inside the coronary arteries that supply blood to heart muscle. This narrows these arteries, reducing the flow of blood and causing angina (pain or pressure in the chest) and heart attacks. More details of CHD are given in unit 14.6. There are a number of links between diet and CHD in addition to obesity:

- **High levels of salt** produce hypertension (increased blood pressure) that increases the risk of CHD. In the UK we consume, on average, twice the recommended salt intake, due largely to its addition to most processed foods.
- **High blood cholesterol** contributes to the formation of plaques in the coronary arteries, especially the form known as low density lipoprotein (LDL). LDLs are used to transport cholesterol from the liver to various tissues, including artery walls where they leave their cholesterol – especially at sites of damage. High density lipoproteins (HDLs) transport cholesterol in the opposite direction, i.e. from the tissues to the liver, ready for excretion. HDLs therefore help guard against CHD.
- **High fatty acid intake**, especially saturated fatty acids, increases the risk of CHD. For example, an increase in trans fatty acids from the recommended 2% of food intake to 4% increases the incidence of CHD by a quarter.

On the other hand, there are a number of aspects of the diet which seem to reduce the risk of coronary heart disease. These include:

- **Eating dietary fibre (non-starch polysaccharides)** protects against obesity and reduces insulin levels in the blood, so reducing CHD.
- **Moderate consumption of alcohol**, especially wine and beer, has been shown by some studies to reduce the risk of CHD.
- **Eating oily fish**, such as mackerel and herring.

1 Proteins are digested into amino acids. Nutritionists consider some of these amino acids to be essential amino acids. A deficiency of protein in the diet can lead to significant damage to health.

a Explain why some amino acids are described as *essential* in the diet. *(2 marks)*

b Describe the consequences for a young child of a **very** restricted intake of protein in the diet. *(3 marks)*

The table shows the daily energy requirements and some of the daily nutrient requirements for a woman aged 25 and a lactating (breast feeding) woman of the same age and body mass.

energy or nutrients	daily requirements	
	woman aged 25	lactating woman
energy / kJ kg^{-1}	145	185
protein / g kg^{-1}	0.75	0.93
vitamin A / μg	600	950
vitamin D / μg	0	10
calcium / mg	700	1250

c With reference to the table, explain why the energy and nutrient intakes are greater for the lactating woman.
(In this question one mark is awarded for the quality of written communication.)

(8 marks)

d State **three** further components of a balanced diet, other than digestible carbohydrate, that are not given in Table 3.1.

(3 marks)
(Total: 16 marks)
OCR 2802 Jan 2001, B (HHD), No.3

2 a List **five** components of a balanced diet. *(5 marks)*

The body mass index (BMI) is calculated using the following formula

$$BMI = \frac{\text{body mass in kg}}{(\text{height in metres})^2}$$

The BMI can be sub-divided into the categories shown in the table.

BMI	category
below 20	underweight
20–25	acceptable
25–30	overweight
over 30	obese
over 40	very obese

A young woman has a body mass of 60 kg and is 1.6 metres tall.

b (i) Use the formula above to calculate her BMI. Express your answer to the nearest whole number. *(1 mark)*

(ii) Use the table to identify this woman's category. *(1 mark)*

Obesity is on the increase in the UK.

An epidemiological study of some people born in the UK in March 1958 showed that fat children had the highest risk of becoming obese in adulthood. It also showed that most obese adults had not been fat in childhood or adolescence.

c List **three** degenerative diseases or conditions that obese people are at particular risk of developing.

(3 marks)

d Suggest strategies that health authorities and groups such as *Weight Watchers* should adopt to reduce obesity in the UK. *(2 marks)*
(Total 12 marks)
OCR 2802 May 2002, B (HHD), No.1

3 a Complete the following passage about Vitamin D.

Vitamin D promotes the strengthening of the skeleton by ensuring a plentiful supply of … ions in the blood. Infants and children who are deprived of vitamin D develop the disease called … which is associated with weak and deformed bones that cannot support their weight. Some women who have many pregnancies and breast feed their babies may suffer from a softening of bones called … . Vitamin D is found in some foods, such as … . However, most of the body's vitamin D is made in the body when the … is exposed to … *(6 marks)*

Dietary Reference Values (DRVs) are based on studies of the British population. The table shows DRVs for protein for women.

DRV	15–18 years	19–50 years	over 50 years
Estimated Average Requirement (EAR)/g day^{-1}	37.1	36.0	37.2
Reference Nutrient Intake (RNI)/g day^{-1}	45.4	45.0	46.5

b Explain the difference between the Estimated Average Requirement (EAR) and the Reference Nutrient Intake (RNI). *(2 marks)*

c State **two** reasons why the protein intake of 19–50 year old women might be greater than the values given in the table. *(2 marks)*

d Explain why women over the age of 50 should maintain a protein intake similar to that of younger women. *(2 marks)*

e Outline **two** advantages of publishing Dietary Reference Values. *(2 marks)*

Children in many developing countries suffer from severe protein energy malnutrition. As a result they show reduced growth and are susceptible to infectious diseases, such as measles.

f Explain why young children with protein energy malnutrition are especially susceptible to measles. *(2 marks)*

Supplementary feeding programmes provide such children with a diet rich in carbohydrate but with limited quantities of high quality protein.

g Suggest reasons for providing these children with this type of diet. *(2 marks)*

(Total 18 marks)

OCR 2802 Jan 2002, B (HHD), No.3

4 Surveys of the British population show that requirements for nutrients, such as iron, show a normal distribution. Three Dietary Reference Values (DRVs) are based on this normal distribution as shown in the figure.

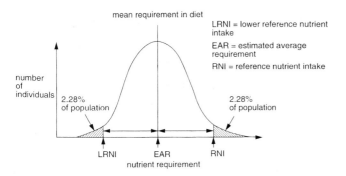

mean requirement in diet

LRNI = lower reference nutrient intake
EAR = estimated average requirement
RNI = reference nutrient intake

number of individuals

2.28% of population

2.28% of population

LRNI   EAR   RNI
nutrient requirement

a With reference to the figure, state the percentage of the population that has a requirement for iron between the LRNI and RNI. *(1 mark)*

The table shows the DRVs for iron for males and females aged 17.

	males			females		
	LRNI	**EAR**	**RNI**	**LRNI**	**EAR**	**RNI**
iron intake /mg day⁻¹	6.1	8.7	11.3	8.0	11.4	14.8

b With reference to the figure and the table, explain why

  (i) nutritionists recommend that people consume the RNI for nutrients, such as iron; *(2 marks)*

  (ii) the DRVs for iron for females are higher than those for males. *(2 marks)*

People with anorexia nervosa are likely to consume much less than the LRNI for iron.

c Suggest the likely consequences for people with anorexia nervosa of consuming less than the LRNI for iron over a long time. *(3 marks)*

(Total 8 marks)

OCR 2802 Jun 2001, B (HHD), No.7

5 a State **three** components of a balanced diet which provide energy. *(3 marks)*

Four investigations of the energy intake of 14 and 15 year old boys and girls have been carried out in the UK since the 1930s. These studies show that the average intake of energy has decreased, while the average body masses of both boys and girls have remained the same. The results of these investigations are shown in the table.

investigation	average energy intake / kJ per day	
	boys	**girls**
1930s	12 873	11 088
1960s	11 739	9534
1970s	10 962	8484
1980s	10 478	8316

b (i) Calculate the percentage decrease in energy intake for boys between the 1930s and 1980s. Show your working. *(3 marks)*

  (ii) Suggest **two** reasons for the fact that the intake of energy has decreased between the 1930s and the 1980s while the average body mass has remained constant. *(2 marks)*

  (iii) Explain why the energy intake of girls is lower than that for boys of the same age. *(1 mark)*

In 1991, the British Government's Committee on Medical Aspects of Food Policy (COMA) published dietary reference values (DRVs). The Estimated Average Requirements (EAR) is the dietary reference value for energy intake. EARs for different age groups are calculated from basal metabolic rates, the amount of energy needed to support growth and the amount of physical activity.

c Explain the value of publishing EARs for dietary energy. *(3 marks)*

d Suggest **two** problems that might be encountered in calculating the EAR for any one age group. *(2 marks)*

The overconsumption of energy-rich food can lead to obesity, which can increase the chances of becoming seriously ill.

e (i) Explain what is meant by *obesity*. *(1 mark)*

  (ii) Suggest **three** ways in which obesity may lead to serious illness. *(3 marks)*

(Total 18 marks)

OCR 2802 2000 Specimen paper, B (HHD), No.4

Gaseous exchange and exercise

Human gaseous exchange system

In this unit and those that follow, we shall explore in detail the effect of exercise on the gaseous exchange and circulatory systems of humans. To help you understand this chapter it is important to have covered the following material that appears earlier in the book:

- Unit 4.2 – Diffusion (sections 4.2.1 and 4.2.2)
- Unit 4.6 – Exchange between organisms and the environment (sections 4.6.1 and 4.6.2)
- Unit 4.7 – Gaseous exchange in mammals (all sections)
- Unit 8.1 – The need for transport systems in organisms (sections 8.1.1, 8.1.2 and 8.1.4)
- Unit 8.3 – The structure and functions of blood (section 8.3.2)
- Unit 8.5 – Obtaining oxygen – gaseous exchange in the alveoli (all sections)
- Unit 8.6 – Transport of oxygen (sections 8.6.1 and 8.6.2)
- Unit 9.1 – The mammalian heart (all sections)
- Unit 9.2 – The cardiac cycle (all sections)
- Unit 9.3 – Cardiac cycle and blood pressure (all sections).

As this chapter shows how gaseous exchange and blood transport are linked, a revision of each of these sections would be helpful even if you have covered all the material listed above.

13.1.1 Distribution of alveoli and blood vessels in lung tissue

The arrangement of alveoli and blood vessels within the human lungs is illustrated in figure 13.1, and information on how gaseous exchange occurs between air in the alveoli and blood within the vessels is covered in sections 4.7.2 and 8.5.1.

13.1.2 Distribution of tissues within the trachea, bronchi and bronchioles

Once air has been drawn in through the mouth or nose, it passes through a region at the back of the mouth called the pharynx, before entering the trachea. The **trachea** is a flexible air-way that is supported by pieces of **cartilage** (Fig 13.1). These pieces of cartilage do not form complete rings, but are horseshoe-shaped. This allows the adjacent oesophagus to expand without obstruction when food passes along it on its way to the stomach. The wall of the trachea is made up of **smooth muscle**, lined with **ciliated epithelium**, which possesses **goblet cells**.

The trachea branches into two bronchi, one **bronchus** entering each of the two lungs (Fig 13.1). The structure of

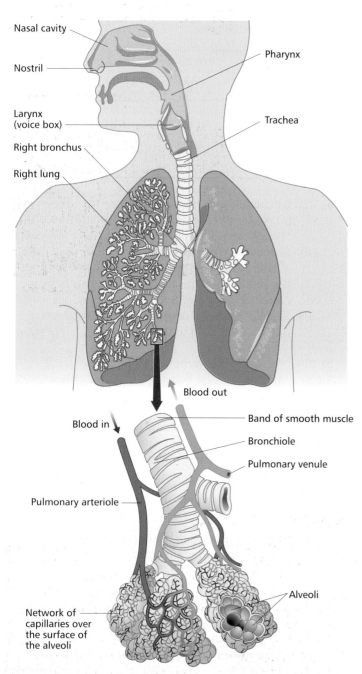

Nasal cavity

Nostril

Pharynx

Larynx (voice box)

Trachea

Right bronchus

Right lung

Blood out

Blood in

Band of smooth muscle

Bronchiole

Pulmonary venule

Pulmonary arteriole

Alveoli

Network of capillaries over the surface of the alveoli

Fig 13.1 *Human gaseous exchange system showing positions of trachea, bronchi and bronchioles as well as the arrangement of blood vessels and alveoli in the lungs*

the bronchi resembles that of the trachea, only they are smaller in diameter. They too are supported by cartilage, have walls of smooth muscle and are lined with ciliated epithelium, which contains goblet cells. The bronchi divide repeatedly to give a series of smaller tubes, called **bronchioles**. These are similar in appearance to the bronchi, although the tiniest ones lack cartilage and terminate in a cluster of alveoli (Fig 13.1).

13.1.3 Functions of tissues within the trachea, bronchi and bronchioles

- **Cartilage** is a rigid, but flexible supporting material. Its incomplete rings support the smooth muscle of these tubes, keeping them in an open position. In this way, it prevents the trachea and bronchi from collapsing when the air pressure inside them is lowered when breathing in. Being flexible and arranged in rings with spaces between each ring, it allows the trachea and bronchi to bend and extend – essential when bending or stretching the neck as a whole.
- **Smooth muscle** is tissue which is capable of contraction, but which is not under voluntary control. Although found in the trachea, bronchi and bronchioles, it is in the bronchioles where it has its effect. As the smaller bronchioles are not supported by cartilage, contraction of the rings of smooth muscle around them causes the bronchioles to constrict. In this way, the flow of air to and from the alveoli can be restricted and therefore controlled.
- **Elastic fibres** are flexible fibres that recoil if stretched. Once the smooth muscle in the walls of the bronchioles has contracted, it cannot itself dilate (open up) the lumen of the bronchioles. This is done by the elastic fibres, which become stretched when the smooth muscle contracts. As the smooth muscle relaxes, so the elastic fibres spring back to their original positions, thereby dilating the bronchioles. They therefore control the air-flow to and from the alveoli by allowing more air to pass through the bronchioles.
- **Goblet cells** are so called because they have long, thin stems and rounded tops, resembling a wine goblet in shape (see photo). They are part of the epithelium of trachea and bronchi and produce mucus that forms a thin layer over the whole inner surface of these structures. Mucus is sticky, and so bacteria, pollen and other dust particles cling to it as the air that contains them moves along these tubes. White blood cells, called macrophages (section 16.1.2), migrate from blood vessels and engulf bacteria and other particles trapped in the mucus. In this way the incoming air is 'filtered' clean and the risk of lung infections is reduced.
- **Ciliated epithelium** is a thin layer of epithelial cells that have hair-like organelles called cilia (section 1.7.2) on one surface. These cilia move in a synchronised manner and so transport the dirt-laden mucus which surrounds them upwards towards the pharynx at a rate of around 10mm min⁻¹. Once at the pharynx, the mucus passes down the oesophagus to the stomach, usually unnoticed. Here the acid conditions kill any bacteria not consumed by the macrophages.

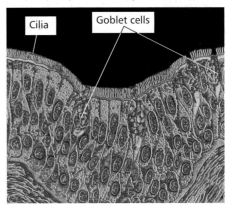

Epithelium of the trachea showing cilia and goblet cells as seen under a light microscope (×800)

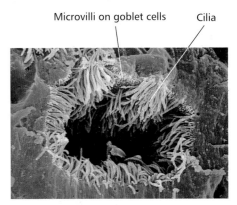

False-colour SEM of epithelium of the trachea showing ciliated epithelium (×3570)

SUMMARY TEST 13.1

Air enters the body through the mouth or nose and enters the trachea, which is supported by (1) shaped pieces of cartilage. The trachea branches into two bronchi. The trachea and bronchi are lined with epithelium, which produces mucus from (2) cells. This mucus is sticky and traps (3) and (4) present in the air. The mucus is moved towards the mouth by (5), present on the epithelium. The bronchi, in turn, divide to form (6), which can constrict when the (7) they contain contracts. These structures are dilated by the action of (8).

Air is moved in and out of the lungs by the movement of the diaphragm (which moves by contraction of its own muscle) and the ribs (which move by the action of the intercostal muscles between them). The volume of air moved during this process of breathing depends upon the extent of these movements and varies considerably depending on whether an individual is exercising or at rest.

13.2.1 Lung capacities

A range of terms is used to describe the changes in the volume of the lungs during breathing. The actual volumes vary a little depending on the size, sex, age, health and fitness of an individual. For this reason, the figures for different lung volumes show variations from textbook to textbook. During normal breathing at rest, you will exchange 0.5dm³ (500cm³) at each breath. This is called **tidal volume**. If you breathe in normally and then, rather than breathing out, continue to breathe in until your lungs are as full of air as you can make them, you will have taken in a further 1.5dm³ (1500cm³) of air (**inspiratory reserve volume**). If you carry out a similar process on breathing out, you will find that you can expel around an additional 1.5dm³ (1500cm³) of air if you force out of the lungs as much as you can (**expiratory reserve volume**). You may think you have completely emptied the lungs of air at this point. However, you would be wrong because around 1.5dm³ (1500cm³), called the **residual volume**, remains in the lungs. Without this residual volume, the moist walls of the alveoli would stick together and it would be impossible to re-inflate the lungs. If you take a maximum breath out, followed by a maximum breath in, you will inhale some 3.5dm³ (3500cm³) of air. This is called the **vital capacity**. These volumes are illustrated in the graph in figure 13.2. The **total lung capacity** is the total volume of air the lungs can hold. This is equal to the vital capacity (3.5dm³) plus the residual volume (1.5dm³). It is therefore 5dm³, although it can vary from 3.5dm³ to 8.0dm³. Although 0.5dm³ of air (tidal volume) is exchanged during normal breathing, only 0.35dm³ of this air actually reaches the alveoli and is hence available for gas exchange. The remaining 0.15dm³ is left in the trachea, bronchi and bronchioles, where no gas exchange occurs. This is known as the **dead space**.

To summarise:

- **Tidal volume** (0.5dm³) – the volume of air normally exchanged at each breath at rest.
- **Residual volume** (1.5dm³) – the volume of air that is always left in the lungs.
- **Vital capacity** (3.5dm³) – the volume of air that can be exchanged between maximum inspiration and maximum expiration.
- **Total lung capacity** (5.0dm³) – the maximum volume of air that the lungs can hold.

VOLUMES AND CAPACITIES

Why are some lung volumes called 'volumes' while others are called 'capacities'? A volume is used to describe a single component, e.g. residual volume, while a capacity is made up of more than one volume. Vital capacity, for example, is made up of the tidal volume, plus the inspiratory reserve volume, plus the expiratory reserve volume (however, the last two terms need not be learned at this level).

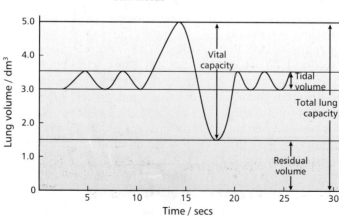

Fig 13.2 *Graph to illustrate lung capacities*

13.2.2 Ventilation rate

Although lung capacities are useful measures, they tell us little about how much gas exchange takes place. What is more useful is how much air is exchanged in a given time. The **ventilation rate** is the total volume of air that is moved into the lungs during one minute. This is equal to the tidal volume multiplied by the number of breaths (breathing rate) and is expressed as $dm^3 \, min^{-1}$. The ventilation rate is also called the **minute volume (V_E)**.

$$\text{Minute volume } (V_E) = \text{tidal volume} \times \text{breathing (ventilation) rate}$$

13.2.3 Measuring lung capacities and volumes

Lung capacities and volumes can be measured using a piece of equipment called a **spirometer**, attached to an instrument called a kymograph, which records changes in volume. The process is as follows:

- A person breathes in air through a tube from a container of oxygen that floats in a tank of water.
- This floating container rises and falls as the person breathes out and in respectively.
- The extent of this rise and fall reflects the volume of oxygen taken out of or into the lungs at each breath.
- The container is attached by an arm to a pen that also rises and falls as the person breathes.
- The pen draws a trace on graph paper attached to the rotating drum of the kymograph.
- Expired air passes through a chamber containing soda lime before it passes back into the oxygen container. This removes carbon dioxide because otherwise it would be re-breathed and high levels of carbon dioxide will increase breathing rates and could lead to the person fainting.
- Because oxygen is used up at each breath, the volume in the chamber slowly reduces and this is shown on the trace produced (Fig 13.3).

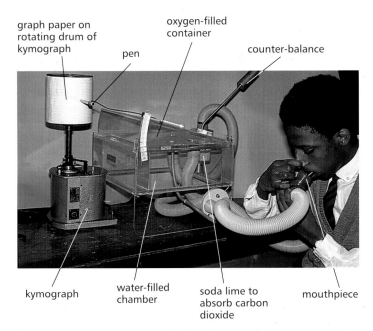

graph paper on rotating drum of kymograph

pen

oxygen-filled container

counter-balance

kymograph

water-filled chamber

soda lime to absorb carbon dioxide

mouthpiece

Student using a spirometer attached to a kymograph

SUMMARY TEST 13.2

The volume of air normally moved in and out of the lungs while at rest is **(1)**dm^3 and is known as the **(2)**. Of this only **(3)**dm^3 actually reaches the alveoli and is therefore available for gas exchange. This is because the remainder, called the **(4)**, remains in the trachea, bronchi and bronchioles. The volume of air exchanged between the deepest breath in followed by the deepest breath out is called the **(5)**, and normally measures 3.5dm^3. Even with maximum effort it is impossible to expel all air from the lungs because a volume of **(6)**dm^3 remains to stop the walls of the alveoli from sticking together. This volume is called the **(7)** volume. The ventilation rate, also known as **(8)**, is the breathing rate multiplied by the **(9)**, measured over a period of **(10)**. Lung capacities can be measured using a **(11)**, which is attached to a recording device called a **(12)**.

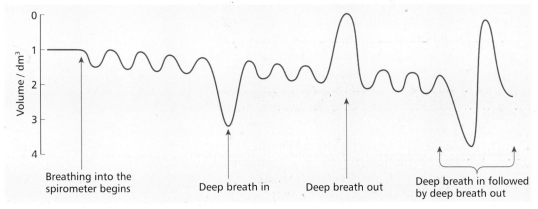

Breathing into the spirometer begins

Deep breath in

Deep breath out

Deep breath in followed by deep breath out

Fig 13.3 *Trace produced on a kymograph as an individual used a spirometer to measure lung volumes*

Pulse rate and its measurement

The pulse rate is the number of times a person's heart beats in one minute.

13.3.1 What is a pulse?

When blood is pumped from the left ventricle of the heart, it enters the systemic circulation via the main artery – the aorta. The surge of pressure created by the contraction of the walls of the left ventricle causes the elastic walls of the aorta to stretch, creating a 'bulge'. This extended portion of the aorta wall then rebounds due to the elastic fibres in its wall. With the semi-lunar valves preventing blood returning to the heart, this recoil action of the aorta wall causes a new pressure bulge to be created in the aorta a little further away from the heart. This process then continues along the aorta and into the rest of the arterial system. It is these pressure bulges, passing in sequence along arteries, that we recognise as the pulse.

13.3.2 Measuring the pulse rate

The pulse rate is normally measured as the number of pulses in one minute. This, of course, is equivalent to the number of heartbeats in a minute. The pulse rate can be measured by putting pressure on an artery in which pulsations can be felt. This is normally any place where an artery runs close to the skin surface with a bone beneath it. The wrist is most commonly used but other places where the pulse can be felt include behind the knee, in the neck or the temple. In practice, pulse rates are normally taken for 30 seconds and doubled to give the pulses per min. This is preferable to measuring for the whole minute because, after exercise, the pulse rate will slow over the course of a full minute and may therefore give a lower value than if 30 seconds is used and the number doubled. A period of less than 30 seconds would mean that too few pulses would be counted for the result to be reliable, i.e. the sample size would be too small.

13.3.3 Variations in pulse rate

Pulse rates are normally regular, i.e. the interval between each pulse is the same. Pulse rates do, however, vary with age and level of fitness. They become slower with age, being 140–150 pulses min^{-1} in the unborn child, gradually reducing to 70–75 pulses min^{-1} in adulthood. Table 13.1 shows the range of pulse rates at various ages.

Other variations occur. For example, the pulse rate:

- rises during excitement
- rises during and following physical exertion
- rises during digestion
- is generally more rapid in females
- is influenced by the breathing rate
- increases by approximately 10 pulses min^{-1} for each 0.5°C rise in body temperature above 36.8°C (normal human body temperature).

Rises in pulse rate normally reflect the need for extra oxygen to provide respiratory energy, e.g. during physical exertion or digestion. Body temperature may rise during fever as a consequence of increased metabolism. The pulse rate also rises in order to increase the circulation of white blood cells and **antibodies** to fight infection and to carry heat away from internal organs to the skin, where it is lost to the environment.

Keeping fit reduces your resting pulse rate

Measuring the pulse rate

Table 13.1 *Average pulse rates according to age*

Age	Number of beats min^{-1}
Fetus	140–150
Newborn infants	130–140
During first year	110–130
During second year	96–115
During third year	86–105
7–14 years	76–90
14–21 years	76–85
21–60 years	70–75
60+ years	67–80

13.3.4 Cardiac output

The pulse rate tells us how many times the heart is beating each minute, but not the volume of blood pumped each minute. To find this we need to know the volume of blood pumped at each beat. This is called the **stroke volume**. By multiplying the stroke volume by the heart (pulse) rate we can calculate the volume of blood pumped each minute. This is called the **cardiac output**.

Cardiac output = heart (pulse) rate × stroke volume

13.3.5 Pulse rate and physical exercise

Cardiac output at rest is around $5\,dm^3\,min^{-1}$ but it can rise to $30\,dm^3\,min^{-1}$ during strenuous exercise. The most efficient means of reaching a given cardiac output is to have a low resting pulse (heart) rate and high stroke volume because this results in less oxygen being needed by **cardiac muscle**. Regular training causes certain changes to the heart.

- **The chambers enlarge** so that a greater stroke volume can be achieved.
- **The cardiac muscle increases** to allow blood to be ejected from the heart more forcefully.

These adaptations of the heart to exercise over a period of time lead to:

- an increased stroke volume
- a reduced resting pulse (heart) rate.

As a result, individuals who are physically fit have larger hearts and a lower resting pulse (heart) rate. The resting pulse rate can therefore give some indication of the level of physical fitness (as shown in table 13.2). It must, however, be remembered that other factors, such as health and age, can influence resting pulse rate.

Table 13.2 Resting pulse rates according to fitness levels

Fitness level	Resting pulse rate / beats min^{-1}
Poor	Above 80
Reasonable	70–79
Good	60–69
Excellent	50–59
Outstanding	Below 50

During exercise, pulse (heart) rate can increase from 70 pulses min^{-1} to 200 pulses min^{-1}. The differences between the pulse rate, stroke volume and cardiac output of individuals who are physically fit compared to ones who are less so, is shown in table 13.3.

We can see from this table that the stroke volume of the fitter individual is always greater than that of the less fit person, and the pulse rate, especially the resting one, is lower.

Table 13.3 Comparison of stroke volume, pulse rate and cardiac output for individuals of different fitness levels

State	Fitness level	Stroke volume / cm^3	Pulse rate / pulses min^{-1}	Cardiac output / $dm^3\,min^{-1}$
At rest	Reasonable	65	73	4.75
	Excellent	90	52	4.68
Strenuous exertion	Reasonable	100	200	20.0
	Excellent	155	190	29.5

SUMMARY TEST 13.3

Pulse rate is a measure of the number of **(1)** in one minute. A pulse is a pressure bulge in an artery that can be felt by gently pressing a superficial artery onto a bone with the fingers. Pulses are normally taken in the **(2)**, but can also be taken behind the knee or in the **(3)**. Pulse rates are usually in the range **(4)** pulses min^{-1} in a fetus but fall to the range **(5)** pulses min^{-1} in an adult. Pulse rates rise during **(6)** and **(7)** or when the body temperature increases. Cardiac output is the volume of blood pumped by the heart and is calculated by multiplying the pulse rate by the **(8)**. Generally, the fitter an individual is, the **(9)** is their resting pulse rate.

Blood pressure

To function normally, mammals must maintain a minimum pressure of blood within their closed double circulation system (unit 8.2). This pressure must then be increased as necessary to meet the demands of exertion. Blood pressure is still measured by the medical profession in millimetres of mercury (mmHg) and at rest this is between 120 and 140mmHg (**systolic**) and 80 and 90mmHg (**diastolic**), depending on age (blood pressure increases with age). The SI unit which should be used is kilopascals (kPa) and the equivalent values are 16.0–18.5kPa (systolic) and 10.5–12.0kPa (diastolic).

13.4.1 Blood pressure in arteries, capillaries and veins

When blood is pumped from the left ventricle of the heart it enters the systemic circulation via the aorta. In the aorta the pressure exerted causes extension of its elastic walls. The pressure falls a little as a result of this extension. The elastic wall of the aorta 'rebounds' in a reaction, called the **recoil action**, that increases the pressure again. This recoil leads to regular fluctuations in blood pressure as it moves along the aorta and other major arteries (Fig 13.4). As the blood moves further from the heart its overall pressure decreases, but only slightly at first. However, when the arteries divide into arterioles, the overall volume of the vessels increases. This leads to a more marked fall in blood pressure and also to a reduction of the pressure fluctuations (pulses). This decrease in pressure is essential to prevent the bursting of the one-cell-thick walls of the capillaries into which the blood next passes. This reduction in pressure continues in the capillaries so that, by the time the blood reaches the venules and veins, it has fallen to almost zero (Fig 13.4). Its return is therefore mostly dependent on the body muscles squeezing the thin-walled veins and the valves within them, thus ensuring that the blood moves towards the heart. The pattern of pressure changes is the same for the pulmonary circulation but the actual pressures are only around 25% of those in the systemic circulation.

Fig 13.4 Blood pressure in arteries, capillaries and veins

13.4.2 What determines blood pressure?

The pressure of blood depends upon a number of factors including:
- how much blood the ventricles pump in a given time (cardiac output)
- the resistance to the flow of blood (e.g. narrowing of blood vessels during vasoconstriction)
- the total volume of the blood.

Blood pressure is increased by:
- increased cardiac output, e.g. during exercise
- increased resistance to blood flow, e.g. caused by the narrowing of the arterioles during vasoconstriction or the reduced elasticity of arteries due to atherosclerosis (section 14.4.1)
- increased blood volume, e.g. due to the retention of more water by the kidneys under the influence of antidiuretic hormone (ADH).

Blood pressure is reduced by:
- decreased cardiac output, e.g. during rest or sleep
- decreased resistance to blood flow, e.g. during the widening of arterioles during vasodilation
- decreased blood volume, e.g. during loss of blood due to haemorrhage or increased loss of water by the kidneys.

13.4.3 Hypertension

Hypertension is the term used to describe sustained high blood pressure when at rest. It is normal for blood pressure to increase temporarily both during and after exercise. Hypertension is not always easy to define as there is much individual variation in 'normal' blood pressure, depending on age, general health

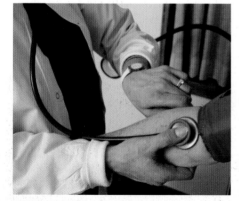
Taking blood pressure

EXTENSION – MEASURING BLOOD PRESSURE

Blood pressure is measured using a device called a **sphygmomanometer**. This is used to measure the blood pressure of the brachial artery (which is found on the inside of the arm) at the elbow. It is measured as follows.

- An inflatable cuff is wrapped around the upper arm.
- A stethoscope is placed on the brachial artery on the inner arm at the elbow to hear the sound of the pulse.
- Air is slowly pumped into the cuff, which constricts the upper arm and restricts the blood supply along the brachial artery.
- When the sound of the pulse stops, the air in the cuff is very gradually released.

- As soon as the pulse is again heard in the stethoscope, the pressure reading in the cuff is recorded. The pressure of the cuff and the systolic blood pressure are almost exactly equal as the blood is just beginning to flow past the cuff at this point. This pressure is therefore the systolic blood pressure.
- The cuff is slowly deflated further until there is no sound in the stethoscope.
- The pressure in the cuff is again recorded – this is the diastolic blood pressure.

1. Cuff pressure = 25 kPa (190 mm Hg)
   Greater than systolic pressure
   Artery closed
   No blood flow
   No sound in stethoscope

2. Cuff pressure = 16 kPa (120 mm Hg)
   Same as systolic pressure
   Artery partly closed
   Intermittent blood flow
   Sound heard in stethoscope

3. Cuff pressure = 10.5 kPa (80 mm Hg)
   Same as diastolic pressure
   Artery open
   Blood flows freely
   Sound in stethoscope disappears

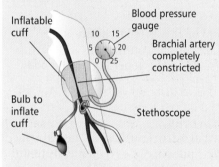
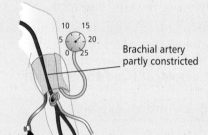
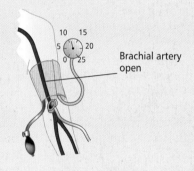

Fig 13.5 *Measurement of blood pressure*

and degree of activity. For example, blood pressure increases as one gets older as a result of the walls of the arteries becoming less elastic. The World Health Organisation defines hypertension as a systolic pressure of 20.5kPa (160mmHg) and a diastolic pressure of 12.5kPa (95mmHg). It is estimated that 10–20% of all UK residents suffer from hypertension. Hypertension is a major risk factor associated with heart disease, as well as increasing the risk of **atherosclerosis**, stroke and kidney failure. It is very much the 'silent killer' because the disease itself has no symptoms to warn of a forthcoming heart attack or stroke. For reasons that are not fully understood, some individuals are prone to hypertension (**essential hypertension**). In others, it is the consequence of some other medical condition, such as kidney failure or hormone imbalance (**secondary hypertension**). There are, in addition, a number of other factors that can contribute to

hypertension. These include:

- smoking
- excessive alcohol intake
- too much salt in the diet
- stress
- obesity
- lack of exercise.

There is also a genetic link because there is a tendency for high blood pressure to run in families. Hypertension can be treated by behavioural changes, such as stopping smoking, reducing alcohol or salt intake etc. and / or medication. Two types of drugs are used.

- **Diuretics** – which cause the kidneys to excrete more water, thus reducing the volume of the blood and hence lowering its pressure.
- **β-blockers** – which reduce cardiac output or relax the smooth muscle in the arteries, thus reducing their resistance to blood flow.

SUMMARY TEST 13.4

The normal systolic blood pressure measured in SI units is in the range (**1**). The equivalent diastolic range is (**2**). As a person gets older, blood pressure usually (**3**). Blood pumped from the left ventricle enters the (**4**) blood system via the main artery, called the (**5**). The recoil action of the walls of this vessel leads to fluctuations in blood pressure called pulses. Blood pressure falls most steeply in the blood vessels called (**6**) and, by the time it reaches the (**7**), it is almost non-existent. If the blood vessels called arterioles dilate, the pressure of the blood will (**8**). If the body retains more water than usual, then blood pressure will (**9**). Sustained high blood pressure when at rest is termed (**10**) and may be the result of medical conditions such as a hormone imbalance or (**11**).

13.5.1 Energy and exercise

The energy required for the muscular movement that is an integral part of exercise is provided by the breakdown of various substances during cellular respiration. These substances include:

- **Glucose** – this is the usual source of energy, and provides a source of **adenosine triphosphate (ATP)**, whether or not oxygen is available.
- **Glycogen** – this is stored in the liver and muscles, and can be quickly broken down to glucose, which in turn can provide energy as above.
- **Triglycerides** – these are stored as fat in **adipose tissue** under the skin and around body organs. A little is stored in muscle cells. Triglycerides provide more ATP than an equivalent mass of glucose or glycogen, and are therefore important as long-term, concentrated stores of energy. This breakdown does, however, require 15% more oxygen and so, when oxygen is in short supply, glycogen stores are preferred.

Whatever the original chemical used, it has to be converted to **adenosine triphosphate (ATP)** before it can be used for muscle contraction. ATP is thus the intermediate source of energy for exercise. It can be generated in the body both **aerobically** and **anaerobically**.

13.5.2 Aerobic exercise

In aerobic exercise, the muscles use up oxygen at a rate less than the rate at which it is delivered from the lungs, via the blood. The presence of oxygen means that the muscles produce ATP by the complete **oxidation** of glucose to carbon dioxide and water. The process releases 40% of the energy available in the glucose molecules. Most moderate exercise such as walking, swimming and jogging is aerobic.

13.5.3 Anaerobic exercise

Anaerobic exercise includes very strenuous exercise that can only be sustained for a short period of time. Running the 100 metres or lifting heavy weights are examples of anaerobic exercise. The muscles use up oxygen faster than it can be supplied, building up an **oxygen debt**. In the absence of oxygen, glucose is broken down, not to carbon dioxide and water, but to lactate (lactic acid). The lactate quickly causes muscle fatigue and cramp if allowed to accumulate. It is therefore rapidly broken down to carbon dioxide and water once the supply of oxygen outstrips its consumption. Although this process releases only 2% of the energy of the glucose molecules, it does allow the muscles to continue to operate despite the lack of an adequate oxygen supply.

13.5.4 Effects of exercise on gaseous exchange

During exercise, the demand for oxygen by muscle cells goes up. To meet the demand, the ventilation rate must be increased by:

- increasing the number of breaths taken each minute, from about 15 at rest to 45 during exercise
- increasing the volume of air exchanged at each breath, from around $0.5dm^3$ at rest to $3.5dm^3$ during exercise.

During moderate exercise such as jogging, the amount of oxygen absorbed by the lungs can increase 13 times, from $0.3dm^3$ min^{-1} to $4.0dm^3$ min^{-1}.

How the increased ventilation rate is triggered is complex, but includes some or all of the following factors:

Jogging is a form of aerobic exercise

- **Lowering of the blood pH** due to an increase in lactate (lactic acid) resulting from the anaerobic breakdown of glucose in muscle cells. The presence of more carbon dioxide in the blood also lowers blood pH.
- **Increased production of carbon dioxide** as a result of aerobic breakdown of glucose in muscle cells.
- **Lowering of the oxygen concentration** of the blood as it is used up by muscle cells.
- **Receptors in muscles and joints** stimulating the inspiratory centre in the brain.

13.5.5 Effects of exercise on the cardiovascular system

The net effect of the following changes to the cardiovascular system during exercise is to supply more oxygen to the muscles and to remove the waste products of respiration, including heat.

- **Cardiac output** increases from around $5\,dm^3\,min^{-1}$ at rest to a maximum of $30\,dm^3\,min^{-1}$ (section 13.3.5)
- **Vasodilation** or **vasoconstriction** of different blood arterioles redistributes the blood towards muscles and away from organs such as the kidney and intestines, whose need for blood is less immediate. The heart needs more blood to maintain the higher cardiac output and the brain must continue to function normally in order to coordinate activities – its supply is therefore largely unchanged during exercise. As exercise generates heat, the blood supply to the skin is increased during exercise to help dissipate this heat to the environment. Table 13.4 describes these changes in the distribution of blood.
- The volume of the blood plasma is reduced due to water loss through sweating, and its glucose level may drop as carbohydrate stores become exhausted. This explains why participants in endurance events, such as marathon running, are encouraged to drink glucose-based liquids.
- Blood acidity levels increase (i.e. pH falls) as more lactate and carbon dioxide are produced.
- There is a rise in **systolic** blood pressure although **diastolic** pressure is largely unchanged.

13.5.6 Effects of exercise on muscles

The increased rate of respiration in muscles during exercise results in the following short-term effects:

- depletion of energy stores (phosphocreatine, glycogen and **triglyceride**)
- reduced oxygen levels as a result of using up the oxygen stored in the myoglobin (unit 8.6)
- raised levels of carbon dioxide and lactate
- increased temperature.

13.5.7 Adrenaline and exercise

Adrenaline is often described as the 'fight or flight' hormone because it prepares the body for action. It is produced in anticipation of exercise, e.g. a race, in order to get the body ready for the forthcoming exertions and it

continues to be produced during exercise. Its effects are summarised in table 13.5.

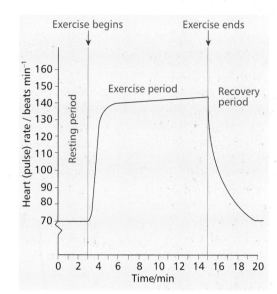

Fig 13.6 *Heart (pulse) rate over a 20-minute period including 12 minutes of exercise*

Table 13.4 *The rate of blood flow to different parts of the body at rest and during strenuous exercise*

Part of the body	Rate of blood flow / $cm^3\,min^{-1}$	
	At rest	During exercise
Liver and intestines	2500	90
Body muscle	1000	16000
Kidneys	1000	300
Brain	750	750
Skin	500	1000
Heart muscle	250	1200

Table 13.5 *Effects of adrenaline in preparing the body for exercise*

Effect	Purpose
Glycogen in the liver is converted to glucose	Makes more glucose available to increase the energy output of muscles
Cardiac output is increased	Increases rate at which glucose and oxygen are delivered to muscles
Vasoconstriction or vasodilation of blood vessels	Blood is redistributed from 'non-essential' organs to muscles, heart and skin
Bronchioles are dilated	Air is more easily inhaled and therefore oxygen is diffused into the blood more rapidly
Sensory perception and mental awareness are increased	Responses to stimuli are more rapid

Fitness is the ability to cope with everyday stresses. It takes many forms, one of which is physical fitness. It is not the same as health, which, as we saw in section 11.1.1, is a state of well-being and freedom from disease. Physical fitness does, however, contribute to a person's well-being and normally makes them less likely to suffer ill health. Physical fitness has many components: agility, power, coordination, strength, speed etc. In this unit we shall look at one such component, namely **aerobic fitness**, also called aerobic capacity.

13.6.1 Aerobic fitness

Aerobic fitness is the ability of an individual to take in and use oxygen. An individual's aerobic fitness depends on three factors:

- effective ventilation of the lungs
- effective transport of oxygen from the lungs to the respiring cells, e.g. muscle cells
- effective use of oxygen within the cells.

Aerobic fitness is therefore affected by the efficiency of the respiratory and cardiovascular systems. Although it is largely determined by an individual's **genes** (e.g. the extent and type of different muscle fibres present), aerobic fitness can be improved by training. It is normally measured by finding out the maximum volume of oxygen that an individual's body can take in and use. This is known as **VO$_2$max** and is usually expressed as dm^3 min^{-1} kg^{-1} or cm^3 min^{-1} kg^{-1}, where kg refers to body mass. Table 13.6 gives some values for VO$_2$max for different age groups and gender.

As individuals cannot work at their maximum aerobic capacity for long periods, a more useful measure is the percentage of their VO$_2$max that an individual can operate at for prolonged periods. A well-trained endurance athlete can work at around 85% of their VO$_2$max for long periods, while a non-athlete would struggle to maintain 65% of VO$_2$max.

Table 13.6 VO$_2$max values for men and women of different ages (cm^3 min^{-1} kg^{-1})

Age	Sex	VO$_2$max / cm^3 min^{-1} kg^{-1}		
		Very poor	Average	Very good
20–29	Male	38	40–41	57+
	Female	28	35–43	49+
30–39	Male	34	40–47	52+
	Female	27	34–41	48+
40–49	Male	30	36–43	48+
	Female	25	32–40	46+
50–59	Male	25	32–39	44+
	Female	21	29–36	42+

13.6.2 Measuring aerobic fitness

Measuring aerobic fitness accurately requires a treadmill and laboratory conditions and is usually carried out as follows.

- The individual runs on a treadmill and all the air they breathe out is collected in a rubber-lined bag, called a Douglas bag.
- The individual continues to run until exhausted.
- The volume and oxygen content of the bag of expired air is measured.
- As we know the percentage of oxygen in the normal atmosphere, we can calculate how much oxygen has been consumed, and hence the VO$_2$max.

As access to this type of technical equipment is limited, it is easier to obtain an estimate of VO$_2$max by measuring heart (pulse) rate. There is a correlation between VO$_2$max and the heart beat at the point at which exercise ends. The greater the aerobic fitness of an individual, the more quickly the heart (pulse) rate returns to its resting value. Figure 13.7 illustrates the typical changes in heart (pulse) rate before, during and after exercise.

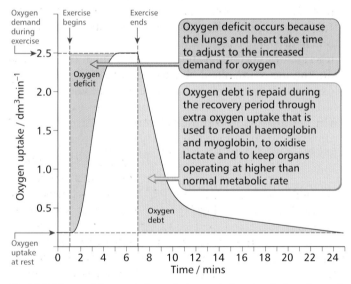

Fig 13.7 Oxygen taken up before, during and after six minutes of exercise

13.6.3 Experiments to investigate the effects of exercise on the body

Before you undertake an experiment involving physical exertion, it is important to be sure that your state of health and physical fitness is appropriate for the particular task. The following precautions should always be taken.

- Practise the exercise for a short period first and do not proceed further if you feel very breathless, sick, dizzy or have any pain, especially in the chest.

- Stop the exercise immediately if you experience any of the above symptoms or any other unusual or unexpected symptoms.
- Check with a doctor first if you suffer from any cardiovascular, respiratory or musculo-skeletal condition. Follow carefully any guidance given.
- Get used to any equipment you are going to use in the experiment.
- Carry out appropriate warm-up exercises.

The effects of exercise on the body are covered in units 13.5 and 13.7. The easiest effect to measure is on the heart rate. This is the same as the pulse rate, whose measurement is described in section 13.3.2. The type of exercise that is undertaken should be appropriate to the abilities and fitness level of the individual concerned. Examples might include:

- jogging / running
- sit ups
- step exercises
- shuttle runs.

Although the process and duration of the exercise will vary depending on the activity chosen, the following is a typical method.

- Take your resting pulse rate for 30 seconds before commencing the exercise.
- Carry out the exercise under the supervision of a trained tutor for 2–5 minutes, or until you can no longer continue.
- Take your pulse rate for 15 seconds on completing the exercise.
- Rest for 45 seconds and take your pulse rate for 15 seconds.
- Continue this pattern for at least 5 minutes.
- Calculate the pulse rate per minute in each case.

13.6.4 Improving aerobic fitness

To improve aerobic fitness a person needs to carry out continuous exercise of the whole body at a level below the maximum possible. Activities which suit this purpose are those which put stress on the respiratory and cardiovascular systems. These include jogging, running, rowing, swimming and cycling. A minimum of 12 minutes exercise is needed because initially the muscles will probably be working **anaerobically** as the body adjusts to the demands being made on it. Time is therefore needed to repay any **oxygen debt**. The activity needs to be continuous because stopping periodically gives the body time to recover rather than putting aerobic systems under stress. 30–40 minutes is ideal for moderately fit individuals; elite athletes will work considerably longer. The intensity at which the activity is carried out will depend on the fitness of the individual. As heart rate increases in relation to the amount of work undertaken, it forms a useful guide to how intense the exercise should be. It is important to overload the aerobic system without pushing it into working anaerobically. One useful guide is to work with your heart rate at the resting heart rate plus 60% of the difference between the resting and maximum heart rate. This is known as the

critical threshold. Maximum heart beat is approximately 220 minus your age.

Therefore, a person aged 30 years with a resting heart rate of 70 beats min^{-1} has a critical threshold of:

$$70 + 60\% \text{ of } (190-70)$$
$$= 70 + 60\% \text{ of } (120)$$
$$= 70 + 72$$
$$= 142 \text{ beats min}^{-1}$$

To improve aerobic fitness in the long term, this form of aerobic training needs to be carried out at least twice a week. Good athletes will train at least five times a week. Although some improvement should be apparent after 4 weeks, the training should be maintained for at least 12 weeks. During this period, the heart rate will decrease for the same workload. It is therefore necessary to increase the workload over the period in order to remain at, or above, the critical threshold.

Regular cycling is an excellent means of improving aerobic fitness

SUMMARY TEST 13.6

Aerobic fitness is normally measured as the maximum volume of oxygen that can be taken in and used. It is abbreviated to the expression **(1)** and is measured in the units **(2)**. A non-athlete can normally only achieve around **(3)**% of their maximum oxygen intake over a prolonged period, whereas top-class athletes might achieve **(4)**%. In the laboratory aerobic fitness can be measured by running on a **(5)** and collecting the exhaled air in a rubber-lined bag, called a **(6)** bag. Alternatively, the **(7)** can be measured as there is a strong correlation between this measurement and aerobic fitness. Improving long-term aerobic fitness requires carrying out aerobic exercise for a minimum period of **(8)** minutes initially, on at least **(9)** occasions each week, for a minimum of **(10)** weeks in total.

Participating regularly in physical sports such as tennis has long-term health benefits

We saw in section 11.1.1 that health is more than simply the absence of disease; it is a state of mental, physical and social well-being. Regular exercise contributes to these states of well-being by helping to keep people physically fit and therefore their bodily systems working efficiently and in harmony with one another. In this state, individuals have the strength, flexibility, coordination and endurance to make them less accident-prone and better able to fight infections and other diseases. Exercise can relax the mind and so contribute to mental well-being. Exercise and sports, especially team sports, bring together people with similar interests, often from different backgrounds, and so improve social skills and integration. Regular exercise cannot change the genes that often determine our overall health, but what it can do is help us maximise the potential we were born with.

13.7.1 Benefits to the heart

Regular exercise over a period of time produces changes to the heart which are designed to improve the rate at which blood transports oxygen from the lungs to the tissues, especially muscles. These changes include:

- **Hypertrophy of the heart** – this increase in size of the heart is often called **athlete's heart**. As the amount of **cardiac muscle** increases so does the force it can generate, allowing it to pump out more blood at each beat.
- **Increased number and size of mitochondria** – this occurs in cardiac muscle fibres and results in an increased energy output.
- **Increase in stroke volume** – the heart not only contracts more forcefully, but also holds more blood, allowing more to be pumped at each beat. The stroke volume for a trained athlete at rest is around 90cm^3, compared to 65cm^3 for an untrained individual. The maximum stroke volumes are 155cm^3 and 100cm^3 respectively.
- **Decrease in resting heart rate** – this is the result of **cardiac output** remaining the same but the stroke volume increasing in a trained athlete. The heart can therefore beat more slowly but still pump the same volume of blood. As a result of aerobic training, the resting heart rate may fall from 73 to 52 beats min^{-1}.
- **Increased maximum cardiac output** – while the cardiac output at rest is largely unchanged, training can increase the maximum output by around 50%.

13.7.2 Benefits to the blood and vascular system

In the same way as it affects the heart, regular exercise can produce adaptations of the blood and vascular system that improve the circulation of the blood. These include:

- **Increased volume of blood** – mostly as a result of an increase in the volume of blood plasma. However, there is also a small increase in the number of red blood cells. This provides more **haemoglobin** and hence a greater quantity of oxygen carried by the blood.
- **Decrease in blood pressure when at rest** – as a result of increased elasticity of the arterial walls.
- **Increase in the number of blood capillaries** – this occurs both in the lungs and in **striated muscle**. This improves the rate of gaseous exchange in both tissues.

13.7.3 Benefits to the lungs

Given that a good supply of oxygen is essential for performance during exercise, it is perhaps surprising that even the best-trained athletes do not show marked differences in lung volumes to those of an average person. The differences that do occur, however, are:

- **A small increase in lung volumes,** such as **vital capacity.**
- **An improvement in the efficiency of the intercostal muscles** (muscles between the ribs) and other respiratory muscles; this increases the volume of air forcibly exhaled and the rate of breathing.
- **Improved network of pulmonary capillaries** around the alveoli, which increases the efficiency of gaseous exchange.
- **An increase in the size of alveoli.**

13.7.4 Benefits to the muscles

As exercise is primarily the use of muscles, it is hardly surprising that there are a number of long-term benefits to the musculo-skeletal system that result from regular exercise. These include:

- **Increased muscle size and strength** – muscles build up in size and strength, a process called hypertrophy, as a result of power-training exercises like weight lifting, rather than endurance exercises like running.
- **Increased levels of myoglobin** in muscle fibres to store oxygen (unit 8.6).
- **Increased number and size of mitochondria** in muscle fibres and therefore an increase in the rate of **aerobic** respiration and hence energy release.
- **Greater enzyme activity** of those enzymes involved in respiration, thereby adding to the efficiency of energy release.
- **Larger stores of glycogen and triglycerides** to act as energy reserves, especially when the duration of exercise is prolonged.
- **Improved capillary network** thereby improving the rate of diffusion of oxygen and carbon dioxide between muscle cells and the blood.

13.7.5 Overall health benefits

Before looking at other ways in which exercise benefits health, it is necessary to appreciate that exercise and sport can also involve health risks, especially if undertaken without adequate training and preparation. These risks include damage to muscles and / or skeleton from over-exertion, falling or physical contact. Examples of injuries include muscle strains, pulled **tendons** and broken bones. By contrast, the benefits of regular and appropriate exercise are considerable. These include:

- **Reduced risk of coronary heart disease** both directly as a consequence of strengthening the heart and increasing the blood flow through it, and indirectly through lowering blood pressure and blood **cholesterol,** and reducing obesity. Exercising three times a week can halve the risk of a heart attack.
- **Reduced risk of hypertension** – as a result of better elasticity of the artery walls and greater dilation of them, which reduces the resistance to blood flow within them.
- **Less obesity** – exercise uses up energy, e.g. one hour of jogging can use 4000 kJ – the equivalent of 140g of fat. Losing weight also reduces the strain put on joints.
- **Lowering of blood cholesterol** and therefore the build-up of fatty **plaques** on the artery linings, which cause **atherosclerosis.**
- **Slowing the process of atherosclerosis** – arteries harden less rapidly due to improved elasticity and lower cholesterol levels.
- **Maintaining sensitivity of cells to insulin** – this helps to keep blood sugar levels within the normal range, and so reduces the risk of type II **diabetes.**
- **Reduced risk of strokes** due to increased blood flow through the brain, lower blood pressure and less atherosclerosis.
- **Lower risk of osteoporosis,** especially in females, because exercise builds up the calcium content of bones, making them harder and less liable to break.
- **Increased resistance to infection,** although this may be lowered for the first 24 hours after vigorous exercise.
- **Greater body flexibility** and therefore less risk of falls and strains.
- **Stronger ligaments and tendons.**
- **Better balance and coordination.**
- **Better mental health** as a consequence of being in a more relaxed psychological state and feeling good about yourself and your performance. Exercise can also improve sleep and reduce stress.

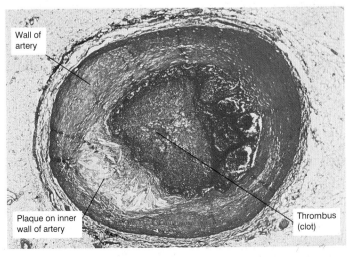

Wall of artery

Plaque on inner wall of artery

Thrombus (clot)

Regular exercise can help prevent coronary heart disease such as this thrombus in a human coronary artery

1 a Explain the term *aerobic exercise*.   (2 marks)

A student investigated the effect of aerobic exercise on his father, using a cross country skiing machine and a digital pulse meter. He began by taking his father's resting pulse several times during the day and calculating a mean resting pulse rate.

The father completed five sessions on the skiing machine each at a different level of difficulty. The student recorded how long it took for his father's pulse to return to its resting value after each session. The investigation was repeated a second time and mean values calculated.

The results of this investigation are shown in the table.

level of difficulty	mean maximum pulse rate/beats per minute	mean time taken to recover/min
1	133	4.0
2	143	5.5
3	174	10.0
4	184	12.0
5	187	12.5

b Explain why the student determined his father's mean resting pulse rate before the exercise began.
     (1 mark)

c Explain why it took several minutes for the father's pulse to return to its resting value after the exercise was finished.   (3 marks)

d Explain how the student could extend the investigation to find out how much exercise on the skiing machine is necessary for his father to achieve a significant improvement in aerobic fitness.   (3 marks)
     (Total 9 marks)
     OCR 2802 May 2002, B (HHD), No.2

2 Figure 1 shows some cells from part of the epithelium of the gaseous exchange system.

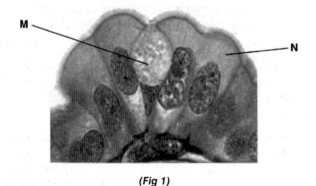

(Fig 1)

a Name the type of cell labelled **M** shown in figure 1 and state its function.   (2 marks)

Four structures found in the gaseous exchange system are listed below.

        alveolus, bronchiole, bronchus, trachea

b Underline the structures which contain cell type **M**.
     (1 mark)

c State the likely effect of heavy smoking on cell type **M**.
     (1 mark)

d Name the type of cell labelled **N** in figure 1.   (1 mark)

e Complete the table below to show the function of each of the following in the gaseous exchange system.

	function
cartilage	
smooth muscle	
elastic fibres	

     (3 marks)

A spirometer measures the volumes of gas breathed in and out of the lungs. Figure 2 shows the results obtained from a 17 year old male who was sitting down while breathing in and out of a spirometer.

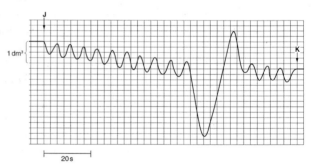

(Fig 2)

f Describe this person's breathing between points **J** and **K** on the spirometer trace.   (4 marks)
     (Total 12 marks)
     OCR 2802 Jan 2002, B (HHD), No.4

3 A survey was carried out in 1998 on the blood pressure of a sample of the population of England. High blood pressure is known as hypertension. In this particular survey, anyone with a systolic blood pressure of 18.7 kPa (140 mmHg) or over, or who was taking drugs to lower blood pressure, was recorded as having hypertension.

Some of the results of this survey are shown in the following table.

The table shows the mean systolic blood pressure for all men and women surveyed and the means for each of the age groups shown.

	all ages	age groups						
		16–24	25–34	35–44	45–54	55–64	65–74	75 and over
men mean systollic blood pressure/kPa	18.2	17.1	17.4	17.5	18.2	18.9	19.7	20.0
women mean systollic blood pressure/kPa	17.7	16.0	16.1	16.5	17.6	18.7	19.9	20.7

a Use the information in the table to find the answers to the following.

  (i) State the mean systolic blood pressure for all men and all women in the sample. *(1 mark)*

  (ii) Calculate the percentage increase in mean systolic blood pressure between the ages of 16–24 and 65–74 in women.
  Show your working and express your answer to the nearest whole number.
(2 marks)

  (iii) Describe the change in mean systolic blood pressure in men with increasing age. *(2 marks)*

The table below shows the percentages of all men and women surveyed who had hypertension and also the percentage for each age group.

	all ages	percentage of people in each age group in the survey who had hypertension						
		16–24	25–34	35–44	45–54	55–64	65–74	75 and over
men	40.8	16.0	20.5	26.1	42.3	59.8	69.9	72.8
women	32.9	4.2	6.9	13.2	30.8	51.6	72.8	77.6

b Use the information in both tables to explain whether there is any evidence to support the suggestion that men are more at risk from hypertension than women.
(3 marks)

c Explain the advantages to health of carrying out this survey. *(2 marks)*

(Total 10 marks)

OCR 2802 Jun 2003, B (HHD), No.6

4 The figure shows a transverse section of a bronchus from the lung of a mammal.

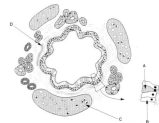

a Name **A** to **D**. *(4 marks)*

b Describe how the cells lining the bronchus protect the alveoli from damage. *(4 marks)*

There are elastic fibres between the cells lining the gaseous exchange surface in the alveoli.

c Describe the function of the elastic fibres in the alveoli. *(3 marks)*

The table shows some measurements of a person's breathing. Ventilation rate is the volume of air breathed in during one minute.

d With reference to the table,
  (i) calculate the ventilation rate at rest; *(1 mark)*
  (ii) explain the meaning of the term *vital capacity*; *(2 marks)*
  (iii) state how the person increased their ventilation rate even though their breathing rate remained constant. *(1 mark)*

tidal volume at rest	500 cm³
vital capacity	4600 cm³
breathing rate at rest	12 breaths per minute
ventilation rate during exercise	20 000 cm³ min⁻¹

(Total 15 marks)

OCR 2802 Jun 2001, B (HHD), No.1

5 Some athletes find that they recover faster after strenuous exercise by continuing to exercise slowly. This is sometimes called 'warming down'. An investigation was carried out to see if this was due to the removal of lactate from the blood.

A cyclist exercised vigorously on an exercise bicycle for six minutes and then rested for 34 minutes. When fully recovered, the same cyclist repeated the procedure, but this time cycled at a slow speed following the six minutes of fast cycling. Blood samples were taken at intervals and analysed for the concentration of lactate. The results are shown in the graph.

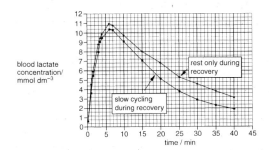

a With reference to the graph,
  (i) describe the changes in lactate concentration in the blood when the cyclist exercised and then rested during recovery; *(3 marks)*
  (ii) explain the changes you have described in (i); *(4 marks)*
  (iii) suggest how cycling at slow speed during the 'warming down' period helps to lower the concentration of lactate in the blood more quickly than resting completely. *(2 marks)*

Trained athletes can exercise at much higher levels than untrained people before their blood lactate concentration starts to increase. This is due to changes in the cardiovascular system and in muscles.

b Describe the changes that occur in the cardiovascular system and in muscles **during training**. *(5 marks)*

(Total 14 marks)

OCR 2802 Jun 2001, B (HHD), No.2

Tobacco smoke and its effects on gaseous exchange

Table 14.1 *Percentage of the adult UK population who smoked tobacco 1974–2001* (source: Office for National Statistics)

Year	Men	Women	Both
1974	51	41	45
1978	45	37	40
1982	38	33	35
1986	35	31	33
1990	31	29	30
1994	28	26	27
1998	28	26	27
2001	28	26	27

Table 14.2 *Daily consumption of manufactured cigarettes per smoker 1949–2001* (source: Office for National Statistics)

Year	Men	Women
1949	14.1	6.8
1959	18.4	11.0
1969	18.9	13.7
1979	21.6	16.6
1990	16.8	13.9
2001	15.0	13.0

Smoking during pregnancy threatens not only the life of the mother but also that of the unborn child

Life insurance companies have calculated that, on average, smoking a single cigarette lowers one's life expectancy by 10.7 minutes – longer than it takes to smoke the cigarette! While this is a statistical deduction rather than a scientific one, there is now clear scientific evidence to support the view that smoking cigarettes damages your health and reduces life expectancy (unit 14.3).

14.1.1 Who smokes and how much?

In the UK around 12 million adults (aged 16 years and over) smoke cigarettes, and a further 3 million smoke tobacco in other forms (pipes and / or cigars). Smoking became popular, especially with men, after the first world war. In 1948, the first year in which statistics were collected, 82% of men smoked tobacco in some form. In women, smoking reached its peak in 1966, when 45% of adult females smoked. Its popularity declined amongst both groups up to the early 1990s but has remained constant since that time (table 14.1). There are now around 11 million ex-smokers in the UK population. There is now little difference between the proportion of men (28%) and women (26%) in the UK population who smoke. The highest proportion of smokers is in the 20–24-year age group, where 40% of men and 35% of women continue to do so. The proportion then declines steadily with age, with only 17% of the over 60-year age group who smoke.

Smoking is more prevalent amongst people in routine or manual occupations than those in managerial or professional ones – 33% compared to 22% in men, and 30% compared to 20% in women (figures for 2001). In terms of health, it is not merely the number of people who smoke but how frequently they do so and, to some extent, the type of tobacco they use. The average daily consumption of cigarettes in 1979 was 21.6 for men and 16.6 for women, but has declined in both groups since (table 14.2).

While smoking is on the decline in much of the developed world, it is increasing in many developing countries, where the World Health Organisation considers it to be a disease reaching **epidemic** proportions.

14.1.2 What is in tobacco smoke?

Tobacco smoke is a cocktail of up to 4000 different chemicals and is released into the atmosphere in two ways:

* **mainstream smoke** from the filter or mouth end of the cigarette/cigar/pipe
* **sidestream smoke** from the burning tip of the cigarette/cigar/pipe tobacco.

Around 85% of tobacco smoke in a room will be sidestream smoke – the form which contains higher concentrations of many toxins than mainstream smoke. Breathing in this smoke is called **passive smoking** and presents a health hazard to people nearby who inhale it. Of the thousands of chemicals in tobacco smoke, three important ones are:

* **Carbon monoxide** – a poisonous gas also found in car exhaust fumes.
* **Nicotine** – a poisonous alkaloid drug that is addictive. 60mg of nicotine placed on the tongue would kill an individual within minutes. It is absorbed by the body very rapidly, reaching the brain in less than 30 seconds
* **Tar** – a sticky, brown substance responsible for the staining of the fingers and teeth of smokers. It appears in tobacco smoke as minute droplets.

14.1.3 Effect of carbon monoxide on gaseous exchange

Once inhaled, carbon monoxide readily diffuses through the walls of the alveoli in the lungs and into the red blood cells. Here the carbon monoxide combines with **haemoglobin** to form a stable compound called carboxyhaemoglobin. Whereas oxyhaemoglobin releases its oxygen at the tissues, carboxyhaemoglobin does not release its carbon monoxide and so the haemoglobin is not available to carry oxygen. To make matters worse, haemoglobin has a higher affinity for carbon monoxide than it has for oxygen. Hence, even a small proportion of carbon monoxide in the lungs is taken up in preference to oxygen. As a result the oxygen-carrying capacity of the blood of a smoker is reduced by some 5–10% and even by as much as 15%. This accounts for the breathlessness experienced by many smokers, especially when taking exercise. It also puts a strain on the heart, which has to pump more blood than in a non-smoker in order to transport the same amount of oxygen around the body. This contributes to certain cardiovascular diseases (unit 14.4).

14.1.4 Effects of tar on gaseous exchange

Tar in tobacco smoke is a mixture of chemicals that enter the respiratory tract as an aerosol of minute droplets. About 70% of this tar is deposited on the air-ways and alveoli. It is an irritant that causes inflammation of the mucous membranes lining the trachea, bronchi and bronchioles, resulting in them producing more mucus. At the same time, the tar thickens the epithelium and paralyses the cilia on its surface (section 1.7.2). As a consequence these cilia cannot remove the mucus secreted by the epithelial lining. The mucus, laden with dust and microorganisms, therefore accumulates in the lungs, leading to infections and damage. The cough, typical of many smokers, is the result of trying to remove this build-up of mucus from the lungs. These various responses, either directly or indirectly, reduce the rate of gaseous exchange in a number of ways.

- Build-up of mucus on the walls of the air-ways reduces their diameter and so limits the rate at which air can reach the alveoli.
- Mucus accumulating in the alveoli limits the air that they can contain and lengthens the diffusion pathway.
- Coughing can cause damage to the air-ways and alveoli; scar tissue builds up which again reduces air movements and rates of diffusion.
- Infections arise because the cilia no longer remove mucus and pathogens – these lead to **emphysema**, which reduces the surface area of the alveoli.
- **Allergens** such as pollen also accumulate, leading to further inflammation of the air-ways, reduced air-flow in and out of the lungs, and possible **asthma** attacks.

In addition to its effects on gaseous exchange, tars also contribute to chronic obstructive pulmonary disease (sections 14.2.1 and 14.2.2), cardiovascular disease (unit 14.4) and lung cancer (section 14.2.3).

It has been calculated that smoking these 20 cigarettes would, on average, reduce your life-expectancy by three and a half hours.

SUMMARY TEST 14.1

Smoking by men and women in the UK declined from the middle of the last century until the early 1990s, since when it has remained stable at or around **(1)**% of the population. Smokers in the UK most frequently come from the **(2)** age group. Tobacco smoke contains thousands of chemicals but three are of particular importance to health: carbon monoxide, tar and a poisonous and addictive alkaloid called **(3)**. Carbon monoxide forms an irreversible bond with haemoglobin to form **(4)**. As a result the oxygen-carrying capacity of the blood can be reduced by **(5)**%. This explains the **(6)** experienced by many smokers. 70% of the tar inhaled is deposited on air-ways such as the trachea, **(7)** and **(8)**. This leads to the paralysis of the **(9)** on the epithelium of these air-ways, causing **(10)** laden with bacteria, dirt and tar to accumulate in the lungs. The bacteria may cause infections leading to a disease called **(11)**, in which the surface area of the alveoli is reduced.

It is estimated that, in the UK, around 120 000 premature deaths occur each year as a result of smoking. About 20% of these are due to two diseases, emphysema and chronic bronchitis, and 25% are a result of lung cancer. Chronic bronchitis and emphysema frequently occur together, in which case the term **chronic obstructive lung disease** is used to describe this combined condition.

14.2.1 Chronic bronchitis

Chronic bronchitis is a long-term, life-threatening disease, unlike acute bronchitis, which lasts only a few days and frequently accompanies common ailments such as colds and influenza. Around 1 million people in the UK suffer from chronic bronchitis and some 300 000 people die from the disease each year. It results from irritation of the lining of the bronchi and bronchioles by dust, fumes, atmospheric pollution and, most significantly, tars in tobacco smoke. This irritation causes:

- swelling of the epithelial membranes of these airways, restricting the passage of air and causing breathlessness
- excessive production of mucus from the goblet cells (section 13.1.3) of the epithelium
- damage to the cilia of the epithelium, causing a slowing or complete halt to their action of moving dirt-laden mucus out of the lungs
- accumulation of 'dirty' mucus in the lungs, reducing air movement and gas exchange
- coughing, as reflex actions cause the forced expulsion of mucus which would otherwise 'drown' the patient
- build-up of scar tissue on the bronchi and bronchioles, as a result of the coughing, leading to even thicker walls and greater restriction of air-flow
- infections, such as pneumonia, due to bacteria accumulating rather than being 'swept away' along with mucus by the cilia.

Chronic bronchitis develops slowly and the main symptoms of coughing and breathlessness only become apparent when the disease is well developed. It is difficult to reverse the damage caused once the disease is established.

14.2.2 Emphysema

One in every five smokers will develop the crippling lung disease called **emphysema** which, together with other related obstructive lung disorders, kills 20 000 people a year in the UK alone. The disease develops over a period of 20 or so years and it is virtually impossible to diagnose until the lungs have been irreversibly damaged. In its early stages the only symptom is a slight breathlessness but, as this gets progressively worse, many people become so disabled that they cannot even get out of bed. People with emphysema usually die of respiratory failure, often accompanied by infection. A small number die of heart failure as the heart becomes enlarged and over-worked trying to pump blood through arteries that have become constricted as a result of lack of oxygen.

Healthy lungs contain large quantities of elastic connective tissue comprised predominantly of the protein elastin. This tissue expands when we breathe in and returns to its former size when we breathe out. In emphysematous lungs the elastin has become permanently stretched and the lungs are no longer able to force out all the air from the alveoli. The surface area of the alveoli is reduced and they sometimes burst. As a result, little if any exchange of gases can take place across the stretched and damaged air-sacs.

The damage is brought about by abnormally high levels of elastase, an enzyme formed in some of the white blood cells, which breaks down elastin. Elastase also degrades other proteins so that, in the latter stages of the disease, breakdown of

Section through a healthy human lung

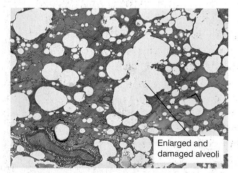

Section through the cancerous lung of a smoker; note the rounded cancer tumours (white) and the widespread cigarette tar deposits (dark areas)

Normal lung tissue as seen under a light microscope

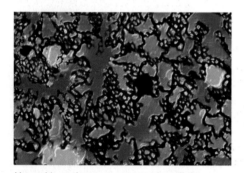

Enlarged and damaged alveoli

Lung tissue damaged by emphysema as seen under a light microscope

lung tissue results in large, non-functional holes in the lung. In healthy lungs elastin is not broken down because a protein inhibitor (PI) inhibits the action of the enzyme elastase. However, in smokers, it has been suggested that the oxidants in cigarette smoke inactivate PI, resulting in greater elastase activity and hence a breakdown of elastin.

Elastase is produced by **phagocytes** which need it so they can migrate through tissue to reach sites of infection. This is part of the body's normal inflammatory response. In smokers, where a large number of phagocytic cells are attracted to the lungs by the particulate materials in smoke, a combination of the release of elastase and a low level of its natural inhibitor lead to a lot of tissue degradation. The only way to minimise the chances of getting emphysema is not to smoke at all, or to give up – the function cannot be restored to smoke-damaged lungs but giving up can significantly reduce the rate of further deterioration.

14.2.3 Lung cancer

Over 80% of all lung cancer deaths are caused by smoking and a quarter of all smokers die from this cause. The additional risk of getting lung cancer is related to the number of cigarettes smoked (Fig 14.1). Tobacco smoke contains a number of **carcinogens** (factors that cause cancer). These cause considerable damage to the **genes** of the epithelial cells which line the lungs and are therefore exposed to these carcinogens. Among these mutant genes are ones that control normal cell division. It is these mutant genes that give rise to lung cancer. One identified carcinogen is benzopyrene (BP) found in the tar of tobacco smoke. A derivative of BP binds directly to the tumour-suppressor gene p53, which normally stops cell division and destroys mutated cells. With this gene inactivated, epithelial cells divide by **mitosis** in an uncontrolled manner leading to the formation of a **tumour**. Growth of this tumour is slow and it may take over 20 years for any symptoms to develop. These symptoms include:

- persistent cough, which is a reflex action to the obstruction to the airways caused by the tumour
- blood in the sputum, resulting from damage to the lung tissues caused by the tumour and coughing
- shortness of breath because the tumour is obstructing the airways and replacing the alveoli

- hoarseness or other changes to the voice due to pressure of the tumour on the airways, larynx or nerves serving the larynx
- wheezing noises as air is forced along airways obstructed by the tumour.

Cancer cells may break away from the original tumour and spread the disease to other parts of the body (metastasis), leading to a wide variety of symptoms as a result of these secondary tumours.

More details about cancer in general are given in unit 6.4.

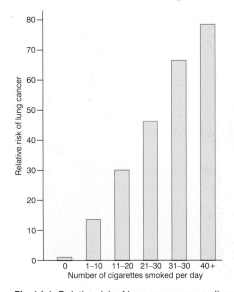

Fig 14.1 *Relative risk of lung cancer according to daily cigarette consumption (Source: Souhami & Tobias, Cancer and its management, 1986)*

EXTENSION – SMOKING AND OTHER CANCERS

Lung cancer is not the only form of cancer where smoking is known to increase risk. Over 90% of patients with cancer of the mouth and throat either smoke or chew tobacco. Heavy smokers are 20–30 times more likely to die from cancer of the larynx (voice box) than non-smokers. The risks are further increased amongst those who regularly drink alcohol and smoke tobacco. Other cancers with a higher incidence and death rates in smokers include cancer of the oesophagus, bladder, kidney, pancreas, stomach and liver.

SUMMARY TEST 14.2

Of the 120 000 premature deaths in the UK caused by smoking, 25% result from lung cancer and 20% from emphysema and chronic bronchitis, which together are called **(1)**. Emphysema's early symptom is **(2)**, although this may take up to 20 years to develop. Emphysema results from white blood cells producing the enzyme **(3)** so that they can migrate through the lung tissues to fight infections and engulf particles from tobacco smoke. The enzyme breaks down the protein **(4)** in the alveoli, resulting in the production of non-functional holes. Chronic bronchitis is due to swelling of the **(5)** membranes of the airways and the production of excess mucus from the **(6)** cells within them. Damage is also caused to the **(7)** on the membranes and, as a result, dirt-laden mucus accumulates in the lungs. Over **(8)**% of all deaths from lung cancer are attributable to smoking. Tar in tobacco smoke contains cancer-causing chemicals called **(9)** to alter certain genes in the lining of the airways. Cells divide to form a **(10)** from which cells may become separated and spread the disease to other parts of the body – a process called **(11)**. Symptoms of lung cancer include **(12)**, **(13)** and **(14)** as well as chest pain, hoarseness and a wheezing noise when breathing.

There are over 50 different illnesses and ailments that arise more regularly amongst smokers than non-smokers, and more than 20 of these can kill. Half of all teenagers who now smoke will die from diseases caused by tobacco if they continue to smoke. It is estimated that, since 1950, 6 million people in the UK and 60 million people worldwide have died from tobacco-related diseases. How can we be sure that tobacco really is responsible for so many deaths and illnesses? There are two main types of evidence that link smoking to these diseases:

• epidemiological evidence
• experimental evidence.

14.3.1 Epidemiological evidence linking smoking to disease

Epidemiology is the study of the **incidence** and pattern of disease with a view to finding means of preventing and controlling it. To do this, epidemiologists collect data on diseases and then look for correlations between these diseases and various factors in the lives of those suffering them. The world's longest-running survey of smoking began in the UK in 1951. This survey and further ones elsewhere in the world have revealed a number of general statistical facts about smokers:

• A regular smoker is three times more likely to die prematurely than a non-smoker.
• The more cigarettes smoked per day, the earlier, on average, a smoker dies. This is illustrated by the graph in figure 14.2.
• Smokers who give up the habit improve their life expectancy compared to those who continue to smoke.
• One in two long-term smokers will die early as a result of smoking.

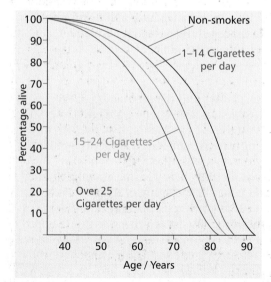

Fig 14.2 *Life expectancy related to the number of cigarettes smoked*

Perhaps the most compelling epidemiological evidence that smoking is damaging to health comes from the correlation between lung cancer and smoking. Figure 14.3 shows how the number of cigarettes smoked in a year and the number of deaths from lung cancer over the last century in the UK follow a very similar pattern. The lung cancer death rate lags behind that of cigarettes smoked because the disease takes 20 years or more to cause death. A similar correlation is shown for the incidence of lung cancer and the number of cigarettes smoked per day by men in the USA (Fig 14.4). Other epidemiological statistics that show a correlation between lung cancer and smoking include:

• A man smoking 25 cigarettes a day is 25 times more likely to die of lung cancer than a non-smoker.
• The longer one smokes, the greater the risk of lung cancer. Smoking 20 cigarettes a day for 40 years increases the risk of lung cancer eight times more than smoking 40 cigarettes a day for 20 years.
• When smoking ceases, the risk of developing lung cancer decreases and approaches that of a non-smoker after around 10–15 years (depending on age and amount of tobacco consumed).
• The death rate from lung cancer is 18 times greater in a smoker than a non-smoker.

Similar correlations exist that link pulmonary diseases to smoking. These include:

• More than 80% of deaths from **chronic obstructive pulmonary disease** are attributable to smoking as the disease is very rare amongst lifetime non-smokers.
• The incidence of chronic obstructive pulmonary disease increases with the number of cigarettes smoked.
• 98% of **emphysema** sufferers are smokers.
• Pneumonia is more common amongst smokers than non-smokers and is twice as likely to cause death.

Cigarette manufacturers and some smokers argue that these epidemiological correlations are coincidental. While the statistics do point very clearly to the likelihood that smoking is hazardous to health, there needs to be a clear causal connection between smoking and disease before we can say that the case is proven. This causal connection is provided by experimental evidence.

14.3.2 Experimental evidence linking smoking to disease

Experiments were carried out in the 1960s in which dogs were made to inhale cigarette smoke. The smoke was either inhaled directly or first passed through a filter tip. Those dogs that inhaled the filtered smoke remained generally healthy. Those inhaling unfiltered smoke developed chronic obstructive pulmonary disease and early signs of lung cancer. It would therefore appear that filter tips do remove some of the harmful constituents of tobacco smoke. The next stage was to use machines to simulate the action of smoking and to collect the harmful constituents that accumulated in the filters. These were then analysed chemically and each constituent was tested in the laboratory for its ability to damage epithelial cells and mutate the **genes** they contain. This was done by adding tar to the skin of mice or to cells that had been grown in culture. As a result of such tests it has been shown that a number of the constituents of the tar found in cigarette smoke act as **carcinogens**. One example is benzopyrene (BP), which has been shown to be a powerful mutagen. It

is absorbed by epithelial cells and converted to a derivative, BPDE, which binds with the tumour-suppressor gene p53 (section 6.4.2) and mutates it to an inactive form. In its inactive state it allows uncontrolled cell division of epithelial cells and hence the growth of a **tumour**. The key evidence linking smoking to lung cancer is that, when the mutations of the p53 gene in a cancer cell are examined, they occur at three specific points on the DNA. When the BPDE derivative from tobacco smoke is used to mutate the p53 gene, it causes changes to the DNA at precisely the same points.

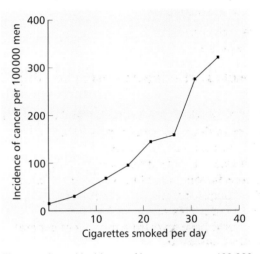

Fig 14.4 Annual incidence of lung cancer per 100 000 men in the USA correlated to the daily consumption of cigarettes

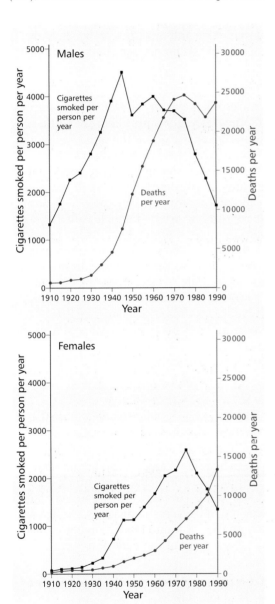

Fig 14.3 Incidence of deaths from lung cancer in the UK correlated to cigarettes smoked per year (1910–1990)

SUMMARY TEST 14.3

Epidemiological evidence is used to link smoking to various types of pulmonary disease. For example a smoker is (**1**) times more likely to die prematurely than a non-smoker. The death rate from lung cancer is related to the number of cigarettes smoked, although there is a time lag of around (**2**) years between smoking and the symptoms of the disease. Evidence shows that you are 25 times more likely to die of lung cancer if you regularly smoke a total of (**3**) cigarettes a day. If you give up smoking the risk typically decreases to that of a non-smoker after (**4**) years. Similar correlations exist for smoking and pulmonary diseases. For example, 98% of people who suffer from (**5**) are smokers and smokers are twice as likely to die from infections such as (**6**). All this evidence shows a clear association between smoking and certain diseases but it is the experimental evidence that shows a (**7**) connection. One such experiment showed that benzopyrene, found as part of the (**8**) in cigarette smoke, acts as a (**9**) by mutating the p53 gene that controls cell division. As a result of this mutation cells divide to form a (**10**).

Cardiovascular disease and smoking

Cardiovascular disease consists of degenerative diseases of the heart and circulatory system. There are many forms, of which the three most significant are:

- atherosclerosis
- coronary heart disease
- stroke.

It is the leading cause of death in many countries of the world, including the UK and the USA. In the UK it is responsible for 40% of premature deaths, killing 140 000 people under the age of 75 years each year. Cardiovascular disease is multifactorial (has many causes), many of which are self-inflicted, including smoking. In particular it is the carbon monoxide and nicotine components of tobacco smoke that increase the risk of cardiovascular disease amongst smokers.

14.4.1 Atherosclerosis

Atherosclerosis is the form of arteriosclerosis in which the walls of the arteries thicken as a result of **cholesterol** and other fatty substances being deposited on the artery lining. It begins with the formation of streaks on the inner wall of the artery which then increase in size to form patches, known as **plaques**, that eventually thicken the wall and narrow the lumen (opening) of the artery (Fig 14.5). These thickenings, called **atheromas**, most commonly occur in the larger arteries and are the result of fibres, dead **smooth muscle** cells and cholesterol being deposited. These may eventually rupture the artery lining, leaving it rough and uneven. This in turn causes a blood clot to form on it which further prevents blood flow and starves the tissues that the artery serves of oxygen. Atheromas are uneven and disturb the flow of blood, which can then gradually form a clot, known as a **thrombus**. This thrombus can become so large that it blocks the blood vessel where it forms, or small pieces, called **emboli**, may break off and block smaller vessels elsewhere. These blockages cause the blood supply to particular tissues to be reduced (**ischaemia**). Over many years, calcium may also become deposited in the atheroma, causing the artery wall to harden. This condition is known as **arteriosclerosi**s and is particularly associated with aging.

14.4.2 Coronary heart disease

Coronary heart disease (CHD) is a disease affecting the coronary arteries which supply blood to the cardiac muscle of the heart (section 9.1.1). It is also called ischaemic heart disease because, as a result of an atheroma or thrombus, the supply of blood to the heart muscle is restricted. This causes a condition called **angina** in which symptoms such as chest pain and difficulty in breathing occur, especially when exercising. If the blockages of the coronary arteries are very large, part of the heart may be completely deprived of oxygen and die. This is called a **myocardial infarction**, or heart attack. In the UK around half a million people suffer a heart attack each year and almost one-third of these die as a result.

14.4.3 Strokes

A **stroke** occurs when the supply of arterial blood to part of the brain is interrupted in some way and, as a result, cells in that region die. It is also called a **cerebrovascular accident** and is of two types.

- **A thrombosis** (blood clot) develops either on an atheroma in an artery supplying the brain with blood, or an atheroma develops elsewhere and a piece of it breaks away to block an artery supplying the brain. In either case the blockage prevents blood, and therefore oxygen, reaching a particular region of the brain.

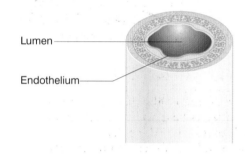

Lumen

Endothelium

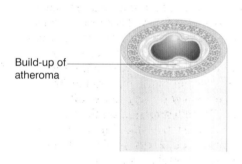

Build-up of atheroma

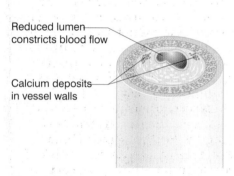

Reduced lumen constricts blood flow

Calcium deposits in vessel walls

Fig 14.5 *Build-up of atheroma*

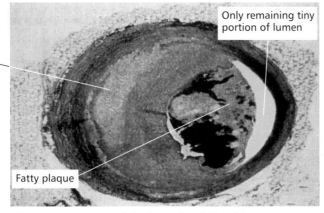

Artery wall thickened with fibres, dead smooth muscle cells and fatty substances

Only remaining tiny portion of lumen

Fatty plaque

Human coronary artery with a fatty atheroma partially obstructing the interior

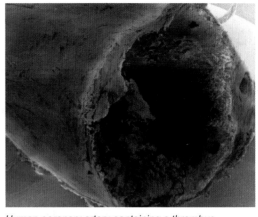

Human coronary artery containing a thrombus

- **An aneurysm** is a weakness in the wall of an artery, often the result of atherosclerosis. These aneurysms frequently burst, leading to a **haemorrhage** and therefore the loss of blood supply to the region of the brain served by that artery.

The effects of a stroke depend upon the region of the brain that is deprived of blood. Most commonly a stroke affects the middle cerebral artery, causing paralysis on one side of the body.

14.4.4 Effects of carbon monoxide on the cardiovascular system

We saw in section 14.1.3 that carbon monoxide combines easily, but irreversibly, with the **haemoglobin** in red blood cells to form carboxyhaemoglobin and thereby reduces the oxygen-carrying capacity of the blood. To supply the equivalent volume of blood to the tissues, the heart must therefore work harder. This can lead to hypertension (high blood pressure), described in section 13.4.3, which increases the risk of coronary heart disease and strokes. In addition, the reduction in the oxygen-carrying capacity of the blood means that it may be insufficient to supply the heart muscle during exercise. This leads to angina or, in severe cases, a myocardial infarction.

14.4.5 Effects of nicotine on the cardiovascular system

Nicotine in tobacco smoke stimulates the production of the hormone adrenaline by the adrenal glands, leading to an increase in heart rate and raised blood pressure (hypertension). In addition, it reduces the ability of arteries to dilate, which further increases blood pressure. As a consequence there is a greater risk of smokers suffering coronary heart disease (2.5 times greater than in a non-smoker) or a stroke (1.5 times greater than a non-smoker or a light smoker, but rising to 3 times for those smoking 20 or more cigarettes a day). Nicotine also makes the red cells in the blood more 'sticky', and this leads to a higher risk of thrombosis and hence of strokes or myocardial infarction (heart attack).

EXTENSION – PASSIVE SMOKING

Non-smokers who have inhaled other people's tobacco smoke for a number of years have an increased risk of contracting certain diseases. The UK's Scientific Committee on Tobacco and Health has estimated that those regularly exposed, e.g. where a member of the family or work colleague smokes, have a 20–30% increased risk of contracting lung cancer and coronary heart disease. Non-smokers exposed to between 1 and 19 cigarettes per day for 10 years or more have an increased risk of coronary heart disease of 23%. Those exposed to over 20 cigarettes over the same period have an increased risk of 31%. Passive smoking therefore causes several hundred deaths in non-smokers each year in the UK. One well-known case was that of the entertainer Roy Castle, a non-smoker who died of lung cancer after spending much of his life performing in smoke-filled clubs. The consequences of passive smoking have led to a ban on smoking in many work places, public buildings, theatres, restaurants and on public transport.

SUMMARY TEST 14.4

Cardiovascular diseases are degenerative diseases of the heart and circulatory system which are the main cause of death in many countries. In atherosclerosis the walls of the arteries thicken due to deposits of **(1)** and other fatty substances. Initially patches called **(2)** develop, which eventually increase in size to form **(3)**, which may block the **(4)** of the artery. As a result the flow of blood is disturbed and so a **(5)** may develop. If this happens in a coronary artery the blood supply to a region of the heart may be cut off, depriving it of **(6)**. This region of heart muscle may die, something called a **(7)**, or heart attack. If, however, the blockage is in an artery serving the brain, then a stroke, also called a **(8)**, is the result. One type of stroke occurs when a weakness in the artery wall, called an **(9)**, bursts, leading to a **(10)**. Some cardiovascular disease is the result of the gas **(11)** in tobacco smoke, which combines irreversibly with **(12)**. Another constituent of tobacco smoke, nicotine, increases the production of the hormone **(13)**, which in turn leads to a rise in **(14)**.

14.5.1 Differences in distribution of cardiovascular disease (CVD)

When you look at the **incidence** of cardiovascular disease (CVD) in different countries around the world, you are struck by the differences that are evident. The World Health Organisation (WHO) set up a global monitoring project called MONICA in 1979 to compare the incidence and mortality from CVD in a wide range of populations throughout Europe. These are some of the facts that emerged:

- Almost half (49%) of all deaths in Europe are a result of CVD (55% of deaths in women and 45% of deaths in men).
- In those European states that are members of the EU, the rate was slightly lower, at 42% (46% in women and 38% in men).The percentage of deaths in men from all causes in the EU are given in figure 14.6.
- In all 49 countries of Europe studied, CVD was the most common cause of death.
- In Bulgaria, CVD causes the death of 58% of women aged under 75 years, whereas in Iceland it is only responsible for 20% of women aged under 75 years.
- The death rate from coronary heart disease (CHD) for men aged 35–74 years is eight times higher in Russia than in France and for women it is twelve times higher.
- The death rates from CHD are falling in most northern and western European countries, but are rising in most central and eastern ones (Fig 14.7).
- The death rate from strokes for men aged 35–74 years living in Italy fell by 41% between 1983 and 1993, but rose by 25% for men in Romania over the same period.

These selected items of information have been presented to illustrate both the scale of the problem of CVD and the differences in its pattern of distribution, even within a single continent such as Europe.

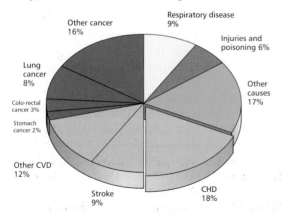

Fig 14.6 *Causes of death in men in the EU (figures from the latest available data for each country in the EU, published by the World Health Organisation in 1999)*

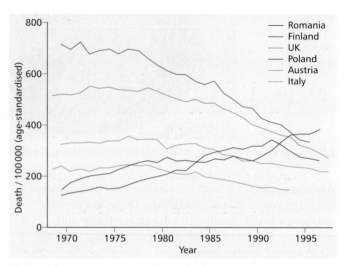

Fig 14.7 *Death rates from coronary heart disease for men aged 35–74 years in six European countries from 1968 to 1997*

Globally too, there are wide variations in the incidence of CVD. In general, developed countries show a higher incidence than developing ones. It is estimated by the WHO, for example, that around 4.5 million people in China suffer from CHD. This is not much more that the 4 million estimated to suffer from CHD in the UK, which has a population at least 25 times smaller. The incidence of CHD is similarly much lower in sub-Saharan Africa and parts of Latin America than it is in countries such as Australia, Canada, Japan and the USA. Even within the developed countries there are large differences in the levels of CHD (Fig 14.8). The differences are not limited to those between countries, but exist within an individual country. For example, in the UK, the incidence of CHD is greatest:

- in men
- in Scotland, northern Ireland and the north of England
- among the less well-off
- among certain ethnic groups.

The reasons for these many differences in the global pattern of distribution of CVD are complex and differ from country to country. Much depends upon the prevalence of a number of recognised risk factors (section 14.5.2), each of which increases the chances of developing CVD. Many of these risk factors, e.g. high salt and saturated fat intake, smoking, large alcohol consumption, stressful lives and lack of exercise, are associated more with developed rather than developing countries. This explains the differences in these two types of countries and why emerging economies, e.g. Cuba and Chile, show an increase in CVD. It is also true that developed countries often have a better standard of health care. Residents of these countries are frequently cured of other diseases which would otherwise have caused death before the onset of CVD in their later years.

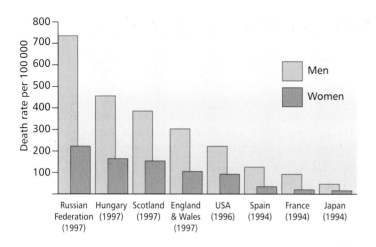

Fig 14.8 *Death rates from coronary heart disease in various countries*

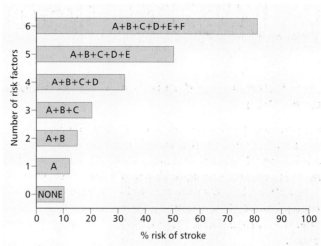

Fig 14.9 *The combined impact of six risk factors on the likelihood of a 70-year-old man experiencing a stroke over the next 10 years*

14.5.2 Factors affecting the incidence of cardiovascular disease

There are a number of factors that individually increase the risk of an individual suffering from CVD. When combined together, four or five of these factors produce a disproportionately greater risk (Fig 14.9). What then are these risk factors? They fall into two main categories:

- factors which are predetermined and over which the individual has no control
- factors over which the individual can exercise control.

Factors beyond an individual's control include:

- **Heredity** – genes determine many factors which affect CVD. For example:
  - blood pressure
  - blood **cholesterol** levels
  - heart disease such as **atherosclerosis**.
- **Age** – the risk of CVD increases steadily with age. For example, 90% of all strokes occur in persons over 55 years old (Fig 14.10).
- **Sex** – men are at greater risk than women of the same age. It is important to realise, however, that, because women live longer, the incidence of CVD is greater overall in women than in men.
- **Diabetes** – although Type I (early onset) diabetes tends to run in families, there is no simple pattern of inheritance and it can arise from other causes. Diabetics are at an increased risk of CVD.
- **Ethnicity** – some races, e.g. black Afro-Americans, are at increased risk of some forms of CVD, e.g. strokes. The reasons may be genetic or due to other factors, such as customs, e.g. those that affect diet.
- **Medical history** – those who have already suffered a stroke or heart attack are at increased risk of doing so again.
- **Air pollution** – this can increase the risk of CVD if it is severe and prolonged.

Factors which an individual can, to some extent, control include:

- **Smoking** – smokers are between two and six times more likely to suffer from CVD. Giving up smoking is the single most effective way of increasing life expectancy.
- **High blood pressure** – excessive prolonged stress, certain diets and lack of exercise all increase blood pressure and hence the risk of CVD.
- **Blood cholesterol level**s – these can be kept lower by eating less saturated fat in the diet.
- **Obesity** – a body mass index of over 25 (section 12.7.1) brings an increased risk of CVD.
- **Diet** – high levels of salt in the diet raise blood pressure while high levels of saturated fat increase blood cholesterol concentration. Both therefore increase the risk of CVD. By contrast, foods that act as **antioxidants**, e.g. vitamin C, reduce the risk of CVD, as does non-starch polysaccharide (dietary fibre).
- **Physical activity** – aerobic exercise (unit 13.6) can lower blood pressure and blood cholesterol as well as help avoid obesity – all of which decrease the risk of CVD.

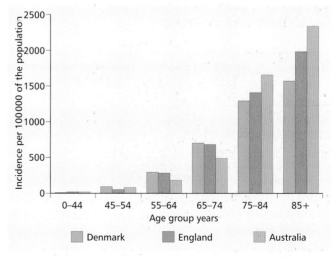

Fig 14.10 *Annual number of cases of stroke by age in three different countries*

207

We saw in unit 14.5 that cardiovascular disease is the cause of 49% of all deaths in Europe. One form of cardiovascular disease is coronary heart disease (CHD) (section 14.4.2), which is responsible for a third of all deaths of people aged 45–64 years in the UK. Clearly it is of utmost importance to try to reduce CHD because:

- it kills a large number of people prematurely
- it leads to disability and a poorer quality of life for sufferers
- it is a major contributor to the costs of the National Health Service (NHS).

14.6.1 Prevention versus cure

The cause of CHD is multifactorial and some of the factors involved can be controlled by individuals (section 14.5.2). It is logical, therefore, that any programme designed to reduce the incidence of CHD should be based on prevention. However, we know that some factors, e.g. heredity, are outside the control of individuals and therefore we must equally have the means to cure CHD. Given that nicotine is addictive, should the government put more resources into helping smokers give up their habit by providing free nicotine replacements in the same way as they provide methadone as a substitute for heroin? Is there a social responsibility on individuals who have, or are prone to, CHD to avoid smoking, keep their weight down, take exercise etc. in order to avoid costly bypass and transplant operations – or should they be free to make their own lifestyle decisions regardless?

14.6.2 Prevention of coronary heart disease (CHD)

Although we can do nothing about our age, sex or inherited **genes**, there are other measures we as individuals can take to avoid CHD. These include:

- not taking up smoking
- avoiding becoming over-weight
- reducing salt intake in the diet
- reducing intake of **cholesterol** and saturated fats in the diet
- taking regular **aerobic** exercise
- keeping alcohol consumption within safe limits
- increasing intake of non-starch polysaccharides (dietary fibre) and **antioxidants** in the diet.

What role can governments and health authorities play in preventing CHD? First and foremost they can develop programmes of health education and so make us aware of the latest scientific evidence linking certain factors to CHD. They can then advise on various means by which we can reduce our risks. Such programmes are clearly more effective if targeted at those most at risk, e.g. those who smoke, are overweight, have a history of heart disease, the poor, males etc. Specific measures include:

- anti-smoking campaigns
- warnings on tobacco products
- health awareness measures, e.g. healthy eating
- establishment of clinics to regularly and routinely monitor blood pressure and blood cholesterol levels
- dietary information including regulations on labelling food, especially processed food
- provision of drugs that lower blood cholesterol levels to people at risk
- encouraging exercise and physical fitness through advertising and providing sports facilities

Cutting down on foods rich in cholesterol and saturated fat helps prevent coronary heart disease

EXTENSION – ANTIOXIDANTS AND HEART DISEASE

Antioxidants are chemicals that prevent **oxidation** via highly reactive intermediate compounds called free radicals. Examples of antioxidants include vitamins C and E and β-carotene and have been promoted as preventing a range of conditions, including heart disease. This claim is based upon the suggestion that the formation of plaques in the process of atherosclerosis is partly the result of the oxidation of the **low density lipoprotein** cholesterol by compounds known as oxidants. Antioxidants have been found to inhibit oxidant formation, to reduce the activity of those already formed and even to repair injury caused by oxidants. There is certainly epidemiological evidence to show that greater antioxidant intake in the diet results in a lower risk of heart disease.

- warning of the dangers of obesity, smoking etc.
- legislation to control content of processed food and cigarettes (the EU has introduced measures from 2002 to reduce tar and nicotine levels in cigarettes).

14.6.3 Treatment of coronary heart disease (CHD)

However good the preventative measures, there will always be cases of CHD, not least because factors such as heredity are not currently preventable. There are four main methods of treating CHD.

- **Drugs** can be taken which reduce blood pressure (e.g. **diuretics**), lower blood cholesterol (e.g. lovastatin) and reduce the formation of blood clots (e.g. aspirin).
- **Angioplasty** can be used to open up coronary arteries that are blocked by **atherosclerosis**. This involves inserting a catheter into an artery in the groin or armpit under local anaesthetic. This is then directed, with the help of an X-ray machine, to the blocked region of the coronary artery. A balloon or small mechanical device attached to the catheter is then used to break up the **plaque** and open up the artery. A stainless steel mesh tube, called a stent, can be inserted if necessary and left in the artery to keep it permanently open.
- **Coronary bypass operations** involve taking a piece of vein, usually from the leg, and grafting it between the aorta and the coronary artery at a point beyond the blockage. In this way blood is re-routed around the blockage and the **cardiac muscle** is supplied with an adequate supply of oxygen. It is sometimes necessary to carry out more than one bypass at the same time – figure 14.11, for example, illustrates a triple coronary bypass.
- **Heart transplant operations** involve transplanting the living heart from a recently deceased person into a patient. First carried out by Dr Christiaan Barnard in South Africa in1967, this proved to be ground-breaking surgery. Initially the operation gave only a limited extension of life because the patient's immune system rejected what, to it, was foreign material. A major breakthrough came with the discovery of cyclosporin – a drug extracted from a soil fungus – which was very effective at suppressing the immune system. Other immuno-suppressant drugs now give heart transplant recipients 5–10 extra years of relatively normal life. Nevertheless, it is still necessary to have a donor with a close tissue match, e.g. a genetic relative. As a result the supply of donor hearts is limited and research is currently taking place into using pigs' hearts or artificially manufactured ones instead. To make the pigs' hearts compatible with the recipient's tissues a breed of pigs with human genes inserted into them has been developed so that their organs will not cause a major immunological reaction if implanted into humans. It may be possible, in time, to inject the specific DNA of the proposed recipient into an embryo pig some months before an operation, so that the heart that develops is an even closer match. Ethical and legal issues are raised by transplantation of animal organs into humans but we are likely to see the first such operation in the not too distant future.

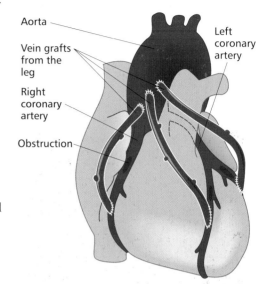

Fig 14.11 *Triple coronary heart bypass*

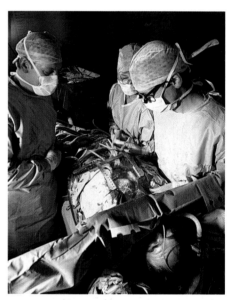

Heart bypass operation

SUMMARY TEST 14.6

In Europe, cardiovascular disease (CVD) accounts for **(1)**% of all deaths and one form of CVD, coronary heart disease (CHD) accounts for a total of **(2)** deaths in the UK. On the basis that prevention is better than cure, individuals can reduce the risk of CHD by reducing the intake of cholesterol, **(3)** and **(4)** in their diet and increasing that of **(5)** and **(6)**. To treat CHD, drugs can be used or an operation can be performed in which a small balloon is inserted with a catheter into a coronary artery and inflated to open the blockage. The operation is called **(7)**. Alternative treatments are **(8)** and heart transplants. To overcome the shortage of donor hearts it may be possible to transplant them from animals such as a **(9)**. These can be bred with human genes incorporated into them. This helps to prevent **(10)** of the heart by the recipient's **(11)** system.

1 Figure 1 shows a section of lung tissue.

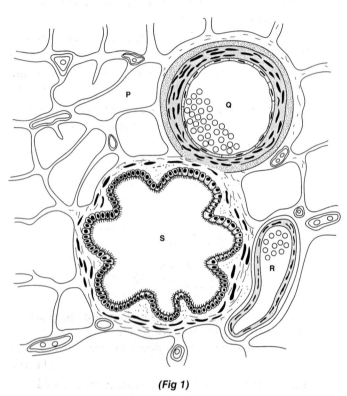

(Fig 1)

a The letters **P** to **S** in figure 1 are in the centres of four different structures.
Name these structures. *(4 marks)*

Figure 2 shows sections of lung tissue at lower magnification than in figure 1.

- **A** is from a non-smoker
- **B** is from a smoker who suffers from emphysema

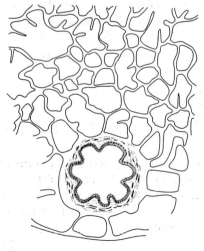

(Fig 2A)

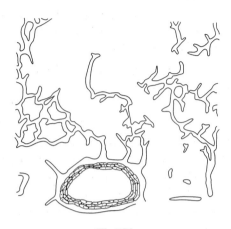

(Fig 2B)

b Describe how the lung tissue in figure 2**B** differs from the lung tissue in figure 2**A**.

Your answer should refer only to features visible in these drawings. *(2 marks)*

c Explain why people with emphysema have difficulty in forcing air out of their lungs when they breathe out. *(2 marks)*

d Describe **one** piece of epidemiological evidence that would suggest a link between emphysema and the smoking of cigarettes. *(1 mark)*

(Total 9 marks)

OCR 2802 Jan 2003, B (HHD), No.2

2 The table shows deaths from lung cancer, coronary heart disease (CHD) and stroke in the UK in 1997.

disease	deaths of men			deaths of women		
	all ages	under 75	deaths under 75 as % of all ages	all ages	under 75	deaths under 75 as % of all ages
lung cancer	22 021	12 822	...	13 234	7387	56
CHD	76 490	38 105	50	64 069	16 090	25
stroke	24 898	8415	34	41 502	7249	17

a (i) Complete the table by calculating the percentage of deaths from lung cancer that occur among men under the age of 75.
Express your answer to the nearest whole number. *(1 mark)*

(ii) State **one** other piece of information that is required to assess how important these diseases are as causes of death in the UK. *(1 mark)*

(iii) 'Cardiovascular diseases, such as CHD and stroke, shorten the lives of men more than they do the lives of women.'

Explain whether the data in the table support this statement. *(3 marks)*

b Smoking is a significant contributory factor to the development of cardiovascular diseases.

Explain how the constituents of cigarette smoke cause cardiovascular diseases, such as coronary heart disease (CHD) and stroke. *(7 marks)*

(In this question, 1 mark is available for the quality of written communication.)

(Total 13 marks)
OCR 2802 Jan 2003, B (HHD), No.5

3 Cigarette smoke contains tar, which is a mixture of many different chemicals. Some of these may stimulate changes to cells in the lining of the bronchi. These changes may lead to cancerous growths in the lung.

a (i) What name is given to chemicals that cause cancer? *(1 mark)*

(ii) Describe briefly the changes that occur **in the cells** lining the bronchi in response to these chemicals. *(3 marks)*

b State **two** symptoms of lung cancer. *(2 marks)*

Studies have shown that some aspects of smoking behaviour increase the risk of developing lung cancer.

Figure 1 shows the changes in the percentage of smokers in the male population of the UK between 1950 and 1998.

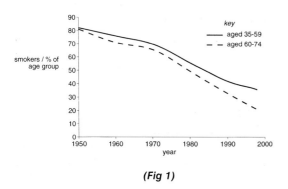

(Fig 1)

Figure 2 shows the changes in mortality from lung cancer in men over the same period of time.

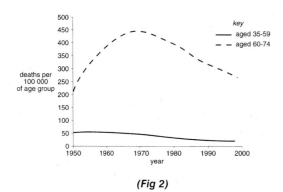

(Fig 2)

c Describe and explain the changes in mortality from lung cancer among men in the UK between 1950 and

1998. Use the information in figures 1 and 2 to help your answer. *(4 marks)*

(Total 10 marks)
OCR 2802 May 2002, B(HHD), No.6

4 Stroke and coronary heart disease are both diseases of the cardiovascular system. They involve blood vessels in different parts of the body.

a State **two** ways in which strokes differ from coronary heart disease. *(2 marks)*

b Describe the effects of *carbon monoxide* and *nicotine* on the cardiovascular system. *(4 marks)*

The figure shows the annual death rates from coronary heart disease (CHD) in men and women in certain countries in the late 1990s.

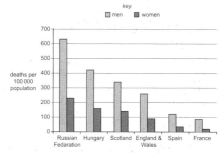

c Use the data from the figure to compare the death rates from CHD in these countries.
Credit will be given if you use figures to support your answer. *(3 marks)*

d State the ways in which health authorities could reduce death rates from CHD. *(3 marks)*

(Total 12 marks)
OCR 2802 Jan 2002, B (HHD), No.2

5 Genetic factors are known to play an important role in the development of coronary heart disease (CHD).

One inherited condition that increases the risk of developing CHD is known as familial hypercholesterolaemia (FH). In FH, the blood cholesterol concentration is raised much higher than normal because of poor metabolism of cholesterol in the liver. Blood cholesterol concentrations greater than 250 mg 100 cm^{-3} are considered to be high.

a Explain why people with FH are especially at risk of coronary heart disease. *(4 marks)*

b State the **dietary** advice that would be given to people with FH. *(2 marks)*

c Explain how coronary heart disease may be treated by surgery. *(3 marks)*

d The Human Genome Project has identified a number of genes that influence human health.
Explain how genetic tests could help people who might be at risk of degenerative diseases, such as coronary heart disease. *(3 marks)*

(Total 12 marks)
OCR 2802 Jun 2003, B (HHD), No.4

Infectious diseases

Cholera

The human body is, in many ways, an ideal environment for the growth of many microorganisms. It provides a warm, constant temperature, a near-neutral pH, a constant supply of food and water in well-balanced proportions for growth and a ready supply of oxygen. It is therefore not surprising that our bodies are colonised by a large variety of microorganisms, many of which have the capacity to cause disease. Those that cause disease are called **pathogens** and the extent to which a pathogen causes damage is known as **virulence**. To be considered a pathogen, a microorganism must:

- gain entry to the host
- colonise the tissues of the host
- resist the defences of the host
- cause damage to the host's tissues.

One such pathogen causes the disease known as cholera.

15.1.1 Cause and means of transmission

False-colour TEM of Vibrio cholerae – the bacterium causing cholera

In 1883, Robert Koch isolated the agent causing cholera – a curved, rod-shaped bacterium called *Vibrio cholerae*, which is characterised by the presence of a flagellum at one end. Cholera is transmitted by ingestion of water and, more rarely, food, that has been contaminated with faecal material containing the pathogen. Such contamination can arise because:

- drinking water is not properly purified
- untreated sewage leaks into water-courses
- food is eaten which is contaminated by those preparing and serving it
- organisms, especially shellfish, feed on untreated sewage released into rivers or the sea.

The main symptoms of cholera are diarrhoea and consequent dehydration. Up to 75% of those with cholera have few if any symptoms. Consequently they act as carriers, unwittingly spreading the disease. Less than 10% of infected patients develop the moderate or severe dehydration associated with cholera.

15.1.2 Worldwide importance

Cholera is most easily transmitted where there is a lack of clean water, sanitation is poor and houses lack basic facilities

In 1817, a **pandemic** of cholera affected the world and since that time there have been six more, the last of which is currently in progress. This seventh pandemic began in 1961 in Indonesia and spread rapidly to Bangladesh (1963), India (1964), the USSR, Iran and Iraq (1965–66). In 1970 it reached West Africa – an area which had not experienced a cholera outbreak for more than a century. It has now become **endemic** to much of the continent of Africa. Cholera spread rapidly around the world, reaching Latin America in 1991 (from where it had also been absent for over 100 years). It has now spread throughout South America.

The strain of *Vibrio* initially responsible for the present pandemic is serogroup 01 and is of the variety called 'El Tor'. In 1992, however, a new strain, serogroup 0139 'Bengal', emerged in Bangladesh and has been responsible for more recent outbreaks of the disease. There is evidence that 'Bengal' is more virulent than 'El Tor'.

Globally, cholera is of great importance, killing, as it does, an estimated 120 000 people each year. In some areas fatality rates may exceed 20% of those contracting the disease; with proper treatment, this should not exceed 1%. In 1999 alone, there were over a quarter of a million cases worldwide, over 80% of which were in Africa.

15.1.3 Prevention and control

To prevent the transmission of cholera, it is clearly necessary to prevent faecal matter containing *V. cholerae* from contaminating food and water. This can be achieved by:

- ensuring water supplies are clean, uncontaminated and treated, e.g. chlorination treatment
- proper sanitation and sewage treatment
- personal hygiene, e.g. washing hands after using the toilet
- food hygiene, e.g. wearing gloves when handling food.

In developing countries, as the population of cities grows rapidly, there is often neither the money nor the resources to provide the necessary infrastructure. As a result, there may be no clean water supply or sewerage system. Housing may be of a poor standard, with no running water or toilet facilities. Without the means to cook, food such as shellfish may be eaten raw, increasing the risk of contracting cholera. Cholera outbreaks frequently arise or worsen following natural disasters, such as earthquakes or war, because:

- services such as water supply and sewage disposal are disrupted
- health services are overstretched
- residents are made homeless
- fleeing refugees spread the disease to neighbouring areas.

Outbreaks of cholera in Basra and Baghdad followed the 2003 war in Iraq.

It is clearly not possible to build water treatment plants and sewerage systems and to re-house millions of people overnight. The main preventative, in the short term, therefore has to be **vaccination**. A vaccine based on inactivated *V. cholerae* 01 has been available for over 40 years but has had limited effect for two main biological reasons:

- it does not prevent transmission of the bacterium
- it only gives **immunity** for a period of about 6 months.

Two oral vaccines, which are easier to administer, are now available:

- a genetically modified form of *V. cholerae* that produces few, if any, symptoms but still induces the body to produce **antibodies** against it
- a mixture of dead *V. cholerae* bacteria and a form of the toxin produced by the bacteria.

Unfortunately, neither vaccine is very effective in children less than 2 years of age and both are currently based on *V. cholerae* 01, although vaccines against *V. cholerae* 0139 are available and being tested. The continuing evolution of new strains of *V. cholerae* makes the development of effective vaccines difficult.

V. cholerae colonises the epithelium of the gut, causing diarrhoea. It is therefore beyond the reach of the body's immune system and so oral **antibiotics** are of little use in controlling the disease because they are expelled from the gut before they can be absorbed. Antibiotics may be given intravenously in severe cases, but misguided use of these has led to the emergence of resistant strains of the bacterium. In the meantime, the main method of treating the disease is to replace the lost fluids through oral rehydration therapy (see extension box). Apart from the social advantages of controlling cholera by providing better housing and sanitation, there are economic ones too. In countries where the disease is endemic, the poor health of workers means that economic output is reduced. Tourism suffers because people are afraid of catching cholera and even food exports are hit because countries do not want to risk importing contaminated produce. Eradication of cholera would reverse these economic disadvantages.

Diseases like cholera cause excessive peristalsis. As a result the contents of the intestines are removed too quickly for normal reabsorption of water to take place in the colon. The patient rapidly dehydrates, sometimes to such a degree that his / her life is threatened. This is especially the case with infants – indeed diarrhoea is one of the leading causes of infant mortality in developing countries. Not only is water lost in such infections but also certain nutrients and salts.

A simple treatment, **oral rehydration therapy**, is cheaply available and much effort is now being expended in educating parents about its benefits to their children. A solution of sugar and salt of a specific concentration is given to the patient by mouth. This replaces the nutrients, salt and water lost in the diarrhoeal fluid. The solution must be given regularly and in large amounts throughout the duration of the illness.

SUMMARY TEST 15.1

Organisms that enter, colonise and damage the tissues of a host organism are called **(1)**, and the extent of the damage they cause is known as **(2)**. One such organism is the bacterium called **(3)**, which causes cholera. This disease is transmitted mostly through **(4)** that is contaminated with faecal material containing the bacterium, although, to a lesser extent, contaminated **(5)** may be responsible. Outbreaks of cholera affect many people in many regions of the world and are therefore called **(6)**. A total of around **(7)** people are estimated to die each year from cholera and the worst-affected continent is **(8)**. Prevention in the long term depends on better housing, clean water and proper **(9)**, as well as better hygiene. In the short term, **(10)** is the best preventative measure. Treatment may involve injecting **(11)** and replacing fluids using **(12)**.

15.2 Malaria

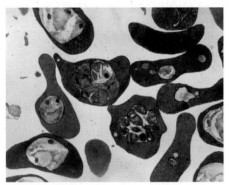

Coloured TEM of the malarial parasite Plasmodium *infecting red blood cells*

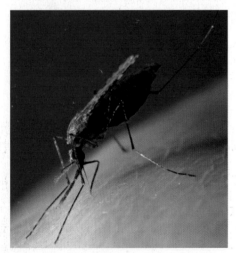

An Anopheles *mosquito, the vector of the disease malaria, feeding*

Most microorganisms that cause disease in humans are prokaryotes (unit 1.8) but a few are **eukaryotes**. One example of a disease caused by a eukaryotic **parasite** is malaria. It affects the liver, red blood cells and brain, causing pain, shivering, sweating and anaemia.

15.2.1 Causes and means of transmission

Malaria is caused by parasites of the protoctist genus *Plasmodium*. Four species of *Plasmodium* can produce the disease in various forms. Of these, *Plasmodium falciparum* is the most widespread and dangerous. It has two hosts: humans, and mosquitoes belonging to the genus *Anopheles*. Of the 380 or so species of *Anopheles*, around 60 are capable of transmitting malaria to humans. Organisms such as *Anopheles* mosquitoes that transfer a parasite to its main (primary) host are known as **vectors**. The life cycle of *Plasmodium* can be summarised as follows.

- After mating, the female *Anopheles* mosquito requires a meal of blood in order for her eggs to mature. The eggs are later laid in water.
- When biting a human, the female mosquito injects saliva to prevent the blood clotting and blocking her needle-like mouthparts.
- If the saliva contains *Plasmodium*, this enters the human blood system and is carried to the liver.
- In the liver, *Plasmodium* multiplies to produce massive numbers that are released into the blood.
- These enter red blood cells, multiply and form gametes which are released into the blood plasma.
- When another female mosquito bites an infected human, these gametes are sucked up, along with the blood.
- The gametes fuse to form a zygote that develops into a cyst on the wall of the mosquito's stomach.
- The cyst releases *Plasmodium* cells which migrate to the salivary gland, ready to be injected into a human next time the mosquito feeds on blood.

15.2.2 Worldwide importance

Around 300 million people worldwide suffer from malaria, with between 1.5 million and 3.0 million dying from the disease each year. It ranks third behind pneumonia and tuberculosis as the major cause of death in the world. There is one death from malaria somewhere in the world every 12 seconds and it is on the increase. The World Health Organisation (WHO) estimates that malaria will increase annually by 16% and may soon become responsible for more deaths than any other disease.

Unlike other diseases such as cholera, tuberculosis (TB) and acquired immune deficiency syndrome (AIDS), which have no vector organism, malaria is confined to regions where the *Anopheles* mosquito is found. Tropical climates provide the best breeding and living conditions for the *Anopheles* mosquito. Malaria therefore occurs most commonly in the 100 tropical countries where 40% of the world's population live. Malaria is **endemic** to 91 of these countries. 90% of all sufferers live in sub-Saharan Africa, with nearly 70% of the remaining cases in just six countries: India, Brazil, Sri Lanka, Vietnam, Columbia and the Solomon Islands.

15.2.3 Prevention and control

In deciding how to control malaria we need to look at the three living organisms involved in its spread:

- humans, who are highly mobile and spread the disease far and wide
- mosquitoes, which are highly mobile as flying adults but more or less stationary as larvae and pupae

- *Plasmodium*, which has developed resistance to drugs and adapts genetically to new circumstances.

Effective prevention and control of malaria involves a complementary range of measures aimed at all three biological elements of the cycle, namely:

- **Control of mosquitoes** – through reducing their populations by:
  - draining marshes and other areas of water where the mosquitoes lay their eggs and the larvae develop
  - introducing fish, which consume mosquito larvae, into marshes, ponds etc. (biological control). However, these fish may also eat beneficial aquatic life
  - spraying fresh-water areas with *Bacillus thuringiensis*, a parasite that kills mosquito larvae but is harmless to other wildlife (biological control)
  - spraying fresh-water areas with insecticides to kill mosquito larvae (chemical control). These insecticides often kill beneficial aquatic organisms as well and may accumulate in food chains, harming other wildlife and even humans
  - spraying oil over fresh-water areas to prevent the mosquito larvae breathing the air they need to survive (chemical control).
- **Control by humans** – through avoiding being bitten by mosquitoes by:
  - keeping doors and windows closed as far as possible
  - using insect repellents on the skin
  - wearing clothing that covers most, if not all, the skin
  - sleeping under mosquito nets.
- **Control of *Plasmodium*** – through using drugs to:

- treat those affected, by administering drugs such as quinine and chloroquine
- protect those not yet affected, by taking preventative (prophylactic) drugs, such as chloroquine and proguanil, for at least a week before exposure to the parasite and for 4 weeks after exposure.

The social and economic consequences of malaria are huge. For example, it is estimated that a single bout of malaria costs a sum equivalent to over 10 working days in Africa. International efforts to control malaria were successful in the late 1950s and early 1960s but the disease has since returned with a vengeance. Reasons for this resurgence may include some or all of the following:

- poor education and a failure to follow anti-malarial measures properly
- wars and political unrest preventing implementation of anti-malarial measures
- migrants, both political and economic, spreading the disease to previously unaffected areas
- poverty in affected countries leaving few resources for medical treatments or education
- ease of movement, e.g. by jet travel, spreading the disease globally
- mosquitoes breeding in untreated water, e.g. puddles
- insecticide resistance in mosquitoes
- banning of certain insecticides, e.g. DDT
- drug resistance in *Plasmodium*
- **global warming** increasing the breeding range and life-span of *Anopheles* mosquitoes
- no vaccine being available as yet (see extension box).

EXTENSION – A VACCINE EFFECTIVE AGAINST MALARIA

For years scientists have sought an effective vaccine to control malaria. Why then should this be proving so difficult when vaccines exist for many other diseases? To begin with, *Plasmodium* is a eukaryotic organism. As such, it has features and systems very similar to our own cells and so it is 'less obvious' to the immune system. It also means that it possesses many surface antigens. *Plasmodium* also 'hides away' inside red blood cells. Surrounded by the membrane of the red blood cell, *Plasmodium* does not cause an antigenic response from the body's immune system (= **antigenic concealment**). However, the disruption it causes to the red blood cells by consuming their **haemoglobin** would normally result in the red blood cells being sent to the spleen

to be disassembled because they are functionally useless. In the process, *Plasmodium* cells would also be destroyed. *Plasmodium* avoids this fate by secreting a protein that extends through the membrane of the red blood cell and anchors it to the wall of the blood vessel so it cannot be carried to the spleen. The presence of this protein, however, 'exposes' *Plasmodium*, and so the immune system starts to destroy the protein. *Plasmodium* has a final trick up its sleeve, however. No sooner has the immune system been sensitised to the protein than *Plasmodium* produces a different protein to anchor the red blood cell to the wall of the blood vessel. This is called **antigen shifting** and is the main reason why a vaccine is proving so difficult to develop.

SUMMARY TEST 15.2

Malaria is caused by four species of the genus *Plasmodium*. *Plasmodium* belongs to the group **(1)** and it has two hosts: humans, and mosquitoes belonging to the genus **(2)**. As mosquitoes spread the disease to the main host, humans, they are called a **(3)**. Around the world some **(4)** million people suffer from malaria and up to 3 million die each year, mostly in **(5)**. Control of mosquitoes can help prevent malaria. This is carried out by draining marshes, where the **(6)** stage of the mosquito lives, or by spraying chemical agents such as **(7)** or **(8)** on the water to kill them or using biological methods like **(9)** or **(10)** to have the same effect. Humans can help to avoid being bitten by mosquitoes by using **(11)** to cover them when sleeping or putting **(12)** on the skin to discourage them. Drugs such as **(13)** and **(14)** may be used to prevent contracting malaria when visiting a country where it is endemic.

Unlike many other infectious diseases, AIDS is a relative newcomer, having first been diagnosed in 1981. It is already one of the most serious diseases in human history. While it currently does not kill as many people as other diseases in one year, it has a fatality rate of 100%. Without treatment, all AIDS sufferers die. Treatment only prolongs life; it does not provide a cure.

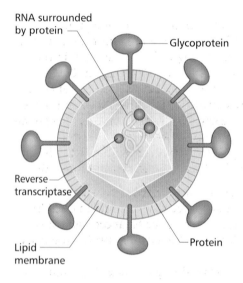

Fig 15.1 *Structure of human immunodeficiency virus*

15.3.1 Causes and means of transmission

AIDS is caused by the **human immunodeficiency virus (HIV)**, a spherical retrovirus whose structure is shown in figure 15.1. Once infected with HIV, an individual is said to be **HIV positive**, a condition which persists throughout life. As the virus remains dormant for about 8 years on average, an HIV-positive person does not show any symptoms during this period but can act as a carrier, often unwittingly spreading the disease. The virus can be detected in virtually all body fluids of an HIV-positive individual. However, since it is only in blood, semen or vaginal fluid that its concentration is high, the virus is usually spread through sexual intercourse or transfer of infected blood from one person to another – as when drug-users share a hypodermic needle – or from mother to baby during childbirth. There is evidence that HIV can be transmitted from mother to baby across the placenta as well as in breast milk. In the past, the transfusion of HIV-infected blood has spread AIDS, although blood is now routinely screened to avoid any risk to patients, such as haemophiliacs, who depend on transfused blood. An infected person has such a low incidence of HIV in the faeces, urine, sweat, saliva and tears that contact with these substances does not present a risk. In any case, the virus quickly dies outside the human body, and therefore even blood, semen and vaginal secretions must be transferred directly.

Having entered the blood, HIV infects white blood cells, such as T helper cells (section 16.4.2), as well as macrophages and brain cells, to which the virus readily binds. It becomes enclosed within the cell membrane and so its **antigens** do not stimulate an immune response from the **lymphocytes** in the blood. Replication of the virus is controlled and it frequently becomes dormant. It is months or years later that replication recommences, so that it is on average 8 years before AIDS develops. Most victims then die within 2 years, usually of opportunist pathogens that take advantage of the impaired immune system. Table 15.1 gives details of the frequency of different forms of transmission in the UK. This shows that sexual transmission accounts for over 80% of HIV transmission and always has. However, heterosexual sex is now almost twice as likely to be the means of transmission than sex between men in the UK.

Table 15.1 *Causes of HIV transmission in the UK*

Form of transmission	% of all cases	
	2002	**1982–2002**
Heterosexual sex	54	32
Sex between men	28	52
Mother to baby	2	2
Injecting drug use	2	7
Tissue/blood transfusion	Negligible	3
Undetermined/other	14	4
Total number of cases	**5338**	**56 108**

15.3.2 Worldwide importance

AIDS is truly a global disease. Unlike malaria, which is confined to regions where the *Anopheles* mosquito is found, AIDS has affected almost every country in the world and certainly all the major regions (table 15.2). The current **pandemic** has spread rapidly since its origins and there are now estimated to be 42 million people infected with HIV worldwide, in addition to almost 25 million who have died from AIDS over the past 20 years. In the year 2002 alone an estimated 5 million people were infected with HIV (table 15.2).

HIV / AIDS is far more prevalent in sub-Saharan Africa than elsewhere. In this region:

- 29.4 million people are estimated to be currently infected with HIV
- 3.5 million were infected with HIV in 2002 alone
- 8.8% of the total population was HIV positive in 2002 alone
- in countries such as Zimbabwe, 25% of the population is HIV positive
- as a result primarily of AIDS, life expectancy in South Africa fell from 65 years to 55 years between 1995 and 1999.

Table 15.2 *Total number of living HIV-infected people by region (December 2002) (source: Joint UN Programme on HIV / AIDS)*

Region	Number (millions) infected with HIV	% of total population (15–49 years) with HIV
Sub-Saharan Africa	29.40	8.80
South and South East Asia	6.00	0.60
Latin America	1.50	0.60
East Asia and Pacific	1.20	0.10
Eastern Europe and Central Asia	1.20	0.60
North America	1.00	0.60
Western Europe	0.57	0.30
North Africa and Middle East	0.55	0.30
Caribbean	0.44	2.40
Australia and New Zealand	0.15	0.10
Total	42.00	1.20

Table 15.3 *Worldwide estimates of HIV / AIDS cases as of the end of 2002 (source: Joint UN Programme on HIV / AIDS*

Infections and deaths	Number (millions)			
	Men (15–49 years)	Women (15–49 years)	Children (under 15 years)	Total
Number of people newly infected in 2002	2.2	2.0	0.8	5.0
Total numbers of people with HIV / AIDS in 2002	19.4	19.2	3.2	41.8
Deaths from HIV / AIDS in 2002	1.3	1.2	0.6	3.1
Total number of AIDS deaths from 1982 to 2002	9.8	10.2	4.9	24.9

Because AIDS sufferers fall victim to other infections; the **prevalence** of the disease in Africa is especially significant because it is this region that also has the highest **incidence** of other diseases, such as cholera and malaria. Also the rise in TB cases worldwide has been linked to an increase in the number of AIDS cases. This has a devastating effect on the economy of African countries because:

- AIDS affects mostly people in the 20–40 years age range and these are usually the most economically productive sector of any population
- scarce financial resources have to be spent on expensive drugs, leaving little for economic development.

15.3.3 Prevention and control

There is, as yet, no cure for AIDS. Much effort is being expended in developing a vaccine but progress is being hampered by the rapid rate at which HIV mutates, the fact that HIV 'hides' itself within the lymphocyte cell membrane and the risk that a vaccine from attenuated HIV could induce cancers. Furthermore, as AIDS affects almost exclusively humans, there are no suitable animals on which to test new drugs. Current approaches to finding a suitable treatment for AIDS involve development of:

- drugs which inhibit HIV
- a vaccine to prevent AIDS
- medicines which boost the immune system of AIDS sufferers
- treatments for the other infections which develop in AIDS sufferers.

Drugs are expensive and often beyond the means of sufferers in Africa and other developing countries. Many of these drugs are now increasingly effective so that, in the UK, deaths dropped from 1516 in 1995 to 235 in 2001. These drugs, however, often have unpleasant side effects, including headaches, diarrhoea and even permanent nerve damage. Nor are they a cure, but rather a means of delaying the onset of AIDS. With neither a cure nor a vaccine available at present, preventive measures remain the best means of containing the disease. Such measures include:

- **Advising HIV-positive mothers not to breast-feed** as the virus can be transferred from mother to child in breast milk. This must however be balanced against the benefits of breast-feeding which may outweigh the risks of HIV.
- **Contact tracing** to avoid the spread of HIV anyone who is found to be HIV positive is asked to contact others who he/she might have infected, e.g. sexual partners, so that they may be tested and treated as necessary.
- **Education** is paramount and involves informing the population of the risks and how to minimise them.
- **Needle-exchange schemes** for drug-users who can exchange used needles for new ones so that needle-sharing becomes unnecessary.
- **Screening of blood from donors** now occurs routinely and it is heat treated to kill HIV so that the risk to haemophiliacs and others using blood products is negligible in developed countries.
- **HIV testing for individuals** at particular risk, e.g. injecting drug-users, prostitutes and male homosexuals.
- **Using condoms** or other barriers, such as femidoms and dental dams, for all forms of sexual contact. These act as a physical barrier and prevent the mixing of body fluids.

The biggest problem with containing the AIDS pandemic is the time it takes for the disease to become apparent. During this period, HIV-positive individuals can spread the disease widely, often unwittingly. Another problem is the social stigma that HIV still carries. HIV-positive individuals may feel ostracised and isolated and therefore are deterred from seeking treatment.

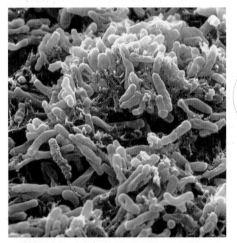

SEM of a cluster of Mycobacterium tuberculosis, *the bacterium causing tuberculosis* (×8500)

Tuberculosis (TB) is an infectious disease that can affect any part of the body but is usually found in the lungs because these are the first site of infection, causing coughing, shortness of breath, fever and sweating. It kills approximately 2 million people each year and the incidence of the disease is growing, even in developed countries where once it was comparatively rare.

15.4.1 Causes and transmission

Tuberculosis is caused by one of two rod-shaped bacteria: *Mycobacterium tuberculosis* or *Mycobacterium bovis*. It is estimated that up to 30% of the world's population have one or other form of the bacterium within their bodies. These people form two groups which differ **epidemiologically**.

- Most do not suffer any symptoms and the infection is controlled by the body's immune system. These individuals cannot pass on the disease. The bacteria may, however, be activated, often after many years, especially when the person's immune system is weakened by other infections, or by AIDS.
- Some develop the disease because the bacteria overcome the body's defences, causing symptoms such as a persistent cough, tiredness, loss of weight and fever. These individuals can transmit the disease to others.

Despite its **prevalence** around the world, it is not all that easy to contract TB compared to many other infectious diseases. It is spread through the air by **droplet infection** when infected individuals cough, sneeze, laugh or even just talk. However, it normally takes close contact with an infected person over a period of time, rather than a casual meeting in the street, to transmit the bacteria. TB is therefore usually spread between family members, close friends or work colleagues, especially in crowded and poorly ventilated conditions. TB can also be spread from cows to humans because *M. bovis* also infects cattle. Meat and especially milk may contain the bacterium. Some groups are at greater risk of contracting TB. These include people who are:

- in close contact over long periods with infected individuals, e.g. living, and especially sleeping, in over-crowded conditions
- suffering from HIV infections
- suffering from other medical conditions that make the body less able to resist the disease, e.g. diabetes, lung disease such as silicosis
- undergoing treatment with immuno-suppressant drugs (e.g. following transplant surgery)
- malnourished
- working or residing in long-term care facilities where relatively large numbers of people live close together, e.g. old people' homes, care homes, hospitals and prisons
- from countries where TB is common
- alcoholic, injecting drug-users and / or homeless.

15.4.2 Worldwide importance

TB is a global disease which, at one time, appeared to be under control, at least in developed countries (Fig 15.2). However, a major resurgence in the disease led the World Health Organisation (WHO) to declare a global emergency in 1993. It is now estimated that:

- Over 8 million people develop TB each year
- TB kills 2 million people in the world annually

SUMMARY TEST 15.4

Tuberculosis (TB) is caused by two types of bacterium. The one found only in humans is **(1)** while the other, called **(2)**, is found in humans and cattle. The disease is transmitted through the **(3)**. Around **(4)**% of the world's population is thought to be infected by tuberculosis bacteria but, of these, only around **(5)**% suffer symptoms and pass the disease on to others. Certain groups of people are more at risk, including those living in **(6)** conditions, those who are malnourished, and those whose immune systems are weakened by medicine and diseases like silicosis or who are **(7)**. TB kills around **(8)** million people annually. It can be treated using the **(9)** vaccine or by drugs. One problem with the drugs is that they need to be taken for **(10)** months and many people fail to complete the treatment, leading to **(11)** strains of the bacteria that cause TB.

- 30% of the world's population is currently infected with *Mycobacterium*
- 5–10% of people infected with *Mycobacterium* will become sick and infectious with TB at some time during their lives
- nearly 1% of the world's population is newly infected with TB each year
- someone in the world is newly infected with TB every second.

In Eastern Europe and Africa the disease is now on the increase after 40 years of decline. The greatest number of cases occurs in South East Asia, although numbers are rising rapidly in sub-Saharan Africa, largely as a result of the AIDS **epidemic** there (section 15.3.2).

Reasons why TB is on the increase include the following.

- The spread of HIV (unit 15.3) throughout the world has compromised the individual's immune system and allowed TB to develop when ordinarily it would be controlled.
- Increased movement of people as a result of global trade and tourism has spread the disease worldwide.
- War and political unrest has led to mass movements of refugees, who are often housed in densely populated camps and shelters. Even where treatment programmes are implemented, refugees often move on before the 6-month treatment is complete.
- Poorly managed TB treatment and prevention programmes lead to incomplete treatment of TB or the wrong dose or drug being administered.
- Drug-resistant forms of *Mycobacterium* have developed throughout the world, largely as the result of incomplete treatment.
- A greater number of people are living 'rough'. They are therefore often in poor health and may temporarily live in crowded sheltered accommodation.
- There is a larger proportion of elderly people in the population and older people often have less effective immune systems.

15.4.3 Prevention and control

The main biological preventative measure for tuberculosis (TB) is **vaccination**. All children in the UK are routinely tested for their **immunity** to TB. Vaccination of those individuals who are already immune is unnecessary and dangerous. Those without immunity are given the **Bacille-Calmette-Guerin (BCG) vaccine**. This is an attenuated (weakened) strain of *M. bovis*, the organism that causes bovine TB. While this form of mass **immunisation** has been the major reason for the fall in cases of TB in the UK, there is no doubt that improved social conditions have also played a part. Better housing, with less over-crowding, less damp conditions and better nutrition, has, until recently, reduced the ease with which the disease can be transmitted.

Other biological means of controlling TB include treatment with drugs, such as isoniazid. The drug must be taken for 6 to 9 months to be effective. One problem with all drug treatment for TB is the long period for which the drugs must be taken. When individuals are ill, they willingly take the drugs as they are keen to recover. The drugs initially destroy the least-resistant strains of *Mycobacterium*. After a number of months, the patients feel better because the vast majority of *Mycobacterium* have been killed. The temptation is for patients to consider themselves cured and to stop taking the drugs. This is almost the worst course of action because the few bacteria that remain are those that are most resistant to the drug. These resistant strains survive, multiply and spread to others. There is therefore a selection pressure that leads to the development of strains of *Mycobacterium* that do not respond to the drug. It is in this way that multiple drug resistant strains of TB (MDR-TB) have developed. To overcome the problem, a 'cocktail' of three or four drugs is used to ensure that at least one will be effective. In addition, in 1991, WHO introduced a scheme for the detection and cure of TB called DOTS (Direct Observation of Therapy). This combines:

- a political commitment from the government of each country involved
- services of laboratories that can test for the presence of *Mycobacterium* in the sputum samples taken from the population
- supplies of a range of anti-TB drugs
- monitoring and recording of each patient's treatment
- direct observation of the treatment by health and community workers and volunteers to ensure that drugs are taken properly and the course of treatment is completed.

Another biological control method is the pasteurisation of milk, which kills any *M. bovis* present. Cattle herds are also regularly checked by veterinary surgeons for any sign of TB or the bacterium causing it.

In addition to the biological means of control discussed above, there are social and economic measures that can be introduced to reduce the number of TB cases. These include:

- better education about TB, particularly the need to complete all courses of drugs
- more and better housing, leading to less over-crowding
- improved health facilities and treatments, including more effective drugs, vaccination programmes and contact tracing
- better nutrition to ensure that immune systems are not weakened by poor diet.

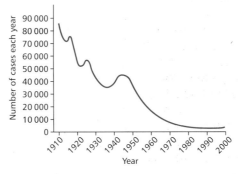

Fig 15.2 *Graph showing the number of TB cases in the UK during the period 1910–2000*

Antibiotics

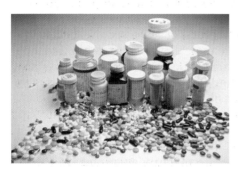

One in every six prescriptions issued by doctors in the UK is for an antibiotic

Antibiotics are substances produced by microorganisms that can destroy other microorganisms or inhibit their growth. Although most are produced by bacteria, such as *Streptomyces*, a few are made by fungi such as *Penicillium*. Strictly speaking therefore, substances produced by plants and animals, or those produced totally synthetically, are not antibiotics. However, the term is often used more loosely and many of the antibiotics in use today are semi-synthetic because chemists have altered the structure of a naturally produced substance to make it more effective. Chloramphenicol, used against typhoid and pneumonia, is made entirely synthetically. First discovered by Alexander Fleming in 1928, but not produced commercially until 1941, penicillin was the first antibiotic to be used to treat infections. One in every six prescriptions issued by doctors in the UK each year is for an antibiotic. The word 'antibiotic' is derived from Greek words meaning 'against life'. In practice, these drugs are rather narrower in their operation and are generally substances made by microorganisms. They are either:

- **biocidal (bacteriocidal)** – killing other organisms (bacteria)
- **biostatic (bacteriostatic)** – preventing the growth and multiplication of microorganisms (bacteria).

15.5.1 How antibiotics work

In general, antibiotics work in two ways – either by preventing bacteria from making normal cell walls or by interfering with the internal biochemistry of the bacteria. More particularly, they operate by:

- **Preventing cell wall synthesis.** The cell walls of bacteria, like those of plant cells, are essential to prevent the cells from bursting when water enters them by **osmosis (osmotic lysis)**. Antibiotics inhibit the synthesis and assembly of important peptide cross-linkages in bacterial cell walls, weakening them and causing them to burst, thereby killing the bacterium. As these antibiotics inhibit the proper formation of cell walls, they are only effective when bacteria are growing. Penicillin is an antibiotic that works in this way.
- **Interfering with nucleic acid synthesis.** Some antibiotics, such as the anthracyclines, inhibit the synthesis of DNA (unit 5.3), while others, like rifampicin, inhibit the formation of **messenger RNA** and so prevent transcription (unit 5.6). In these cases, the bacterium is unable to grow or multiply.
- **Interfering with protein synthesis.** Antibiotics like streptomycin will bind to ribosomes of bacteria, but not to those of mammalian cells. Other antibiotics, such as tetracycline, bind equally well to both mammalian and bacterial ribosomes, but the drug is taken up only by bacterial cells. In both cases, protein synthesis (unit 5.6 and 5.7) is prevented, and the bacterial cells cannot grow and multiply.
- **Damaging cell membranes.** Amphotericin B is an antibiotic which distorts the lipid bilayer of the cell surface membrane of fungi. The damaged membrane allows the cell contents to escape, killing the fungus.

Figure 15.3 summarises these various actions of antibiotics.

15.5.2 Role of antibiotics in the treatment of infectious disease

No antibiotic is effective against all forms of infection, although two main groups are recognised:

- **Broad-spectrum antibiotics** are effective against a large range of microorganisms.

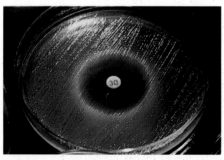

Action of antibiotic on bacterial cells. The clear region around the disc containing antibiotic indicates that the antibiotic has killed the bacteria

- **Narrow-spectrum antibiotics** are effective only against a selected type of microorganism.

An ideal antibiotic is totally effective in killing pathogenic cells without harming any host cells or interfering with the host's metabolism. These rarely occur and more often the antibiotic produces some side effects. Streptomycin, for example, was successful in treating tuberculosis but led to nerve damage, with the result that some patients became deaf. Choosing an appropriate antibiotic for a given disease involves screening different ones against different strains of the microorganism causing the disease (pathogen). To do this, the pathogen, e.g. a strain of bacterium, is grown on an agar plate to give an evenly distributed colony known as a bacterial lawn. A series of small pads, each containing a different antibiotic, is placed on this lawn. If the antibiotic is effective in killing or preventing the growth of the bacteria, the area around it will be clear of bacteria (see photo). This is known as an **inhibition zone**. The diameter of this zone gives a relative indication of the effectiveness of the antibiotic. If the antibiotic is not effective at all, bacteria will grow up to the edge of the pad. Antibiotics have had a major impact in controlling a wide range of infectious diseases, including TB, pneumonia and cholera. Their effectiveness has, however, been reduced due to the emergence of strains of pathogens that are resistant to particular antibiotics.

tolerance to the antibiotic, but rather from a chance mutation within the bacteria. This mutation resulted in bacteria that could produce an enzyme, penicillinase, which broke down antibiotics such as penicillin before it was able to kill them. Thereafter, whenever penicillin is used, only the non-resistant forms of the bacteria are killed. There is therefore a selection pressure that favours the resistant form when the bacteria are exposed to penicillin, and this gradually becomes predominant in the population. The medical implications were obvious and, by 1950, many staphylococcal infections were already completely penicillin resistant. More significantly, the **allele** for **antibiotic resistance** is carried on **plasmids**, and these loops of DNA can be transferred from cell to cell by natural as well as artificial means. Resistance can therefore find its way into other bacterial species. Over-use of antibiotics, e.g. for minor infections that present no danger, increases the likelihood of selection of resistant strains over strains that are susceptible to the antibiotic. A failure to complete the full course of antibiotics also increases the chances of this happening, for the reasons explained in section 15.4.3. Resistance to antibiotics is an increasing problem. A survey of hospitals in five European countries showed that, in 40% of samples taken, bacteria were resistant to one or more antibiotics. In New York City, one-third of all TB cases are resistant to one antibiotic. Some strains of *Staphylococcus aureus* in hospitals in the UK are resistant to at least four different antibiotics.

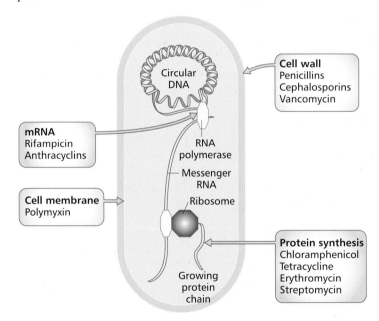

Fig 15.3 *Sites of antibiotic action*

15.5.3 Antibiotic resistance

The commercial production of antibiotics in the 1940s was heralded as a major breakthrough in the control and possible elimination of diseases caused by bacteria. Unfortunately, it was not long before it was realised that some antibiotics no longer killed bacteria in the way they had previously. It was found that these bacteria had developed resistance, not as the result of a cumulative

---

SUMMARY TEST 15.5

Antibiotics are substances produced by **(1)** and they operate in two general ways. Those called **(2)** prevent pathogens such as bacteria from multiplying and growing by interfering with the bacterium's ability to make either **(3)** or **(4)**. Those called **(5)** kill the pathogens by preventing them making **(6)**, with the result that the cell bursts due to the influx of water. Some antibiotics are effective against a wide range of microorganisms and are called **(7)** antibiotics; others are effective only against a selected type of microorganism and are called **(8)** antibiotics. Antibiotic resistance is becoming increasingly common. In one example a mutation has allowed bacteria to produce an enzyme called **(9)**, which breaks down the first commercially produced antibiotic.

1 Cholera is spread mainly by contaminated water and food. In areas where it is endemic it is mainly a disease of young children, although infants who are breast-fed are rarely affected.

In 1970, cholera spread to West Africa, which had not experienced the disease for more than 100 years. The disease spread quickly and eventually became endemic in most of Africa.

a (i) Name the organism that causes cholera. *(1 mark)*

   (ii) Explain how cholera is transmitted from one person to another through the water supply.
   (3 marks)

   (iii) Define the term *endemic*. *(1 mark)*

   (iv) Suggest why infants who are breast-fed are rarely affected by cholera. *(2 marks)*

b Describe **three** public health measures that can be taken to control the spread of cholera. *(3 marks)*

c Explain why malaria does not show the same worldwide distribution as cholera. *(3 marks)*

(Total 13 marks)
OCR 2802 Jan 2003, B (HHD), No.4

2 Tuberculosis (TB) is one of the world's greatest killers. There is a pandemic of TB and this poses great threats to the world's population. It is a disease that is proving very difficult to eradicate.

a Name the organism that causes tuberculosis (TB).
(1 mark)

b Explain what is meant by the term *pandemic*. *(1 mark)*

c Discuss the problems that are involved in eradicating tuberculosis (TB) from the world.
(In this question, 1 mark is awarded for the quality of written communication.) *(9 marks)*

(Total 12 marks)
OCR 2802 Jun 2003, B (HHD), No.5

3 The following passage is adapted from the World Health Organisation's *World Health Report – 1999*

a State **three** ways in which HIV is transmitted.
(3 marks)

b Explain the **problems** that are encountered by governments, in places such as South-East Asia, in controlling the spread of HIV/AIDS.

(In this question, 1 mark is available for the quality of written communication.) *(7 marks)*

HIV/AIDS control in South-East Asia: the challenge of expanding successful programmes.

The human immunodeficiency virus (HIV) was slower to emerge in South-East Asia than in other parts of the world, but is now a serious public health problem and a threat to development. The first patient from South-East Asia with AIDS was identified in 1984. Since then a total of 92 391 cases of the disease have been reported in this region up to 1st July 1997. However, because of under-reporting and under-diagnosis the reported cases only reflect a proportion of the true problem. The World Health Organisation estimates that there are currently more than 5.5 million people in South-East Asia who are infected with HIV – 18% of the global total.

(Total 10 marks)
OCR 2802 Jan 2002, B (HHD), No.6

4 Over 40% of the world's population live in areas where there is a risk of malaria. The disease causes widespread suffering and death especially among children.

a Name the organism that causes malaria. *(1 mark)*

b State **two** geographical areas of the world where there is a high risk of malaria. *(2 marks)*

Malaria is transmitted by a vector, the mosquito *Anopheles*. The figure shows the life cycle of *Anopheles*.

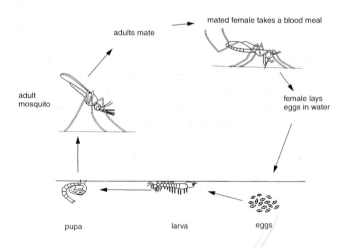

c With reference to the figure,

   (i) describe the role of *Anopheles* in the transmission of malaria; *(3 marks)*

   (ii) outline the biological factors that make malaria a difficult disease to control.
   (In this question, 1 mark is awarded for the quality of written communication.) *(7 marks)*

(Total 13 marks)
OCR 2802 Jun 2001, B (HHD), No.5

5 a (i) Name the organism that causes tuberculosis (TB). *(1 mark)*

(ii) Describe how tuberculosis is spread from infected to uninfected people. *(3 marks)*

The figure shows the number of cases of tuberculosis recorded by health authorities in England and Wales between 1913 and 1998.

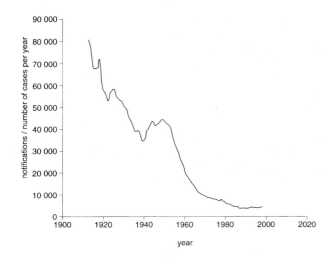

b With reference to the figure, describe the changes in the number of recorded cases of tuberculosis between 1913 and 1998 in England and Wales. *(4 marks)*

c Explain how **social** and **economic** factors have been important in reducing the number of cases of tuberculosis in developed countries such as England and Wales. *(4 marks)*

(Total 12 marks)

OCR 2802 Jan 2001, B (HHD), No.7

6 The bacterium, *Vibrio cholerae*, is the causative agent of cholera. The El Tor strain of *V. cholerae* originally occurred only in Indonesia. In 1961, this strain began to spread replacing existing strains in other parts of Asia. El Tor is now widespread throughout Asia, the Middle East, Africa and parts of Eastern Europe, but has never established itself in Western Europe.

El Tor is hardier than the strain it replaced and the bacteria may continue to appear in the faeces for up to three months after patients have recovered. The bacteria may persist in water for up to fourteen days.

a State **two** ways in which *V. cholerae* is transmitted from infected to uninfected people. *(2 marks)*

Some people infected with cholera have mild symptoms, or none at all, and are carriers of the disease.

b Suggest how laboratory tests could identify carriers of cholera. *(2 marks)*

c Suggest **four** reasons why El Tor has not become established in Western Europe. *(4 marks)*

The United Nations, recognising that most of the outbreaks of cholera were the result of polluted water supplies, set up a 'Decade of Water' in 1981. Its aim was to provide safe water for everyone. Over the decade 1981/1990, the number of people lacking a safe water supply in developing countries dropped from 1800 million to 1200 million.

d Explain why cholera continues to be a worldwide problem, in spite of the 'Decade of Water' campaign.

(In this question, 1 mark is available for the quality of written communication.) *(8 marks)*

The antibiotic tetracycline is sometimes used as a treatment for cholera.

e (i) Suggest **two** ways in which tetracycline can affect *V. cholerae*. *(2 marks)*

(ii) Explain why tetracycline should not be used routinely for all cases of cholera. *(1 mark)*

(Total 19 marks)

OCR 2802, 2000 Specimen paper, B (HHD), No.3

We have seen in chapter 15 some examples of infectious diseases and the damage they can do. Tens of millions of humans die each year from such infections. Many more survive and others appear never to be affected in the first place. Why then are there these differences? Any disease is, in effect, a battle between the **pathogen** and the body's various defence mechanisms. Sometimes the pathogen wins completely and death is often the result. Sometimes the body's defence mechanisms win and the individual recovers from the disease. Having defeated the pathogen, however, the body's defences seem to be better prepared for any second attack from the same pathogen and so repel it before it can cause any harm. This is known as **immunity** and is the main reason why certain people are unaffected by certain pathogens. There is a complete range of intermediates between the stages described above. Much depends on the overall state of health of an individual. A fit, healthy adult will rarely succumb to an infection. Those in ill health, the young and the elderly are usually more vulnerable.

16.1.1   Defence mechanisms

The human body has a range of defences to protect itself from infection. They fall into three main categories, which can be likened to the defences of a typical castle hundreds of years ago:

- **A physical barrier**, e.g. the skin, to prevent entry. Like the walls of a castle, this is the first line of defence.
- **A patrolling 'army' of white blood cells**. Like the soldiers in the castle, these will seek out and kill any invaders.
- **A group of 'guard' cells** that can instantly recognise intruders. Like sentries in the castle, they alert the soldiers to any future danger.

The first line of defence, the physical barrier, takes a number of different forms:

- **A protective covering** – the skin covers the body surface, providing a physical barrier that most pathogens find hard to penetrate. Some pathogens, such as the malarial parasite *Plasmodium*, use a vector, the mosquito, to penetrate this covering and so gain entry to the body (section 15.2.1).
- **Epithelia covered in mucus** – as the body needs to obtain certain substances by diffusion, e.g. oxygen and digested food, there are parts of the body which cannot be covered by a thick layer like the skin. Exchange regions, such as the alveoli in the lungs and the lining of the gut, are covered only by a thin epithelial layer. While

this still acts as a barrier to pathogens, it is more easily crossed than the skin. Many epithelial layers therefore produce mucus, which acts as a further defence against invasion. In the lungs, pathogens stick to this mucus, which is then transported away by cilia (section 1.7.2), up the trachea, to be swallowed into the stomach where the acidic conditions kill the pathogens.

- **Hydrochloric acid in the stomach** – this provides such a low pH that the enzymes of most pathogens are denatured (section 3.2.3) and therefore the organisms are killed.
- **Blood clotting** – any break in the body's coverings is a possible means by which pathogens can get into the body. Blood clotting not only prevents blood loss but also seals off these potential points of entry.

Despite these various precautions, pathogens still frequently gain entry and therefore the body has a second line of defence – a series of specific cellular and chemical defences designed to:

- neutralise any toxins produced by the pathogen
- prevent the pathogen multiplying
- kill the pathogen
- remove any remains of the pathogen.

16.1.2   Phagocytosis

Phagocytosis is the process by which large particles are taken up by cells, in the form of vesicles formed from the cell surface (plasma) membrane. In the blood, two types of white blood cells carry out phagocytosis – **monocytes** and **neutrophils** (section 8.3.3). These cells are known as **phagocytes** and are produced in the marrow of the long

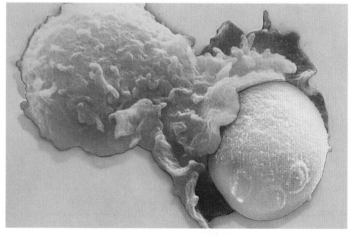

False-colour SEM of phagocytosis of a yeast cell

bones. The process is illustrated in figure 16.1 and is summarised as follows:

- **Antibodies** attach themselves to **antigens** on the surface of the bacterium.
- Proteins, found in the plasma, attach themselves to the antibodies.
- As a result of a series of reactions, the surface of the bacterium becomes coated with proteins called **opsonins**. This process is called **opsonisation**.
- Complement proteins and any chemical products of the bacterium act as attractants, causing neutrophils to move towards the bacterium.
- Neutrophils attach themselves to the opsonins on the surface of the bacterium.
- Neutrophils engulf the bacterium to form a vesicle, known as a **phagosome**.
- Lysosomes (section 1.6.4) move towards the vesicle and fuse with it.
- The enzymes within the lysosomes break down the bacterium into smaller, soluble material.
- The soluble products from the breakdown of the bacterium are absorbed into the cytoplasm of the neutrophils.

Phagocytosis causes **inflammation** at the site of infection. This swollen area contains dead bacteria and phagocytes, which are known as **pus**. Inflammation is the result of the release of **histamine**, which causes dilation of blood vessels in order to speed up the delivery of antibodies and white blood cells to the site of infection.

SUMMARY TEST 16.1

The first line of defence against disease is to prevent the entry of pathogens. The skin provides the main physical barrier, but the thinner coverings, such as the **(1)** layer in the lungs, also secrete **(2)** to provide a further barrier. Other means of preventing entry by pathogens include **(3)** and the secretion of **(4)** by the stomach. Pathogens that do invade the body may be engulfed by cells which carry out **(5)**. These cells are of two types, known as **(6)** and **(7)**, both of which are produced in the **(8)** of the **(9)** bones. In the process the pathogens are coated with proteins called **(10)**. Once engulfed the pathogen is broken down by enzymes released from organelles called **(11)**.

1. The neutrophil is attracted to the bacterium by chemoattractants. It moves towards the bacterium along a concentration gradient

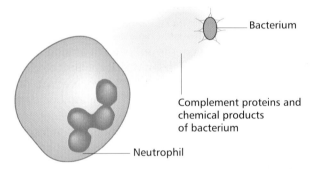

2. The neutrophil binds to the bacterium

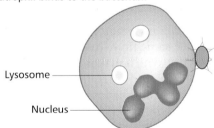

3. Lysosomes within the neutrophil migrate towards the phagosome formed by pseudopodia engulfing the bacterium

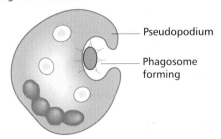

4. The lysosomes release their lytic enzymes into the phagosome, where they break down the bacterium

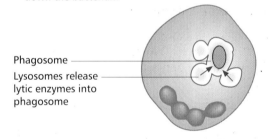

5. The breakdown products of the bacterium are absorbed by the neutrophil

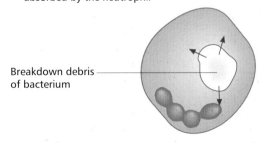

Fig 16.1 *Summary of phagocytosis of a bacterium by a neutrophil*

Principles of immunity – antigens and antibodies

Immunity is the ability of organisms to resist infection by protecting against disease-causing microorganisms that invade their bodies. It involves the recognition of foreign material and the production of chemicals that help destroy it.

16.2.1 Antigens

An **antigen** is any organism or substance that is recognised as non-self (foreign) by the immune system and provokes an immune response. Antigens are usually proteins that make up the cell surface membranes of invading cells, such as microorganisms, or diseased ones, such as **cancer** cells. The presence of an antigen triggers the production of an antibody as part of the body's defence system.

16.2.2 Antibodies

Antibodies are also known as **immunoglobulins (Ig)**. They are proteins synthesised by cells in the blood called **B lymphocytes** (unit 16.3). When the body is invaded by non-self (foreign) material, a B lymphocyte produces antibodies, which react with antigens on the surface of the foreign material by binding to them precisely, in the same way as a key fits a lock. They are therefore very specific, each antigen having its own separate antibody. This massive variety of antibodies is possible because they are made of proteins – molecules that occur in an almost infinite number of forms. Antibodies are made up of four polypeptide chains. The chains of one pair are long and are called **heavy chains**; the chains of the other pair are shorter and are known as **light chains**. The chains are held together by disulphide bridges (section 2.7.1) but, to help the antibody fit around the antigens, they can change shape by moving as if they had a hinge at the fork of the Y-shape. Antibodies have two sites, called **binding sites**, which fit very precisely onto the antigen (Fig 16.2). The binding sites are different on different antibodies and are therefore called the **variable region**. They consist of a sequence of amino acids that form a specific three-dimensional shape which binds directly to a single type of antigen. The rest of the antibody is the same in all antibodies and is known as the **constant region**. This binds to receptors on phagocytes, making **phagocytosis** of **pathogens** easier. There are a number of different groups of antibodies (table 16.1). Each group functions in one or more of the following ways.

The variable region differs with each antibody. It has a shape which exactly fits an antigen. Each antibody therefore can bind to two antigens.

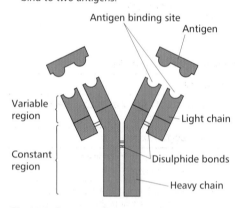

Fig 16.2 *Structure of an antibody*

- **Agglutination** – some antibodies have many binding sites and so can attach to a number of different antigens at the same time. In this way, antibodies cause foreign material to clump together, which makes it more vulnerable to attack from phagocytes.
- **Precipitation** – some antigens are soluble and so antibodies precipitate them out, so that they can be destroyed by phagocytes (section 16.1.2).
- **Neutralisation** – foreign cells often produce toxins, which are chemicals that cause many of the symptoms of a disease. Antibodies can bind to these toxic molecules and so neutralise their harmful effects.
- **Lysis** – having attached themselves to antigens on foreign cells, antibodies then attract other compounds, which bind to them. These include enzymes which help to break down the foreign cells.

These methods of defending the body from infection are illustrated in figure 16.3.

16.2.3 The concept of self and non-self

To be able to defend the body from invasion by foreign material, B lymphocytes must be able to distinguish the body's own cells and chemicals (self) from those that are foreign (non-self). If they could not do this, B lymphocytes would

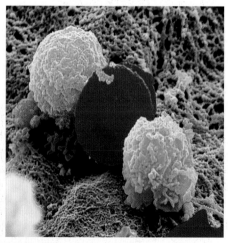

SEM lymphocytes (orange)

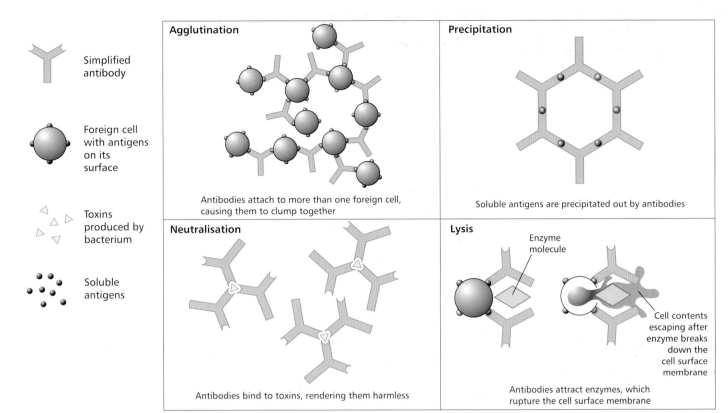

Fig 16.3 *Ways in which antibodies defend the body*

produce antibodies that would destroy the organism's own tissues. How then do B lymphocytes recognise their own cells?

- Each antibody is produced by a different lymphocyte.
- There are more than 10 million different lymphocytes, each capable of recognising a different chemical shape.
- In the fetus, these lymphocytes are constantly colliding with other cells.
- Infection in the fetus is rare because it is protected from the outside world by the mother and, in particular, the placenta.
- Lymphocytes will therefore collide almost exclusively with the body's own material (self).
- Some of the lymphocytes will have receptors that exactly fit those of the body's own cells.
- These lymphocytes either die or are suppressed.
- The only remaining lymphocytes are those that fit foreign material (non-self), and therefore the only antibodies produced are those that attack foreign material.

Table 16.1 *Groups of antibodies (immunoglobulins)*

Immunoglobulin group	Functions
IgM	Serve as receptors on the lymphocyte surface. Promote agglutination
IgG	Promote agglutination
IgD	Serve as receptors for antigens on the B lymphocyte surface
IgA	Prevent bacteria forming colonies on the mucous membrane where they are found Provide early immunity to the newborn when they are passed on in the mother's milk
IgE	Promote the release of histamine to aid the attack on pathogens. Sometimes cause release of histamine in response to harmless antigens, thereby stimulating allergic response

SUMMARY TEST 16.2

An antigen is any organism or substance that is recognised as **(1)** by the immune system. These antigens trigger the production of antibodies, which are made of **(2)** and are also known as **(3)**. Antibodies are synthesised by **(4)** and are made up of a pair of long chains called **(5)** chains and a pair of short ones called **(6)** chains. The two pairs of chains are held together by **(7)**. Antibodies have two sites called **(8)** that fit a specific antigen very precisely. There are a number of groups of different antigens which function in one or more different ways. Some antibodies precipitate out soluble antigens, while others neutralise the **(9)** produced by pathogens, and others break down foreign cells in a process called **(10)**.

Plasma cells differ from the B lymphocytes, from which they develop, in a number of ways. These differences relate to plasma cells' function of producing antibodies at a phenomenal rate – as many as 2000 per second. Compared to B lymphocytes they have a:

- larger size, to accommodate the many organelles needed to manufacture antibodies
- more numerous mitochondria, to provide the energy (section 1.5.3) needed to make antibodies
- more extensive rough endoplasmic reticulum, to make and transport the proteins (section 1.6.1) that make up the antibodies
- more extensive Golgi apparatus, to sort, process and compile the proteins (section 1.6.3) into antibodies and to secrete the finished antibodies.

Immune responses such as phagocytosis (section 16.1.2) are non-specific and occur whatever the infection. The body also has specific responses that fight individual forms of infection. These are slower in action at first, but they can provide long-term immunity. This specific immune response depends on a type of white blood cell called a **lymphocyte**. There are two types of lymphocyte, each with its own immune response:

- B lymphocytes (B cells) – humoral immunity (involves antibodies which are present in body fluids or 'humour')
- T lymphocytes (T cells) – cell-mediated immunity (involves cells).

Both types of lymphocyte are formed from stem cells found in the bone marrow. Their names, however, indicate where they develop and mature:

- B lymphocytes mature in the **B**one marrow
- T lymphocytes mature in the **T**hymus gland.

The maturation process takes place in the fetus and, in the case of B lymphocytes, results in more than 10 million different types, each capable of responding to a different antigen.

16.3.1 Humoral immunity

We saw in section 16.2.2 that, when the body is invaded by foreign (non-self) material, this material possesses antigens that stimulate B lymphocytes to produce antibodies. These antibodies are soluble in the blood and tissue fluid of the body. Another word for body fluids is 'humour' and hence the production of antibodies in this way is known as **humoral immunity**. There are many different types of B lymphocytes, possibly as many as 10 million, and each type produces a different antibody which responds to one specific antigen. When an antigen, e.g. a protein on the surface of a **pathogen** cell, enters the blood or tissue fluids, one type of B lymphocyte will have antibodies on its surface that exactly fit it and therefore it attaches to the antigen. This type of B lymphocyte divides by mitosis (unit 6.2) to form a clone of identical B lymphocytes, all of which produce an antibody that is specific to the foreign antigen. In practise, a typical pathogen, e.g. *Mycobacterium tuberculosis*, has many different proteins on its surface, all of which act as antigens. Pathogens, such as the bacterium that causes tetanus, also produce toxins, each of which will act as an antigen. Therefore many different B lymphocytes make **clones**, each of which produces its own type of antibody. This is known as **polyclonal activation**. For each clone, the cells produced develop into one of two types of cell:

- **Plasma cells**, which secrete antibodies. The plasma cells survive only a few days, but each can make around 2000 antibodies every second during its brief life-span. These antibodies destroy the pathogen, and any toxins it produces, in the ways described in section 16.2.2. The plasma cells are therefore responsible for the immediate defence of the body against infection. This is known as the **primary immune response**.
- **Memory cells** live considerably longer than plasma cells – often for decades. These memory cells do not produce antibodies directly, but rather circulate in the blood and tissue fluid until they encounter the same antigen at some future date. When they do so, they rapidly divide and develop into plasma cells and more memory cells. The plasma cells produce the antibodies needed to destroy the pathogens, while the new memory cells circulate in readiness for a further infection at some time in the future. In this way, memory cells provide

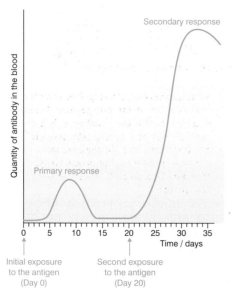

Fig 16.4 Primary and secondary responses to an antigen

long-term immunity against the original infection. This is known as the **secondary immune response**. It is both more rapid and of greater intensity than the primary immune response and ensures that the new infection is repulsed before it can cause any damage – usually with individuals being totally unaware that they have ever been infected. Figure 16.4 illustrates the relative amounts of antibody produced in the primary and secondary immune responses. Figure 16.5 summarises the role of B lymphocytes in immunity.

The way that memory cells function explains why most of us only develop diseases like chickenpox and measles once during our life. The pathogens causing each of these diseases are of a single type and so are quickly identified by the memory cells when they invade the body on subsequent occasions. Cold viruses, by contrast, have over 100 different strains, which are constantly changing. New infections are therefore highly unlikely to be the same as a previous one. With no appropriate memory cells to stimulate antibody production, we have to wait for our slower, less intense, primary response to overcome the infection – during which time we suffer a sore throat and runny nose.

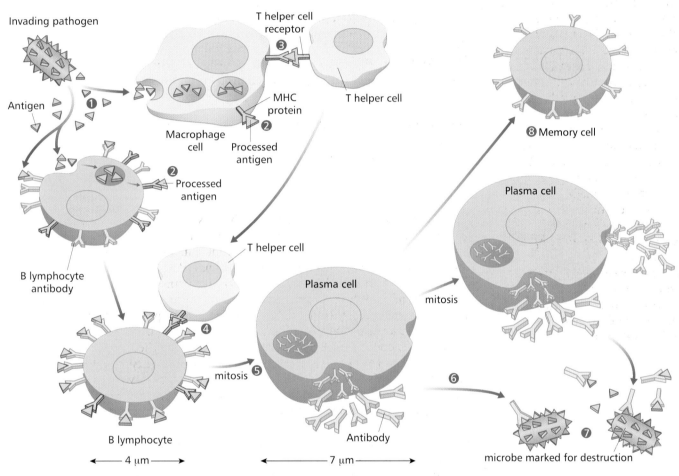

1. Invading pathogen produces antigens that are taken up by both B lymphocytes and macrophage cells by mean of phagocytosis.

2. Both the macrophage cells and the B lymphocytes process the antigens and bind them to a MHC protein (MHC = major histocompatibility complex). The MHC protein presents the processed antigen on the surface of the cells, which are therefore referred to as **antigen-presenting cells**.

3. A T helper cell attaches to the processed antigen on the macrophage cell and becomes activated, making it capable of interacting with B lymphocytes.

4. T helper cells attaches to the MHC proteins with antigens on the surface of the B lymphocyte.

5. The B lymphocyte is activated to divide by mitosis to give a clone of plasma cells.

6. The cloned plasma cells produce antibodies that exactly fit antigens on the pathogen's surface.

7. The antibodies attach to antigens on the pathogen, causing agglutination and lysis of the pathogen, thereby destroying it (= **primary response**).

8. Some plasma cells develop into memory cells that survive for long periods. Future invasions by the same pathogen lead to rapid division of memory cells, some of which develop into plasma cells that produce antibodies. The process is then repeated from stage 6 (= **secondary response**).

Fig 16.5 *Summary of the role of B lymphocytes in the immune response (humoral immunity)*

16.4 T lymphocytes and cell-mediated immunity

SUMMARY TEST 16.4

T lymphocytes respond to their own cells which have been infected by pathogens such as (1). This type of response is called (2) immunity. There are a number of different types of T lymphocytes. One type, called (3) cells, secretes chemicals called (4), which stimulate (5) cells to engulf pathogens by (6) and also stimulate (7) cells to divide to form antibody-producing cells called (8) cells. They also stimulate another type of T lymphocyte called (9) cells, which destroy pathogens.

16.4.1 Cell-mediated immunity

While B lymphocytes respond to non-self (foreign) cells and the foreign products, e.g. toxins, that they produce, T lymphocytes respond to an organism's own cells that have been invaded by non-self material, e.g. a virus or a **cancer** cell. They also respond to transplanted material, which is genetically different. How then can T lymphocytes distinguish these invader cells from normal ones? It is made possible because:

- macrophage cells that have engulfed a pathogen and broken it down, present some of the proteins produced on their own outer surface
- body cells invaded by a virus also manage to present some of the viral proteins on their own cell surface membrane, as a sign of distress
- cancer cells likewise display non-self proteins on their cell surface membranes.

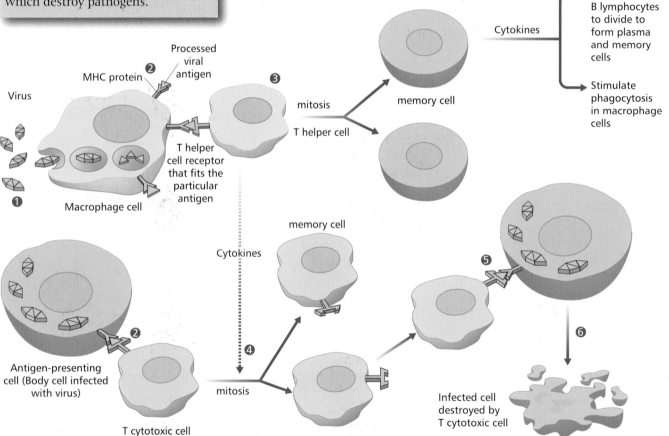

Fig 16.6 *Summary of the role of T lymphocytes in the immune response (cell mediated immunity)*

1. *Viruses both invade body cells and are taken in during phagocytosis by macrophage cells.*

2. *Both the body cells and the macrophage cells process the viruses and bind antigens from them to a MHC protein. The MHC protein presents the antigen on the surface of the cells.*

3. *A T helper cell attaches to the antigen on the surface of the macrophage cells and is thereby stimulated to divide by mitosis. Some of the new T helper cells develop into memory cells that survive for long periods and respond immediately to any new infection by the same virus. Other T helper cells produce cytokines that stimulate B lymphocytes and macrophage cells.*

4. *The cytokines also cause T cytotoxic cells to divide by mitosis. Some of these T cytotoxic*

cells form memory cells that survive and respond immediately to any new infections by the same virus.

5. *Other T cytotoxic cells attach to any body cell presenting the viral antigen (i.e. those that are infected by the virus).*

6. *The attached T cytotoxic cells produce perforins to make holes in the cell membrane and so destroy the cell, along with the viruses it contains.*

The non-self materials on the surface of all these cells act as **antigens** and therefore the term **antigen-presenting cells** is used to describe them. There are many different versions of the two main types of T lymphocytes in the body, each of which has a different receptor protein on its surface. Although these receptors function in a similar way, they are not **antibodies** because they remain attached to the cell rather than being released into the blood plasma.

As T lymphocytes will only respond to antigens that are attached to a body cell (rather than ones that are within body fluids), this type of response is called **cell-mediated immunity**. Figure 16.6 summarises how T lymphocytes respond to a viral infection.

16.4.2 Types of T lymphocytes

There are two main types of T lymphocyte, both of which attach to antigen-presenting cells as described in section 16.4.1. These two types are:
* **T helper cells**, which play a key role in immunity. When they attach to an antigen-presenting cell, T helper cells secrete chemicals called **cytokines (lymphokines)**. These cytokines:

– stimulate macrophage cells to engulf pathogens by phagocytosis
– stimulate B lymphocytes to divide and develop into antibody-producing plasma cells
– activate T cytotoxic cells (T killer cells).
* **T cytotoxic cells (T killer cells)**, which kill body cells that are infected by non-self (foreign) material. They kill not by phagocytosis but by making holes in the cell surface membrane using proteins called **perforins**. These holes allow water to rush into the cell, causing it to burst. As viruses need living cells in which to reproduce, this sacrifice of body cells prevents viruses multiplying.

Both T helper cells and T cytotoxic cells produce their own type of **memory cells**, which circulate in the blood in readiness to respond to future invasions by the same pathogen. Another type of T lymphocyte is the **T suppressor cell**. This, as its name suggests, turns off the actions of the various other lymphocytes once the pathogens have been eliminated from the body. Figure 16.7 summarises the origin and roles of lymphocytes in immunity.

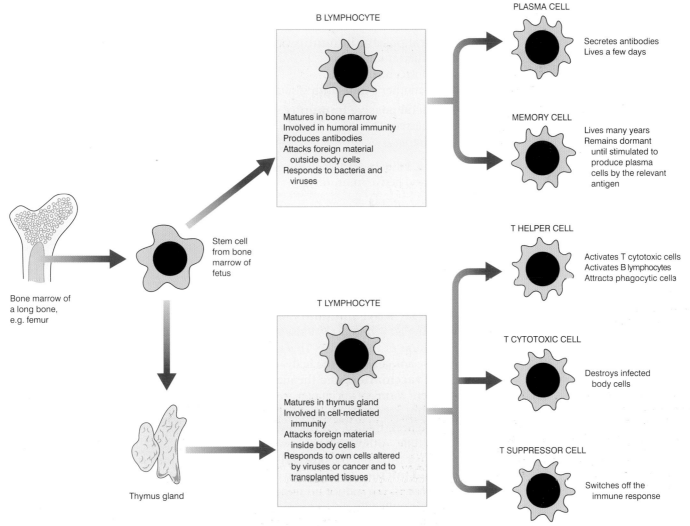

Fig 16.7 *The origin and roles of lymphocytes in immunity*

Immunity is the ability of an organism to resist infection (unit 16.2). This immunity may be naturally acquired or artificially induced. The process of artificially inducing immunity is known as **immunisation**.

16.5.1 Types of immunity

- **Natural immunity** is immunity which is either inherited, or acquired as part of normal life processes, e.g. as a result of having had a disease.
- **Artificial immunity** is immunity acquired as a result of the deliberate exposure of the body to **antibodies** or **antigens** in non-natural circumstances, e.g. **vaccination**.

Both natural and artificial immunity may be passively or actively acquired.

- **Passive immunity** is immunity acquired from the introduction of antibodies from another individual, rather than one's own immune system. It is generally short-lived.
- **Active immunity** is immunity resulting from the activities of an individual's own immune system, rather than an outside source. It is generally long lasting.

16.5.2 Passive immunity

- **Natural passive immunity** occurs when an individual receives antibodies from their mother via
  - the placenta as a fetus
  - the mother's milk during suckling.
- **Artificial passive immunity** occurs when antibodies from another individual are injected. This takes place in the treatment of diseases such as tetanus and diphtheria.

Figure 16.8 illustrates the changes in antibody concentration over time with passive and active immunity.

16.5.3 Active immunity

- **Natural active immunity** results from an individual becoming infected with a disease under normal circumstances. The body produces its own antibodies, and may continue to do so for many years. It is for this reason that many people suffer diseases such as measles only once in a lifetime. The immunity results from the activities of B lymphocyte memory cells (section 16.3.1).
- **Artificial active immunity** forms the basis of immunisation. It involves inducing an immune response in an individual, without them suffering the symptoms of the disease. This is achieved by introducing the appropriate disease antigens into the body, either by injection or by mouth. The process is called **vaccination,** and the material introduced is called **vaccine**. There are different forms of vaccine:
  - **Living attenuated microorganisms** are living microorganisms which have been treated, e.g. by heat, so that they do not cause symptoms, but still multiply. Although harmless, they stimulate the body's immune system. Measles, TB and poliomyelitis can be vaccinated against in this way.
  - **Dead microorganisms** have been killed by some means. Again, they are harmless, but induce immunity. Typhoid, cholera and whooping cough can be controlled by this means.
  - **Genetically engineered microorganisms** can be produced in which the genes for antigen production are transferred from a harmful organism to a

Vaccination in progress. Programmes of vaccination for children have considerably reduced deaths from infectious diseases in developed countries.

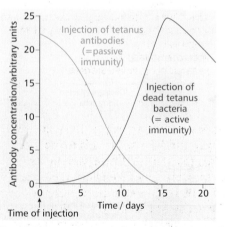

Fig 16.8 *Graph showing the concentration of tetanus antibodies, over time, with active and passive immunity*

harmless one. These are then grown in fermenters and the extracted antigen is separated and purified before injection. Hepatitis B vaccine is of this type.

16.5.4 Immunity in children

As most immunity has to be acquired, either naturally or artificially, it follows that children, and especially newborn babies, are particularly at risk from infection. It takes many years for an individual to build up immunity to a wide variety of diseases. Why then is it that deaths from infections in young humans are not as common as might be expected? There are a number of reasons:

- Even before birth, a fetus has immunity to certain diseases. This is because the placenta allows some antibodies from the mother to pass into the fetus (natural passive immunity). This immunity only lasts for a short period after birth – in the case of measles for around four months.

- The milk formed from the breasts of the mother during the first few days after birth contains antibodies. This early milk, called **colostrum**, has a high concentration of antibodies belonging to the class Immunoglobulin A (IgA). These antibodies both remain in the intestines and are absorbed into the blood providing temporary immunity to a variety of pathogens both inside the intestines and within the body (natural passive immunity).

- In many countries, including the UK, there are programmes of vaccination designed to artificially induce immunity at the most appropriate stage of a child's development. Table 16.2 in the extension, outlines a typical vaccination programme for children in the UK.

EXTENSION – VACCINATION PROGRAMMES

Any programme of vaccination has to strike a balance between providing immunity before the individual is most at risk of contracting a disease and yet avoiding any adverse effects the vaccine may have on normal development. The table below illustrates a typical immunisation programme for children in the UK.

Table 16.2 Typical programme of immunisation in children

Approximate age	Vaccination
2 months	DTP – Diphtheria, Tetanus and Pertussis (whooping cough) – combined vaccine Poliomyelitis Meningitis type C
4 months	DTP combined vaccine – 2nd dose Poliomyelitis – 2nd dose
6 months	DTP combined vaccine – 3rd dose Poliomyelitis – 3rd dose
15–18 months	MMR – Measles, Mumps and Rubella (German measles) – combined vaccine
4–5 years	DTP combined vaccine – 4th dose MMR combined vaccine – 2nd dose Poliomyelitis – booster
11–13 years	Rubella (German measles)
13 years	Tuberculosis (BCG)
15 years	Tetanus and Diphtheria – booster Poliomyelitis – booster

---

Table 16.3 Summary of different types of immunity

	Natural	Artificial
	Inherited or acquired naturally, not deliberately	Acquired deliberately by exposure to causative agent
Passive Results from the introduction of antibodies from another organism's immune system, rather than one's own Short lived	Antibodies pass from mother • to fetus via placenta • to baby during suckling	Antibodies from a different individual or organism are injected
Active Results from the activities of an individual's own immune system Long lasting	Antibodies acquired as a result of a previous infection producing B lymphocyte memory cells, which are reactivated on the second infection	Antigens are injected or given by mouth as a vaccine. They induce the body to produce its own antibodies to the disease. Vaccine may contain • dead pathogen • attenuated pathogen • genetically engineered antigens

---

SUMMARY TEST 16.5

Where immunity is inherited or acquired naturally, it is referred to as (1) immunity. It may be acquired from the introduction of antibodies from someone else rather than an individual's own immune system, in which case it is said to be (2) immunity. Where immunity is given by deliberately exposing the body to antibodies, it is known as artificial immunity. Artificial active immunity forms the basis of immunisation, where antigens are introduced into the body as part of the process known as (3). There are three main forms of microorganism that can be used to produce immunity. These are (4), (5) and (6). Of the various forms of immunity, the full name of each of the following is: via the placenta as a fetus, called (7) immunity; injecting microorganisms weakened by heat treatment, called (8) immunity; and injecting antibodies, such as those from the tetanus bacterium, called (9) immunity.

16.6 Control of disease by vaccination

The development of new vaccines is a highly technological process requiring an extremely high standard of hygiene

The programme of vaccination against various diseases has had considerable success in controlling them. In cases such as smallpox, such programmes have eliminated the disease altogether (section 16.6.3). Yet in other instances similar measures have had less success. To explore why there are these differences we need first to look at what is necessary for effective control of a disease through vaccination and then examine why such measures give different results with different diseases.

16.6.1 Features of a successful vaccination programme

To be effective, a programme of vaccination depends upon:
- **A suitable vaccine** being economically available in sufficient quantity to immunise all of the vulnerable population.
- **Few, if any, side effects** from vaccination. Unpleasant side effects may discourage individuals in the population from being vaccinated.
- **The mechanisms to produce, store and transport the vaccine**. This normally involves technologically advanced equipment, hygienic conditions and refrigerated transport.
- **The means of administering the vaccine properly at the appropriate time**. This involves training staff with appropriate skills at different centres throughout the population.
- **The ability to vaccinate the vast majority (all, if possible) of the vulnerable population**. This is best done at one time so that the transmission of the **pathogen** is interrupted because for a certain period there are no individuals in the population with the disease. This is known as **herd immunity**.

16.6.2 Why vaccination does not eliminate a disease

Even where these criteria for successful vaccination are met, it can still prove extremely difficult to eradicate a disease. The reasons for this include:
- Vaccination fails to induce immunity in certain individuals, e.g. ones with defective immune systems that do not produce the necessary **clones** of B and T lymphocytes (units 16.3 and 16.4).
- Individuals may develop the disease immediately after vaccination, but before their immunity levels are high enough to fight it. These individuals may harbour the pathogen and reinfect others.
- The disease-causing agent (pathogen) may mutate frequently, so that its **antigens** change suddenly (**antigenic shift**) as opposed to gradually (**antigenic drift**). This means that vaccines suddenly become ineffective as the **antibodies** they induce the immune system to produce no longer recognise the new antigens on the pathogen. This happens with the influenza virus, which changes its antigens frequently.
- There may be so many varieties of a particular pathogen that it is all but impossible to develop a vaccine that is effective against them all. The common cold has over 100 types, for example.
- There are, as yet, no effective vaccines against pathogens that are **eukaryotes**. Diseases such as malaria and sleeping sickness cannot therefore be tackled in this way (see extension in unit 15.2).
- Certain pathogens 'hide away' from the body's immune system, either by concealing themselves inside cells, or by living in places out of reach, such as within the intestines, e.g. the cholera pathogen (section 15.1.1).
- Some pathogens suppress the body's immune system, so stimulating it through vaccination is ineffective, e.g. human immunodeficiency virus (section 15.3.1).

- There may be resistance to vaccination for religious, ethical or medical reasons. For example, concerns over the Measles, Mumps and Rubella (MMR) triple vaccine has led a number of parents to opt for separate vaccinations for their children, or to avoid vaccination altogether.

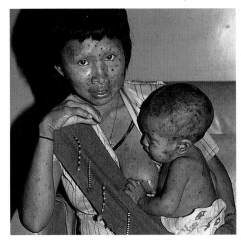

Measles is an infectious disease

16.6.3 Smallpox - the last Jenneration!

The first ever vaccinations were carried out by Edward Jenner in 1794, against smallpox, and this very same disease was the first to be completely eradicated from the world, by the same process, in 1977. Why have we been able to remove smallpox when many other diseases such as measles, tuberculosis, malaria and cholera are still around? There are many reasons, which include:

- There was a simple, safe, easily stored vaccine against smallpox.
- The smallpox vaccine was easily and economically produced and simply administered.
- As a live vaccine, it was especially effective in producing immunity because the pathogen that is introduced into the body can reproduce and therefore persists in the body longer, enabling lasting immunity to build up.
- The smallpox virus was genetically stable, and so it did not mutate or change its antigens and so the vaccine was always effective. This also meant that the same vaccine could be used anywhere across the globe.
- The lethal nature of the disease (up to 30% of the victims died) encouraged people to be involved in the vaccination programme.
- The symptoms of the disease were easily recognised, and so infected patients could be isolated and treated before they spread the disease further.
- The virus did not remain in the body after an infection and so did not form a potential source of future infections.
- No other host organism was involved, making it easier to break the transmission cycle.
- There was a concerted worldwide vaccination programme, coordinated by the World Health Organisation.

16.6.4 The problems of controlling measles by vaccination

In contrast to smallpox, the control of measles by vaccination has proved much more difficult because:

- To eradicate measles effectively it is necessary to ensure almost all of the population (95%) is immune at the same time. This herd immunity interrupts transmission of the pathogen, so preventing susceptible individuals from acquiring it. As the vaccine for measles is only 95% effective, this means virtually everyone in the population must be immune at any one time to eradicate the disease. As this is almost impossible (section 16.6.2), measles is unlikely to be eradicated.
- The measles antigens are complex and it is therefore difficult to make a wholly effective vaccine.
- Some children do not respond effectively to the measles vaccine and therefore need a series of boosters. It is difficult, especially where the population is shifting, to ensure these boosters are administered.
- Parental concern over the side effects of the MMR vaccine has led to a reduced level of vaccination in some countries, including the UK.

16.6.5 The problems of controlling cholera and tuberculosis by vaccination

Control of cholera by means of vaccination is difficult because:

- Cholera is an intestinal disease and therefore not easily reached by the immune system. Any oral treatment, e.g. antibiotics, rarely has time to be effective as it is too rapidly flushed from the intestines by the diarrhoea that is symptomatic of the disease.
- The antigens of the cholera pathogen change rapidly (antigen shifting), making it difficult to develop an effective and lasting vaccine.
- Mobile populations as a result of global trade, tourism and refugees fleeing wars spread cholera and make it difficult to ensure that individuals are vaccinated.

Control of TB by vaccination is difficult because:

- The increase in HIV infections has led to more opportunistic infections of TB in people with impaired immune systems.
- Increasing poverty in many cities has led to overcrowding and more unhealthy conditions that favour the spread of TB.
- Wars and political unrest have created refugees who are often housed in densely populated accommodation.
- Mobile populations as a result of tourism, global trade and refugees have spread the disease worldwide and made it difficult to ensure that individuals are vaccinated.
- The proportion of elderly people in the population is increasing and they often have less effective immune systems and so vaccination is less effective at stimulating immunity.

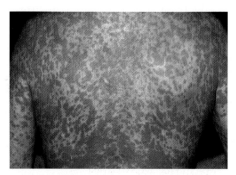

Extensive rash over the back and arms due to an allergic reaction to the antibiotic drug trimethoprim

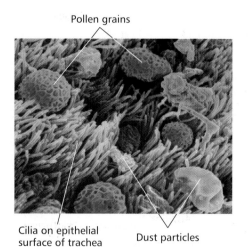

Pollen grains

Cilia on epithelial surface of trachea

Dust particles

SEM of allergens such as dust (pale blue) and pollen (pink) on the surface of the trachea. These allergens may cause asthma and hay-fever

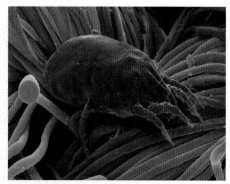

SEM of a dust mite, a common allergen causing asthma

Probably each one of us knows someone who is allergic to something: people who cannot eat peanuts or milk products, cannot wear jewellery with nickel in it, or suffer from eczema, hay-fever or asthma. The list is almost endless. For some people the problem is no more than a minor irritation; for others it makes normal living impossible and can be life-threatening. How is it that so many different things can affect so many people in this way?

16.7.1 Allergens and allergies

We have seen in the previous units how our immune systems protect us from disease by detecting and removing non-self (foreign) matter that gets into our bodies. Sometimes, however, our immune systems respond to a harmless substance in much the same way. This is called an **allergy** and the substance that stimulates the response is called an **allergen**. In most people, these substances cause no response, but in those with an allergy to them, there is an over-reaction of the immune system that can lead to damage to the body tissues. These individuals are said to be **hypersensitive** to the allergen. Just as the immune system builds up a 'memory bank' of previous diseases in order to respond more quickly to subsequent infections, so it does to an allergen. When allergic individuals are first exposed to an allergen, their immune system recognises it as non-self. The B lymphocytes then produce **antibodies** – more specifically antibodies belonging to the group immunoglobulin E (IgE) – with the help of **cytokines** produced by T helper cells (section 16.4.2). These antibodies bind to the surfaces of cells called **mast cell**s – amoeboid-type cells, with a large nucleus, found in all tissues. The individuals are now said to be **sensitised**. When these individuals are further exposed to the same substance, the allergen binds to the IgE antibody on the mast cells, causing them to produce a chemical called **histamine**, as well as other substances, such as cytokines. These cause dilation of blood vessels and the release of lymph from them. The body is, in effect, increasing the supply of white blood cells to fight what it mistakenly perceives to be a harmful infection. The area therefore becomes hot and swollen, in the way that it would if there were a genuine infection. The symptoms may be relieved by administering one of a range of drugs called **antihistamines**. The symptoms of allergies can be divided into two groups:

- a generalised response which affects many body systems and can kill an individual in a short time
- a localised response which affects a more specific region of the body, e.g. asthma and hay-fever affect the lungs.

In a generalised response, allergens, such as penicillin and chemicals in nuts and insect stings, may cause a widespread reaction, including:

- constriction of **smooth muscle** cells in the bronchioles, which restricts the air-flow in and out of the alveoli (section 13.1.3)
- dilation of arterioles leading to a drop in blood pressure

These changes are so sudden and dramatic that they can cause death by asphyxiation and / or lack of adequate blood circulation. Such attacks can be so severe that adrenaline needs to be administered by injection to counteract them.

16.7.2 Asthma and hay-fever

Asthma and hay-fever are two examples of localised allergic reactions. Asthma is now one of the most common diseases in many countries. It affects up to 10% of the world population and accounts for 2000 deaths each year in the UK alone. Some of the most common allergens that stimulate asthma are pollen, animal fur and the faeces of the house dust mite. It can also be triggered or made worse by a range of factors, including air pollutants (e.g. sulphur dioxide, nitrogen oxides and ozone), exercise, cold air, infection, anxiety and stress. One or more of these allergens causes mast cells on the linings of the bronchi and bronchioles to release histamine. This in turn causes:

- the lining of these airways to become inflamed
- the **goblet cells** of their epithelial lining to secrete larger quantities of mucus than normal
- fluid to leave blood capillaries and enter the air-ways
- smooth muscle surrounding the bronchioles to contract and so constrict the air-ways.

Overall, there is a much greater resistance to the flow of air in and out of the alveoli. It is therefore difficult to breathe, and there is a wheezing sound when the sufferer does so. The chest feels tight and there may be coughing. The events of an allergic reaction in an asthma sufferer are summarised in figure 16.9. The condition can be controlled by means of drugs called bronchodilators, which open up the air-ways by relaxing the smooth muscle. To provide immediate relief, these bronchodilators are inhaled using a nebuliser. Research is being carried out to develop a vaccine to prevent asthmatics producing antibodies in response to certain antigens but, as yet, there is no cure. Genetics appears to play a role, as asthma tends to run in families. The number of asthmatics continues to rise, and many explanations have been put forward for this. These include an increase in air pollution, more stress and an increase in the variety of chemicals used in our food and manufactured products. There is also recent evidence that our 'cleaner' lifestyles mean that we are now exposed to fewer allergens as children and therefore become more easily sensitised to them in later life.

Hay-fever, also known as **allergic rhinitis**, gets its name because it is often caused by exposure to pollen, especially that of the grass family. Other allergens, such as dust mite faeces, also lead to the condition. Hay-fever affects the upper respiratory tract, especially the nose. This becomes inflamed, leading to a runny nose and eyes, and a sore throat. While not life-threatening, hay-fever causes misery and severe inconvenience to sufferers, especially during the summer months.

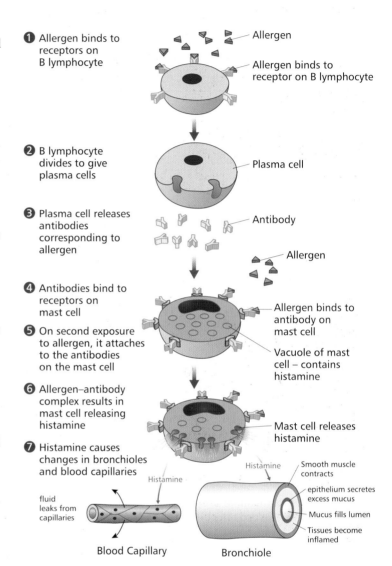

① Allergen binds to receptors on B lymphocyte
- Allergen
- Allergen binds to receptor on B lymphocyte

② B lymphocyte divides to give plasma cells
- Plasma cell

③ Plasma cell releases antibodies corresponding to allergen
- Antibody
- Allergen

④ Antibodies bind to receptors on mast cell
⑤ On second exposure to allergen, it attaches to the antibodies on the mast cell
- Allergen binds to antibody on mast cell
- Vacuole of mast cell – contains histamine

⑥ Allergen–antibody complex results in mast cell releasing histamine
- Mast cell releases histamine

⑦ Histamine causes changes in bronchioles and blood capillaries
- Histamine
- Smooth muscle contracts
- epithelium secretes excess mucus
- Mucus fills lumen
- Tissues become inflamed

fluid leaks from capillaries

Blood Capillary   Bronchiole

Fig 16.9 *Summary of the sequence of events in an asthma allergic reaction*

SUMMARY TEST 16.7

Substances which cause an allergy are given the name **(1)**. When exposed to these substances on the first occasion, susceptible individuals form antibodies from their **(2)** cells. These antibodies are called **(3)** and they surround the surfaces of **(4)** cells found throughout the body. The person is now said to be **(5)**. The next time they are exposed to the same substance, it binds to the antibodies on these cells, causing them to release a chemical called **(6)** and other substances such as **(7)**. Certain chemicals, such as the antibiotic **(8)**, can cause a generalised response that may be lethal. More often substances like pollen cause inflammation of the mucous membranes of the nose. This ailment is called **(9)**. Another such localised response is asthma, in which **(10)** cells of the epithelial linings of the airways produce excess mucus. **(11)** leaves the blood capillaries and enters the air-ways, and the smooth muscle in the walls of the **(12)** contracts, causing them to **(13)**. This makes breathing difficult – a condition that can be relieved by the use of drugs, known generally as **(14)**.

1   Many people suffer from allergies that are unnecessary and exaggerated responses by their immune systems to allergens.

Allergens are harmless and do not cause disease.

The figure shows the stages in an allergic response to an allergen. (Not drawn to scale.)

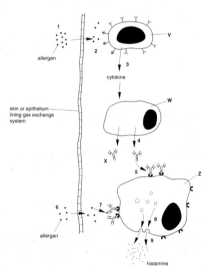

a   Name **two** allergens.   *(2 marks)*

b   With reference to the figure, name

    (i)   the cells labelled **V, W** and **Z**;   *(3 marks)*

    (ii)   the molecule **X**.   *(1 mark)*

c   Explain how molecule **X** attaches to cell **Z**.   *(2 marks)*

When histamine is released during an asthmatic attack, it has a very severe effect on bronchioles.

d   Describe the effect of histamine on bronchioles.

                    (3 marks)

                  (Total 11 marks)

       OCR 2802 May 2002, B (HHD), No.3

2   Tetanus is a disease caused by a bacterium. When the tetanus bacteria enter the body they release a toxin which causes muscular rigidity and extreme pain. Children in the UK are routinely vaccinated against tetanus at an early age.

Figure 1 (opposite) is a diagram that shows three B lymphocytes (**P, Q** and **R**) and the events that occur during an immune response to the tetanus toxin.

a   (i)   Explain what is happening at stages **X** and **Y** in the immune response to tetanus toxin.   *(3 marks)*

     (ii)   Name the cells labelled **S**.   *(1 mark)*

A study investigated active and passive immunity to tetanus toxin. One person, **G**, was injected with the antibodies to tetanus toxin. Another person, **H**, was injected with the vaccine for tetanus and produced antibodies as a result. Blood samples were taken from both people at regular intervals over the following few weeks and analysed for antibodies against tetanus.

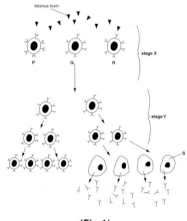

(Fig 1)

The results of the study are shown in figure 2.

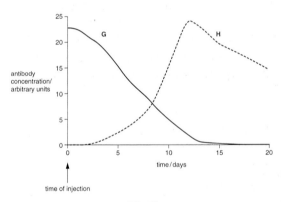

(Fig 2)

b   (i)   Explain why the type of immunity gained by **G** is described as passive immunity.   *(2 marks)*

     (ii)   Describe the advantages of receiving passive immunity to diseases, such as tetanus, compared with active immunity.   *(2 marks)*

c   Explain why there is a slow increase in antibody concentration in the curve for **H**. You may find it helpful to refer to figures 1 and 2 before you write your answer.   *(4 marks)*

d   Explain why person **H** may not need further vaccinations against tetanus.   *(3 marks)*

                  (Total 15 marks)

       OCR 2802 Jan 2002, B (HHD), No.7

3   The following image is a scanning electron micrograph of a phagocytic white blood cell engulfing a cell of the pathogenic yeast, *Candida abicans*.

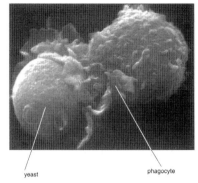

yeast   phagocyte

a State the site of origin of phagocytes in the body.
(1 mark)

b Describe what happens inside the phagocyte after pathogens, such as *Candida albicans* have been engulfed.
(4 marks)

c The following figure shows the structure of an antibody molecule.

Explain how the structure of an antibody is related to its function.

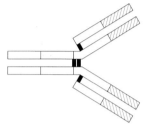

(2 marks)

d There are four different types of immunity.
   • natural active   • natural passive
   • artificial active   • artificial passive

Complete the table below by indicating the type of immunity that is gained in each example given.

example	type of immunity
receiving an injection of a serum containing antibodies, e.g. against tetanus	*passive artificial*
taking an oral vaccine for polio	*active artificial*
catching and recovering from a disease, such as measles	*active natural*
receiving an injection of a weakened strain of a disease-causing bacterium	*active artificial*
babies feeding on breast milk	*passive natural*

(5 marks)
(Total 12 marks)

OCR 2802 Jun 2003, B (HHD), No.3

Question 3 Electron micrograph reproduced by permission of Science Photo Library

4 B lymphocytes are found throughout the body in the blood and in the lymphatic system. When an antigen enters the body, some of these B lymphocytes respond and change into plasma cells that secrete antibodies.

a State the part of the body where B lymphocytes originate.
(1 mark)

b Explain the difference between an antigen and an antibody.
(4 marks)

The figure shows a plasma cell and a B lymphocyte.

plasma cell   7 μm   B lymphocyte   4 μm

c With reference to the figure,
   (i) state **three** ways in which the structure of the plasma cell differs from the B lymphocyte;
   (3 marks)
   (ii) explain the reasons for the differences you have described.
   (3 marks)

d Explain why the production of antibodies is much faster when the same antigen enters the body on a second occasion.
(4 marks)
(Total 15 marks)

OCR 2802 Jan 2001, B (HHD), No.2

5 People may gain active and passive immunity to disease, such as measles.

a Describe **two** different ways, one active and one passive, in which babies may become immune to measles.
(4 marks)

b State **two** reasons why measles has not been eradicated by vaccination.
(2 marks)

Unlike measles, smallpox has been eradicated. This was achieved by a worldwide vaccination programme.

c Explain how it was possible to use vaccination to eradicate smallpox throughout the world.
(In this question 1 mark is awarded for the quality of written communication.)
(7 marks)
(Total 13 marks)

OCR 2802 Jan 2001, B (HHD), No.5

6 a Complete the table below to show **two** differences between active and passive immunity.

	active immunity	passive immunity
1		
2		

(1 marks)

b Complete the table below by describing how each type of immunity is acquired.

type of immunity	how acquired
natural active	*infection from pathogen*
artificial active	*e.g. vaccinations*
natural passive	*e.g. colostrum, placenta*
artificial passive	*e.g. serum injections*

(4 marks)

c Explain, in terms of the structure of antibody molecules, how the immune system is able to produce large numbers of different types of antibodies.
(In this question, 1 mark is available for the quality of written communication.)
(7 marks)
(Total 15 marks)

OCR 2802, 2000 Specimen paper, B(HHD), No.2

17

17.1

Study and examination skills for AS Biology

Moving on from GCSE

Whether it is a football team moving from the first division into the premiership, an employee gaining promotion or a student changing from GCSE to AS Biology, progressing to a higher level can be both demanding and daunting. The introduction of Advanced Subsidiary (AS) courses in September 2000 was, in part, designed to make the transition from GCSE to Advanced level less traumatic. In effect, it was two small steps, rather than one big one.

17.1.1 The Advanced Subsidiary (AS) course

The Advanced Subsidiary (AS) will help you make the necessary adjustments to A level from GCSE because it is assessed at a standard appropriate for candidates who have completed the first year of study of a 2 year A level course, i.e. between GCSE and A level. It may be taken as a 'stand alone' qualification or, when combined with the second half of an A level course (known as A2), the AS forms 50% of the assessment of the total A level. There are three modules at AS and a further three at A2. The full A level therefore comprises six modules. The normal pattern envisaged is for candidates to complete AS at the end of their first year of study and A2 at the end of their second year. Alternatively AS and A2 may be taken together at the end of the second year. Whether you have studied Biology as a separate subject or part of Dual Award Science at GCSE, you will find many

Table 17.1 *Content of OCR Advanced Subsidiary (AS) Biology specification and the mode of assessment*

Module number	Module title	Material within module	Mode of assessment	Length of examination	Proportion of total AS	Proportion of total A level
2801	Biology foundation	Cell structure Biological molecules Enzymes Cell membranes and transport Genetic control of protein structure and function Nuclear division Energy and ecosystems	Written examination	60 minutes	30%	15%
2802	Human health and disease	Introduction to health and disease Diet Gaseous exchange and exercise Smoking and disease Infectious diseases Immunity	Written examination	60 minutes	30%	15%
2803	Transport	The mammalian transport system The mammalian heart Transport in multicellular plants	Written examination	45 minutes	20%	10%
	Experimental skills	EITHER Coursework **or** Practical examination	Teacher assessment Practical examination	 1.5 hours	20% 20%	10% 10%

of the topics studied in AS Biology familiar. It is not the content so much as the depth and detail that you will find different. You will study topics to a depth that enables you to have a much greater understanding of the material. Table 17.1 lays out the main topic headings for each of the three modules on the OCR specification, and gives details on how the material will be assessed.

17.1.2 The different study skills needed for AS

For the reasons given in section 17.1.1, it is unusual for students to find the differences in content at AS a problem. The same cannot always be said for the study methods and the skills required, however. You may find it takes you a little while to adjust to the different style at AS. The course is modular, with a choice of whether to take some modules in the winter (January) and others in the summer (May / June), or to take all the modules in the summer. However, if you are studying at a school or college, you may have little option but to follow the institution's policy in this matter. If you are taking examinations in each module as you complete it, then it is especially important that you come to grips with the new techniques involved as quickly as possible, because your first examination will be upon you quicker than you might imagine. Always ask for help if you are having trouble – you will probably not be the only one in your group to need additional guidance. At AS you will be studying fewer subjects than at GCSE, probably up to five subjects. This means that you will be spending at least twice as long studying each subject at school or college. What many students fail to recognise is that the same is true of the time spent outside the classroom. This, too, must increase for each subject, 4 hours per week being a basic minimum for each AS. Some of the differences to which you will need to accustom yourself are as follows.

- There is a larger body of factual information that you will need to learn. This must be done as you proceed; there is simply too much to be left to learn at the last minute.
- Your understanding must be thorough because, compared with GCSE, greater emphasis is placed on the ability to apply general biological knowledge and principles to novel situations, rather than simply being able to recount a previously learned set of facts.
- The ability to interpret data in a variety of forms is required in order that examiners can test your understanding and not just your rote learning.
- More open-ended questions will appear and you may be required to argue a case for and / or against a particular view rather than give specific answers.

- At AS, the ability to analyse experiments, data and information critically, and to evaluate the accuracy of results and theories, is an extension of the basic training given in these skills at GCSE.
- In order to develop the necessary attributes, you will need to read widely on all aspects of biology (unit 17.5). It is this, above all, which is essential if you wish to obtain a higher grade at AS, where a greater range of sources of information should be used. Scientific journals, for example, are invaluable as a means of keeping up to date in a subject that is developing as rapidly as biology. Do not forget that important information can also be gleaned from newspapers and television programmes, as long as you take a questioning and critical approach to them. Take notes (unit 17.4) as you read or listen – you never know when they might come in useful.
- Practical skills that you developed at GCSE will be an important starting point for the more refined and extensive skills required at AS (unit 17.2), and you may be required to carry out one or more investigations as part of your practical assessment (unit 17.3).
- There is an increasing trend towards the use of mathematical methods, especially statistics, in biology; these demand greater mathematical skill than required at GCSE.

Perhaps the greatest difference, however, is the need for you to take greater responsibility for your own work – to plan and organise for yourself. This requires much self-discipline. Studying for AS Biology can be rewarding and stimulating, but it is always demanding and time-consuming, requiring a sustained effort throughout the course and not just in the final few weeks before an examination.

Practical skills acquired at GCSE will become more extensive and refined at AS

Biology is a practical subject and so you will undertake a number of practical activities during your course. At first you may carry out experiments following directions from your teacher or a worksheet but gradually you will get used to deciding on a problem that you wish to solve and designing your own experiments to do this. Other skills that you will acquire involve using a microscope to look at thin sections of plant or animal organs, as well as learning to make temporary preparations of cells and tissues yourself, and then examining them under the microscope. Whatever form your practical work takes, it is important that you should be able to communicate the results clearly and unambiguously to other people. If you have been using a microscope to discover the structure of, for example, a plant stem, you will need to be able to make clear drawings to illustrate what you have seen. Once your practice sessions are complete you will have to prove to an examiner that you have acquired these practical skills. This assessment may take one of two forms at AS:

- The practical examination tests your ability to carry out instructions and to evaluate the procedure used, together with an exercise using the microscope. In addition you will receive guidelines for an investigation and you will have to submit a plan of the method to be used – this will be done during the days prior to the examination.
- One or more pieces of coursework will give evidence of your ability in all four skill areas to be assessed. These will be marked by your teacher and then samples will be sent to OCR for moderation.

Whether you decide to undertake coursework or do the practical examination, the same standard will be expected from all students.

If you are entered for the examination, rather than the coursework option, you will:

- have to carry out a planning exercise in a limited amount of time prior to the examination

- not be able to choose the topic yourself, but will be informed of it by the examination board

- be told in outline what you need to do (but the advice given in the planning section should be used as further guidance)

- be asked to carry out an experiment during the examination period

- be tested on your ability to follow instructions and work safely – in much the same way as when doing the 'implementing' part of the coursework option

- have to tabulate and prepare tables and graphs – like the 'analysis' section

- be set a series of questions that will enable you to analyse and evaluate the experiment – the guidelines already given will help you to prepare.

If you take the examination option you will also have to demonstrate your ability to use a microscope and to draw biological specimens viewed through it. You may be provided with prepared microscope slides, but you may be asked to make temporary slides of specimens for yourself.

17.2.1 Using the microscope

You will be given instructions from your teacher on how to use the particular microscope that you are provided with. The one illustrated in figure 17.1 is fairly standard. Whatever model of light microscope you use, however, there are a number of things you can do to ensure that you obtain a clear image of what you are looking at. Make sure that:

- the lenses are clean – wipe with lens tissue as necessary
- the objective lens is in place – i.e. it is clicked into position
- the diaphragm is open
- the stage is properly illuminated – if there is a mirror, adjust it so that the plane side reflects the maximum light from below the specimen
- the condenser is correctly adjusted
- the specimen is in the centre of the stage
- low-power focusing is carried out before turning to a higher power objective.

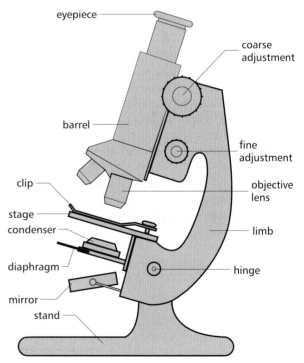

Fig 17.1 *The light microscope*

17.2.2 Staining and preparing temporary slides

A wide range of stains is used to colour different tissues and organelles in a variety of ways. It is therefore important to realise that, in most prepared slides, the colours observed are not the natural colours of the material, but the result of staining. To make a temporary slide the procedures below should be followed.

- Take a clean, dry microscope slide and coverslip.
- Put the material to be observed in the centre of the slide.
- Place a drop of water, or other mounting fluid, over the specimen.
- Place the coverslip at one edge of the slide and at an angle of 45°.
- Pull the coverslip along the slide, using the forefinger and thumb, until it reaches the drop of liquid containing the material.
- Place a mounted needle under the upper edge of the coverslip to support it (Fig 17.2).
- Slowly lower the coverslip, using the needle, until it is lying flat on the slide. Only by lowering it slowly and carefully will air have time to escape and bubbles be avoided.
- Draw off any excess liquid from around the coverslip.

A well-prepared temporary mount should have:

- the material central on the slide and more or less central under the coverslip
- the coverslip (if square) with its sides parallel to the sides of the slide
- no air bubbles under the coverslip
- no excess liquid on the slide or coverslip.

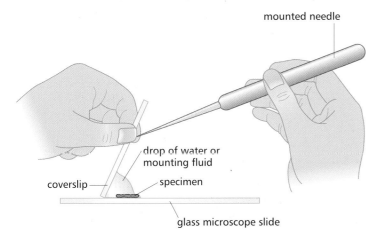

Fig 17.2 Preparing a temporary microscope slide

17.2.3 Drawing specimens

Specimens to be drawn for assessment may be:

- living or dead
- an entire organism or part of it
- microscopic or visible with the naked eye
- real or photographed.

Whatever its form, the specimen can be accurately drawn, whether or not you are a good artist, by following these guidelines.

- Use a good-quality, sharp pencil that is neither so soft that it smudges nor so hard that it leaves an imprint on the paper that you cannot remove; an HB is usually best.
- Draw on good-quality plain paper.
- Choose a suitable scale so that the drawing, labels and annotations will fit comfortably on the page.
- Make large, clear line drawings without the use of ink or coloured pencils.
- Keep the drawing simple by providing only an outline of all the basic structures.
- Draw accurately and faithfully what can be seen. Never draw anything you cannot see, even if it is expected to be present. Never copy from books.
- Draw individual parts of a specimen in strict proportion to each other.
- Provide suitable headings that clearly indicate the nature of the drawing. For microscope drawings, the section (TS / LS) should be stated.
- State the magnification, scale or actual size of the specimen.
- Label fully all biological features, keeping labels away from the diagram, and never label on the actual drawing.
- Avoid crossing label lines and, if possible, arrange labels vertically one beneath the other.
- Use annotations to indicate the functions or biological significance of the labelled parts. Annotations are short explanatory notes written below or alongside the label.

In a practical examination you may not be expected to include all the details given above. They are merely a guide and you should remember to read carefully exactly what the examiners are asking you to do. An example of how to draw a low-power plan of a section through a dicotyledonous leaf is illustrated in section 1.9.4.

The four experimental and investigative skill areas to be assessed are designed to follow on from those you used for Sc1 at GCSE and so you should find them quite familiar. However, this is now AS and not GCSE and so the examiners will be looking for progression. They will expect you to work on practical tasks that are more demanding, both in terms of the practical skills needed to manipulate apparatus and take precise readings and also in terms of the scientific explanations needed to justify your prediction (hypothesis) and to explain the results you obtain. You will no doubt be guided towards suitable investigations by your teacher, as well as by the apparatus available in your school or college. However, if there is a choice, it is always better to work on an area of the specification that interests you.

You may achieve all four skill areas within a single investigation, or carry out a number of individual tasks, each designed to achieve one or more of the skill areas.

As you work through the skills, it is vital to refer constantly to a copy of the criteria against which your work will be marked. Descriptions are provided for marks of 1, 3, 5 and 7. In order to achieve a mark you must fulfil all the criteria for that mark **AND** the ones that precede it. Intermediate marks can be awarded if, for example, you achieve all that is required for 3 marks and half of what is need for a mark of 5; this would be scored as 4 marks.

17.3.1 Planning

This is a vital part of any piece of coursework because it forms the basis of the rest of your investigation. Once you have decided on, or been given an area of study, the following steps are advisable.

- Define clearly the nature of the problem you are going to investigate.
- What do you think will happen? Write a prediction, also known as a hypothesis.
- Explain clearly, in scientific terms appropriate to this level, why you have made this prediction. This will need to be detailed in order to achieve the highest marks so do use a variety of textbooks and the internet as well as any class notes that you might have. Always write clearly and concisely and use correct scientific terminology.
- Consider different methods to investigate this hypothesis.
- Choose one method and justify your choice – include reference to precise volumes, concentrations and times etc.
- Carry out a preliminary experiment to test your hypothesis and to find out whether the method you propose is going to produce sufficient, quantitative results to enable you to accept or reject your hypothesis. It may well be necessary to adjust precise details of your original method.

- Record the results of your preliminary experiment. Carry out further preliminary investigations until you are satisfied with your method.
- Write a full and detailed method for your investigation avoiding the use of personal pronouns, e.g. 'I' and 'we'. This should include:
  - The procedure followed, written in continuous prose and with every step in precisely the correct order. The general rule is that, if your instructions are followed by another person, they would end up doing an experiment that was identical to yours.
  - You should justify the apparatus and materials used. This is often best done by reference to your preliminary experiment.
  - All precise volumes, concentrations and times must be given; never refer to 'about'. You should use SI units throughout.
  - In your experiment you should investigate the effect of a single variable and so you must explain how and why all other factors were kept constant, e.g. the temperature was maintained at 40°C throughout the experiment because this is the optimum temperature for the enzyme; if the temperature had risen too high it may have denatured the enzyme and reduced the rate of reaction. In other words, do not just list the factors to be kept constant and do not just make some vague reference to a 'fair test'.
  - Make it clear how many readings you are intending to take in order to achieve reliable and valid results. A single set of readings is not sufficient – aim to do two repeats.
  - Write a clear risk assessment, including any aspect that may be considered unsafe, however slight. Use Hazcards to help you with the risks associated with any chemicals used and do not forget to make some reference to ethical considerations in relation to living organisms or the environment.

17.3.2 Implementing

If you have thought about your method carefully and have everything well organised in advance then this should be straightforward. Make sure that all the readings you take are done as carefully and as precisely as possible. To get high marks in this section you will have to be using some apparatus that requires a certain level of skill and that will provide you with precise readings – a ruler measuring the length of potato chips is unlikely to be considered sufficiently technically demanding to give you access to the highest marks. If you want high marks, make sure that you have thought about this before starting your experiment. During this section your teacher is likely to be watching you to make sure that you are really carrying out your experiment carefully and that you do not just 'make up' a set of results that looks as though you have been careful. You will also be watched to see if you are working safely – so do not forget your lab coat / apron and do not work with your goggles on top of your head! After collecting all your results, you then have to present them clearly. You will be expected to produce a clear table of all your results, not just the summary ones. This means spending some time designing a table that can clearly display a large data set. Do not forget that all units (e.g. mm^3) must be recorded, but they must appear in the headings to rows and columns and not in every space in the table. All results in one set should be presented to the same number of decimal places. Do not round up and down; use one or two decimal places as appropriate to show that you have taken accurate readings. Do not forget that temperatures should be given precisely too. Just because a water bath was set at 30 °C it does not mean that it actually stayed at that temperature – read the thermometer and record the exact temperature.

17.3.3 Analysing evidence and drawing conclusions

This section is testing a number of different skills. You should first organise all your results into one clear summary table, remembering the rules about units – headings of rows and columns only, and use SI units throughout. You will probably need to do a little bit of maths at this point, even if only to calculate means or rates of reaction. But think carefully about the best way to present your results, bearing in mind that they should be providing evidence for the hypothesis you were testing – do not lose sight of this in the mass of data you may have collected. You then need to decide on the most appropriate graphical representation of your results so that the trends can be analysed. Clearly indicate your independent and dependent variables. At least one of your graphs should be hand-drawn. You may find it useful to mark on your table and / or graph any results that are anomalous, i.e. which do not appear to fit the trend, because you will need to make precise reference to them later.

You will then need to carry out the written part of the analysis. It is often useful to start with a very basic description of what the results show, before dealing with the details. Although you will need to explain the results using scientific knowledge gleaned from a number of sources, nevertheless it is **your** results that you are analysing and you must make clear reference to them throughout. It is generally advisable to 'read' your graph from left to right, describing what it shows and accounting for every major change of slope by reference to the relevant science you know at this level. Similarly, if you have plotted a number of curves on the same axes you should be able to account for any similarities and differences between them. Remember to keep referring to your hypothesis and every reference to your results must be precise, e.g. use coordinates to refer to precise positions and calculate gradients to compare rates of reaction.

17.3.4 Evaluating evidence and procedures

You probably found this a little difficult at GCSE level and it continues to be an area that students find taxing at AS. Look carefully at the criteria and remember that detail is needed for the highest marks. You will need to look at the method you used and consider its suitability for the investigation. Comment on any aspects that could have been improved and what you might have done instead. However, you should not be writing about your own shortcomings. Do not, for example, suggest that you did not read the time accurately or that you forgot to add a particular reagent at the right time. Choose aspects that you could not reasonably have been expected to control. Remember also to look at your results critically. Do they show a trend that supports your hypothesis? Are there any results that do not seem to fit the trend? If so, point out exactly which ones they are and try to give reasons for them not fitting the trend. Even if there are no anomalous results – **state** that this is the case. This is very important and may help you to decide how reliable your results are. Look at the range of results that you processed to produce a mean. Were they very similar to each other, or did they vary widely – another indication of how reliable your results are. Give a numerical value on your sources of error. There is no doubt that the reliability of most results may be improved by increasing the number of samples taken but are there other things you could have done? Could you carry out the same experiment, but using a different method, to see if the results achieved are similar. Suggest, and give details, of another such experiment. When put together, all these aspects should enable you to make a **critical** assessment of how valid acceptance or rejection of your original hypothesis is – do not be too ready to simply accept or reject; **give reasons**.

While for many students notes will be taken largely from lessons, lectures and textbooks, it is important not to under-estimate the valuable contribution that can be made by notes from periodicals, television, films and radio. Whatever the source, there are a number of ways in which note-taking can be made effective.

17.4.1 Distinguishing the important points

In the early stages of your course, check with your tutor that your note-writing is up to standard

Most of what you hear, see or read as part of your AS Biology course will have some relevance. The problem is that there will rarely be time to write it all down. The important thing is for you to understand the main idea being put across and to note this down along with any supporting evidence. Try to appreciate the main theme of the lesson, article or programme; distinguishing this is easier said than done. The title may give you some clue, although the increasing tendency for writers and programme makers to compose a clever play on words sometimes obscures the nature of what follows. It is advisable to check with your tutor in the early stages of your course to ensure that you are on the right lines. Ask yourself whether the information seems to support what you think is the main message – if not, you may be on the wrong lines so try an alternative source that fits the facts better.

Having decided what the theme is, keep relating the rest of the information to it. Learning a fresh topic can be like visiting a town for the first time – you may quickly become lost in the seemingly endless detail of the side streets unless you keep in sight some central, recognisable point that allows you to regain your bearings and make sense of the pattern of the surroundings.

Having recorded the main points, you will be able to fill in the detail later. To do this, make sure you read over your notes as soon as possible after writing them, preferably the same day. In doing so add some of the detail you were not able to write down at the time. Not only will you be re-visiting your notes to help you remember the information, it will allow you to seek clarification from your tutor or another source should you not understand everything you have written.

17.4.2 Writing your notes

While not everyone is capable of speed-writing, almost all of us can develop techniques that permit us to get down a considerable amount of information in a short time. These include:

- Writing in note form as grammatical prose is too time-consuming, e.g. 'insulin from pancreas (islets of Langerhans β cells) glucose $\rightarrow$ glycogen – stored in liver'.
- Using standard abbreviations of biological words, e.g. DNA, ATP, GM.
- Using abbreviated forms of words, e.g. temp., sec., 2° (secondary).
- Developing your own shorthand, especially for commonly occurring biological words, e.g. P/S (photosynthesis), R/P (respiration). These must be restricted to your notes and not used in examinations.
- Avoiding the use of rulers, coloured pencils etc. unless essential for diagrams. Do not waste time underlining with a ruler or giving headings a different colour, it is more important to get down as much information as possible.
- Avoiding the use of correction fluid – if you make a mistake, cross it out. Do not waste precious time using correction fluid and waiting for it to dry.

17.4.3 Organising your notes

Little is more daunting than page after page of unbroken prose. Imagine this book, or even this section, written as one long essay with no subsections or headings. Remember that, when you come to revise, you will probably not feel much like the task anyway. If you are then faced with a seemingly endless collection of continuous notes, you may well give up after a few minutes or even be put off starting altogether.

If possible try to devise a scheme of heading values, where there are three or four levels of subsections. An example is given below:

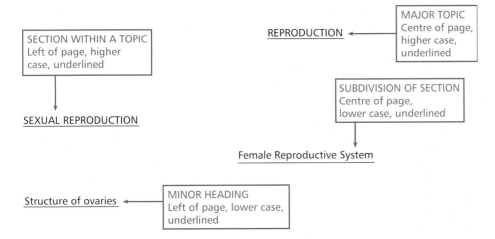

While this arrangement might seem unnecessary and time-consuming, it will prove valuable when you want to access information quickly at a later date and will help you to divide up your revision into manageable portions.

Much of the hard work expended in writing and sub-dividing your notes will be wasted if they are not safely and logically stored. Hardback files with labelled dividers make access easier and save time spent thumbing through a heap of jumbled papers for information on a particular topic.

An index at the start of each file is often useful, although page numbering is not essential because the addition of new material quickly makes this obsolete and necessitates re-numbering. Answers to questions set on a topic can be added at the end of the relevant section of notes. Do not be afraid to re-organise your notes from time to time; it is not always desirable to keep them in chronological order. The order in which topics are taught may depend on many factors, such as the seasonal availability of materials, the availability of apparatus and the preferences of your tutor. Be prepared to arrange your notes in the order that suits you best.

17.4.4 Adding to your notes

Notes should not be static, but should be constantly updated and adapted to suit your needs and those of the specification. As you pick up new information from books, television documentaries, science periodicals and newspapers, add these to the relevant sections of your notes. Loose-leaf files are more or less essential for this purpose. It also helps to leave a number of lines at the end of each subsection so that relevant information can be added at a later date. Always keep notes accessible and add new information immediately it comes to hand. It is good practice to keep the relevant notes at your side when reading a book or article on a particular topic. Adding material over the length of your course helps you to keep up to date. Remember that some notes will be nearly a year old by the time you sit the AS examination, and 2 years old if you continue to study the subject at A2 level. In a subject that changes as rapidly as biology, an awareness of recent developments is a distinct advantage.

Making relevant and concise notes is an important skill

Reading as widely and as variedly as possible on biological topics is one way to increase your chances of a higher grade at AS.

17.5.1 Newspapers and periodicals

Magazines and newspapers often have articles of current biological interest, especially on human related issues, such as health, disease, pollution, GM foods, diet and genetics. While these articles do not always delve into detail or necessarily provide a balanced view, they can help to keep you abreast of recent developments or stimulate an interest in new issues. Read them critically and use them as a source of ideas, arguments and up-to-date information.

Periodicals relevant to AS Biology include *Scientific American*, *New Scientist* and *Nature*, but perhaps the most relevant is *Biological Sciences Review*, a topical and interesting magazine written specifically for students of AS and A level biological subjects. All these contain a wealth of information on the latest discoveries and ideas. Many of the articles go beyond AS and may be difficult to understand. It will not matter if the fine detail is incomprehensible, providing the general principles are clear. Reading these periodicals is the best way of keeping up to date, and some notes, or at least a reference, should be added to the relevant section of your lesson notes.

17.5.2 Books

Books remain the major source of biological information and are used for a number of purposes:

- as an aid to learning, helping you to recall factual information
- for reference as a source of information for the writing of essays or for answering problems
- as a means of improving understanding and providing a greater depth of knowledge
- as a means of generating new ideas
- for enjoyment.

The traditional large textbook is not intended to be read from beginning to end in sequence. Rather it should be used as a reference book, where specific pages or chapters are read in conjunction with the current topic being taught or the current week's homework questions.

It is difficult to recommend to someone a biology book as so much depends on an individual's taste, not to mention the specification being studied. The best answer is to read, initially at least, any book on a topic in which you have a major interest. It may be genetics or pollution or something more specific, like DNA or immunity. You may have a passion for learning how animals adapt to life in deserts, the territorial behaviour of certain fish or migration in birds. The nature of the topic hardly matters, provided you are interested. After all, reading is not intended as a form of punishment, but something to be enjoyed. The enforced reading of particular books is unlikely to engender a love of biology; it is far more likely to have you applying for a change of course. In any case you are unlikely to complete a book you do not enjoy and so you will have gained little, if anything, from the exercise.

Do not feel that all reading has to be directly relevant to your specification. Any reading has some value. It will improve your English through the constant exposure to sentence construction and grammar. It will improve both your biological and non-biological vocabulary as you meet new words (provided you use an appropriate dictionary to discover their meaning). In this way reading will help you to become a

Reading relevant periodicals will increase your understanding of biology

more efficient communicator and to put across your ideas more coherently. As you meet different styles and approaches you will be able to adapt your own writing to include those you find effective. At the same time, you will doubtlessly learn new biological information as well as reinforcing what you already know. You may also be stimulated to do further reading, possibly on topics you had not previously considered. Make brief notes during your reading and add these to your other notes. Jot down page references as you go along; you may need to refer back to these when writing some future essay. It is important to distinguish between reading notes during revision and general reading. The former requires careful attention to each word because, if the notes have been taken properly, they will contain little or no superfluous information. In addition, as you are revising, it may be helpful to read parts a number of times to commit the information to memory. General reading must be much more rapid. Too many students read out loud. That is, they effectively say each word, if not audibly, at least to themselves. You should try to perfect a technique whereby the eyes skim rapidly across the lines, much faster than if the words were spoken. The brain is well capable of taking in the subject content at a speed faster than that necessary to speak the words. This speed-reading is essential if enough books are to be completed to have a significant influence on your performance. Reading speed can normally be increased without any loss of comprehension. Try it and attempt to perfect it through practice.

17.5.3 The internet

The internet is something of a mixed blessing. On the one hand it is a rich source of information on pretty well every topic under the sun. On the other hand the quantity of information can be overwhelming and its accuracy can be doubtful in certain cases. If you have access to the worldwide web, do not hesitate to use it in order to obtain biological information, but be aware that anyone can set up a website where they post information of their choice. There is no guarantee that it is accurate and, very often, it is the views of an individual or group, rather than a scientifically accepted fact. Increasingly, websites on biological issues, such as health and disease, are sponsored by commercial concerns, such as pharmaceutical companies. While the information provided is normally accurate, it is not always balanced, but tends to be biased in favour of the product being promoted. Nevertheless, such information can be useful and provides good practice at selecting relevant information. If possible, visit websites on both sides of a divide, e.g. for and against genetically modified crops. This will help you develop the ability to debate both sides of an argument. Some websites, such as the World Health Organisation site, are a useful source of facts and statistics. They have scientific credibility but are large and not easy to negotiate. Publishers usually provide study support material on the web for each of their biology texts. This can be a most useful source of tests and exercises to help you improve your knowledge and performance.

17.5.4 Teachers and tutors

None of us are so perfect that we are incapable of improvement. Constructive criticism and comment from others more experienced in a particular field is one sure way of upgrading performance. Books can provide information and guidance and help understanding but they cannot alone test how well the information is interpreted and communicated by an individual student. This is where your teachers and / or tutors come in.

Not everyone studies AS Biology full time in a school or college; an increasing number of candidates do so part time and without regular formal lessons. Most, however, still have access to a tutor as part of a distance-learning or supported self-study course. Whichever category you are in you should seek your tutor's advice and guidance throughout the course. Questions set during the course will normally be marked. When these are discussed it is vital to analyse your own shortcomings. Isolate where you have weaknesses. Is it that you do not read the question carefully? Having read the question accurately did you misinterpret what was required? Was the subject content inaccurate or inadequate? Was the detail lacking? Was there an absence of supporting evidence? Did you include too few examples? Did you fail to appreciate the underlying principles or did you get bogged down with unnecessary detail? Did you use the right references? Having determined where you went wrong, think of methods by which you can avoid these problems in future work. If necessary seek your teacher's advice.

Where individual comments are made on your work, take heed of them and be determined not to make the same errors in the future. A number of students ignore criticisms and simply repeat their failings time and time again. Apart from exasperating their teacher, this approach is hardly likely to effect an improvement. Learn to see all criticisms as positive – a means of isolating errors and providing an opportunity for putting them right. After all, if your work is beyond criticism you may as well take the AS without further delay.

Ask for clarification of any comments if you are not sure to what they refer. Seek guidance on the best way to improve. Do not expect sudden or dramatic changes, as this is unlikely. Aim to make slow, steady progress in partnership with your teacher.

The internet can be a useful source of biological information if used selectively

Let us be clear from the outset – there is no universal method of revising that can guarantee success, or that suits everyone. We are all individuals, with all the various differences that our genes and our environment have bestowed on us; a method of study that suits one person may be totally ineffective for another. We do not seek to tell you how you should study, nor to suggest that one method is superior to another, but merely to provide some ideas that you can experiment with. It is up to you to select those methods that suit you best and that bring you the most success. The best advice we can give is to try as many variations as possible, and then to select those that you are most comfortable doing and also that bring the best results. It follows that you need to try out these methods before the final examinations. As these may occur within a few months of starting your course, it is important to use tests set by your teachers as an opportunity to evaluate different methods of revision. You should also set yourself mini examinations, by trying questions such as those at the end of each chapter in this book.

17.6.1 When to revise

Success at AS is achieved by the gradual accumulation of knowledge and understanding throughout the course. The volume of knowledge is too great for it all to be absorbed during a few weeks' revision prior to the examinations. Try to read over notes every day and, if possible, read through the whole weeks' work during the weekend. Prepare for interim tests and examinations thoroughly – only by doing this can you effectively test the efficiency of the revision methods you are using. Students who are disappointed by their final grade often admit to not having revised adequately for practice tests and examinations. How then could they have expected to know how much revision was needed for their actual examinations or modular tests? Always revise thoroughly for all examinations and analyse your results. If they were poor, change your revision methods.

17.6.2 How to revise

Below are 10 guidelines that can help you revise more effectively.

- Choose a place to work where distractions are unlikely. If possible sit at a desk or table, alone in a quiet room. Some students prefer to work with background noise, such as music. This, however, may be distracting. For one thing there is the almost irresistible urge to sing along with the music – something that requires a degree of concentration, meaning that less attention can be paid to revising. In addition, the periodic need to change the radio station, cassette or CD will interrupt your concentration.
- Do not work for too long at a single session. The ability of the brain to concentrate, and so absorb information, diminishes rapidly after a while. The actual time varies with each individual but for most one hour is a reasonable maximum.
- Take breaks of 10–15 minutes between sessions. It helps if you leave the room you are using for study and do something different – take the dog for a walk, watch television, exercise, play cards – anything provided it gives you a complete break from revision. Do things you really enjoy so that the breaks can be seen as a reward for your hard work.
- Be aware that, consciously or subconsciously, you will be seeking ways of avoiding doing revision. Fight the urge to give up. Let someone else answer the door; do not leap up when the telephone rings; better still, keep your mobile

Appropriate revision is essential to success at AS. Sometimes revision can be undertaken in a group – for example by getting others to test your knowledge and understanding

phone turned off. Make it clear to those around you that you have no wish to be disturbed.

- Be aware of 'displacement activities'. These are various activities that you carry out in order to bring relief from work. Revision is often boring and tedious, and you will subconsciously be looking for a means of escape. The sudden craving for coffee or something to eat is no more than an excuse to stop work. You will hardly starve or die of thirst before your next break! There is no need to get up and look out of the window every time there is a noise, or to see whether or not it is raining. Re-organising your CDs is not urgent and the dog can wait for a walk until later. Always be conscious of the dozens of trivial matters that suddenly assume great importance – and ignore them!

- Bring variety to your revision – vary the topics, subjects, times and place of revision and what you do in your breaks.

- Get feedback on the effectiveness of revision by testing yourself or getting others to do so. Closing the book and immediately writing down what you can remember has little value as it only tests short-term memory. Success may depend on remembering information revised days or even weeks earlier. One useful technique is to write short questions as you revise and then try to answer them in a day or so. Even single words can be jotted down, the significance of which you can later try to remember. Go back immediately and re-revise those questions you got wrong or could not answer.

- Practise questions, preferably from past papers. Do them within the allocated time so you can practise working within time limits. Study past OCR papers to make yourself familiar with the styles of questions. Be aware of any recent changes in the format or style of papers and questions. Predicting questions is a risky business and not worth the gamble.

- Use spare minutes for revision. It should be possible to read notes or test yourself during the many spare moments in a day. The 10 minutes spent on a bus or train can be better occupied revising than reading the advertisements around you. The 5 minutes spent waiting for your food to cook or for someone to call can all be put to good use. In themselves they may not be that much, but together they can make a significant contribution to the total revision. They could be the difference between a particular grade and a higher one – in some cases the difference between a pass and a fail.

- Organise revision by making a written timetable well in advance of examinations or modular tests. Be realistic. Do not make it so difficult that you fall behind schedule within the first week. Choose times to revise when you are less likely to be distracted. If you have an 'unfavourite' relative who visits on Wednesday evenings, put down at least 3 hours for this time – you have a better chance of achieving it than on an evening when your favourite television programme is on, or your friends go out to a local club. Leave yourself at least one day a week with no revision, and leave one week in four completely free. This means that, if an unexpected event arises, for example illness, you can use the 'free' days to compensate for lost time. If it does not prove necessary, then either use the time for additional revision or for rewarding yourself with a day off. The break may prove more valuable than revising because it will refresh you for future revision. Do not make a timetable that completely fills your revision time. You will be extremely lucky if you do not have some situation that prevents you revising for at least some of the days in the weeks prior to the examinations.

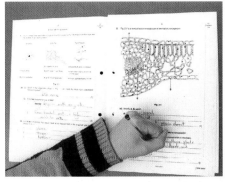

Practising past OCR examination papers is an important part of revision

Having revised thoroughly and prepared yourself well for the examinations, it would be a travesty to waste all this effort by making a hash of the actual papers. Good examination technique alone will not bring success, but it will help ensure your preparation brings its deserved reward.

17.7.1 Preparation for examinations

It is most important to use tests, trial examinations and practice questions to perfect your methods – the final examination is no time to test their efficiency. Students who perform well during a course, but do poorly in examinations, are often over-dependent on books and / or teachers. While both are crucially important to success, over-reliance on them can mean you are less able to think for yourself. Neither will be available to you during the written examinations and so, unless you have learnt to work independently, you may do worse than expected. Such failure is often blamed on 'not being good at exams', but is, in fact, the fault of bad preparation. Indeed much of the blame for poor results is wrongly excused as poor technique rather than the true culprit – inadequate preparation.

Perfecting a sound approach to examinations begins early in your course. It includes developing the ability to read critically, to think independently, and to answer questions, preferably from past papers and under time constraints. The latter will give you practice at interpreting questions and responding in concise, relevant terms. Most of all, to improve you must learn from your mistakes.

17.7.2 Answering examination questions

When it comes to answering examination questions, there are a number of common-sense points which you should follow:

6th-form students sitting examinations

- You can normally use an electronic calculator in OCR AS examinations, so remember to take one with you.
- Read carefully the general instructions given on the front page. All too many candidates ignore these in their rush to get started on the first question. Follow the instructions precisely.
- Read each question with great care. Do not be in a hurry to get started; it pays to be sure you understand what the question requires before attempting it.
- Calculate the time you can afford to spend on each question. At present this is straightforward as both the Foundation module and the Human Health and Disease module have 60 marks each and allow you 1 hour; the Transport module has 45 minutes for the 45 marks available. However, to allow time for reading instructions, reading questions and thinking about your answer, and, to avoid a rush at the end, it is best to allow a little less than 1 minute per mark. Give yourself 9 minutes for every 10 marks, or thereabouts.
- Try to isolate key words in a question and answer in accordance with them. Help in understanding the meaning of commonly used words in questions is given in the chart elsewhere in this unit.
- Make your responses relevant, clear and concise. Do not ramble, repeat yourself or try to disguise inadequate knowledge by waffling or straying from the point.
- If you are stuck for an answer, move on to the next part. If time allows at the end, go back and try again any questions you left blank.
- At present all questions are compulsory on OCR AS Biology papers, so ensure you attempt each one.

17.7.3 Structured questions

The majority of questions on the three OCR theory papers have structured parts, i.e. a series of short, but connected questions on a particular topic.

- In their simplest form they require a single word answer, e.g. 'Name the organ where T lymphocytes mature?' (1 mark)
- In some cases single words are needed to fill in blank spaces in a prose passage or a table.
- Sometimes a short explanation is called for, e.g. ' Explain what is meant by the term tissue?' (2 marks)
- A longer description or explanation may be required, e.g. 'Outline the roles of membranes **within** cells' (6 marks)
- Be guided by the mark distribution as to the amount of detail wanted.
- The lines available for an answer are a guide only – do not feel obliged to fill all the space or worry if you write a few more lines than are provided.
- However, unless you write exceedingly small, you would normally use at least half the lines provided or, unless your writing is excessively large, you would not normally exceed the lines provided. In most cases you are unlikely to fill all the space given.
- In structured questions it is content rather than style that is being tested and so answers should be clear, concise and to the point. There is no need to construct sentences; note-style responses are acceptable.

17.7.4 Questions requiring extended answers

One part of one question on each of the three theory papers (Foundation, Human Health and Disease, and Transport) requires an extended answer. There is an additional mark on each of these for the way in which you communicate your answer. These questions begin with the statement *'In this question, 1 mark is available for the quality of written communication'*. In answering, in addition to giving accurate and relevant biological information, you should:

- Use properly constructed sentences and accurate grammar and punctuation.
- Write legibly and spell words correctly.
- Use an appropriate style and form of writing to suit the question.
- Organise the information in a clear and understandable way.
- Use specialist biological terms and other vocabulary as appropriate.

GLOSSARY OF WORDS USED IN QUESTIONS

Calculate a numerical answer is required and all working should be shown.

Comment recall or infer relevant points of interest in an open-ended way. When applied to graphs and tables of data, usually requires both a description and an explanation.

Compare point out differences and similarities.

Complete fill in the spaces or blanks in a prose passage, table or list.

Deduce make logical and coherent connections between pieces of information, often provided in the question.

Define give only a formal statement.

Describe give an account of the main points, using diagrams if appropriate. This often refers to visual observations, such as describing an experiment or a particular event. It is also used when you need to describe the trend shown by a graph or a table of data. You should not give an explanation within this account.

Discuss debate, giving the various viewpoints and arguments in a balanced, reasoned and objective manner

Distinguish / Contrast make distinctions between; recognise comparable differences.

Estimate give a reasonable approximation or order of magnitude.

Explain show how and why; give reasons for with references to theory.

Give reasons show how and why, give explanations.

List catalogue, often as a sequence of words one after the other. No additional detail is needed.

Name give a specific example, sometimes from information provided in the question.

Outline give the most essential and important points, briefly. No supporting detail is needed.

Select choose appropriately, usually from a list, table or data provided in the question.

Sketch where applied to a diagram this means a simple, free-hand drawing in the correct proportions and showing the main features. Where applied to a graph it means draw the shape and position of the line / curve on labelled axes.

State set down concisely, with little or no supporting evidence.

Suggest put forward ideas, thoughts or hypotheses either to questions which have no single right answer or where you need to apply your knowledge to a new situation.

Sometimes the technical terms used in AS Biology may seem complicated. There is, however, usually a logic to the way many of these words are formed. Biology, as with other branches of science, uses Latin and Greek roots to construct words. Erythrocyte, for example, is the alternative name for a red blood cell. This word is made up of *erythro* meaning 'red' and *cyto* meaning 'cell'. Likewise, disaccharide is made up of *di* (two) and *saccharo* (sugar), while lysosome comprises *lyso* (splitting) and *some* (body), which describes the role of this organelle. The word gluconeogenesis means the formation of new glucose (e.g. from fatty acid), in the liver. The word is made up of *gluco* (sugar) – *neo* (new) – *genesis*(formation). Even if you have never met a word before, you can often work out its meaning if you know some of the Latin and Greek roots given below. For example, use the list to work out what is meant by isopodous and phyllophagous.

a, an	not, without	*deci*	one tenth
ab, abs	away from	*demi*	a half
abdomen	belly	*dendro*	tree, branch
ad	towards	*denti*	tooth
adipi	fat	*dermo, dermato*	skin
aero	air	*di*	two
agri	field	*dictyo*	net
allo, allelo	different	*diplo*	double
amylo	starch	*dorso*	back
andro	male		
ante	before	*eco, oeco, oiko*	dwelling
anthropo	human	*ecto*	outside
anti	against	*encephalo*	brain
arthro	joint	*endo*	inside
astero, astro	star	*entero*	gut, intestine
auto	self	*erythro*	red
auxo	growth	*ex, e, ec*	out of
bacilli	small stick, rod	*fibro, fili*	thread
bi, bin	two	*flagelli*	lash, whip
bio	life	*foli*	leaf
branchio	gill	*fructi*	fruit
bronchio	throat		
bucca	mouth	*galacto*	milk
		gamo	fusion
cardio	heart	*gaster, gastro*	stomach
carni	flesh	*geno*	forming
cata	down	*geo*	earth, gravity
centri	one hundredth	*glyco, glycy, gluco*	sweet, sugar
cephalo	head	*gyno*	female
chloro	green		
chromo, chromato	colour	*haemo, haem*	blood
cili	eyelash	*halo*	salt
coelo	cavity	*haplo*	single
col, com, con	together with	*helix, helico*	coiled
cortex, cortico	covering	*hemi*	half
costa	rib	*hepato*	liver
cuti	skin	*hetero*	different
cyclo	circle	*hexa*	six
cysto	bladder	*holo*	complete
cyto	cell	*homo, homeo*	same
		hydro	water
dactylo	finger, digit	*hygro*	moisture
de	down, away from	*hyper*	excess
		hypo	below

in	inside	*poly*	many
infra	below	*pori*	perforated
inter	between	*post*	after
intra	within	*pre, pro*	before
iso	the same	*proto*	first
		proximi	near
kilo	thousand	*pseudo*	false
		pulmo	lung
lacti	milk		
latero	side	*re*	again
leuco	white, colourless	*reni*	kidney
lipo	fat	*reti*	net
loci	place	*rhizo, rrhizo*	root
logo	study		
luteo	yellow	*saccharo*	sugar
lyso	breakdown	*sacci*	sac
		sapro	decaying
macro	large	*sarco*	flesh
medulli	pith	*schizo*	split
mega	large	*sclero*	hard
melano	black	*scroti*	pouch
meso	middle	*sepsis, septo*	decay
micro	small	*septi*	wall
milli	thousandth	*somo, somato*	body
mito	thread	*spiri*	breathing
mono	one	*sporo*	spore, seed
morpho	form	*squamo*	scale
multi	many	*steno*	narrow
myco, myceto	fungus	*stoma, stomata*	mouth
myo, mys	muscle	*sub*	under
		suberi	cork
nano, nanno	dwarf	*super, supra*	above
neo	new	*sym*	with
nephro	kidney		
neuro	nerve	*telo*	final
		terri	land
oculo	eye	*tetra*	four
oligo	few	*thalamo*	chamber
omni	all	*thermo*	heat
oo, ovo, ovi	egg	*thorax, thoracio*	chest
ops, opto	sight, eye	*toxi*	poison
osmo	push	*trans*	across
osteo	bone	*tri*	three
		tricho	thread
paedo	child	*tropho*	feeding
palaeo	ancient	*turgo*	swollen
patho	disease		
pectori	chest	*uni*	one
pedi, podo	foot, leg	*uter, utero*	sac
penta	five		
peri	around	*vaso*	vessel
phago	feeding	*veno*	vein
pharynx	throat	*ventro*	belly
philo	liking	*villi*	shaggy
phobo	hating	*visci*	sticky
photo	light		
phyllo	leaf	*xantho*	yellow
phyto	plant	*xero*	dry
plankto	wandering	*xylo*	wood
pleio, pleo	many		
pneumo	lung	*zoo*	animal
podo, pes	foot, leg	*zygo*	fusion

Glossary

abiotic an ecological factor that makes up part of the non-biological environment of an organism. Examples include temperature, pH, rainfall and humidity. See also *biotic*.

actin filamentous protein which is involved in contraction within cells, especially muscle cells.

activation energy energy required to bring about a reaction. The activation energy is lowered by the presence of enzymes.

active immunity resistance to disease resulting from the activities of an individual's own immune system whereby an *antigen* induces plasma cells to produce *antibodies*.

active site a group of amino acids that makes up the region of an enzyme into which the substrate fits in order to catalyse a reaction.

active transport movement of a substance from a region where it is in a low concentration to a region where it is in a high concentration. The process requires the expenditure of energy.

acute illness a sudden, but short-term, ailment from which the patient recovers quickly.

adhesion attraction between the molecules of different types. See also *cohesion*.

adipose tissue a form of connective tissue that is made up of cells storing large amounts of fat.

aerobic connected with the presence of free oxygen. Aerobic respiration requires free oxygen to release energy from glucose and other foods. See also *anaerobic*.

aerobic exercise forms of physical exertion, such as swimming and cycling, that involve the lungs and heart in providing oxygen to the muscles for aerobic respiration.

aerobic fitness a measure of the ability of an individual to take in and use oxygen. It depends upon the effective ventilation of the lungs, transport of oxygen by the blood and its use by cells.

allele one of a number of alternative forms of a *gene*. For example, the gene for the shape of pea seeds has two alleles, one for 'round' and one for 'wrinkled'.

allergen a normally harmless substance that causes the immune system to produce an immune response. See also *allergy*.

allergy the response of the immune system to an *allergen*. Examples include hay-fever and asthma.

anaerobic connected with the absence of free oxygen. Anaerobic respiration releases energy from glucose or other foods without the presence of free oxygen. See also *aerobic*.

anion negatively charged *ion* that is attracted to the anode during electrolysis. See also *cation*.

anorexia nervosa a psychological disorder in which patients, frequently young girls, deliberately starve themselves in order to lose weight.

antibiotic a substance produced by one kind of microorganism that inhibits the growth of another.

antibiotic resistance the development in microorganisms of mechanisms that prevent *antibiotics* from killing them.

antibody a protein produced by *lymphocytes* in response to the presence of the appropriate *antigen*.

antigen a molecule that triggers an immune response by *lymphocytes*.

antioxidant chemical which reduces or prevents *oxidation*. Often used as an additive to prolong the shelf-life of certain foods.

apoplast pathway route through the cell walls and intercellular spaces of plants by which water and dissolved substances are transported. See also *symplast pathway*.

arteriosclerosis general term for any degenerative disease of the arteries, e.g. *atherosclerosis*.

artificial immunity resistance to disease acquired as a result of the deliberate exposure of the body to *antibodies* or *antigens* in non-natural circumstances, e.g. *vaccination*.

asthma a *chronic illness* in which there is resistance to air-flow to the alveoli of the lungs as a result of the air-ways becoming inflamed due to an allergic response to an *allergen*.

atheroma fatty deposits on the walls of arteries, often associated with high *cholesterol* levels in the blood.

atherosclerosis narrowing of arteries due to thickening of the arterial wall caused by fat, fibrous tissue and salts being deposited on it.

ATP (adenosine triphosphate) *nucleotide* found in all plants and animals, which is produced during respiration and is important in the transfer of energy.

atrioventricular node (AV node) area of muscle between the atria and ventricles of the heart that plays an important role in coordinating the heartbeat.

basal metabolic rate (BMR) the minimum amount of energy needed to maintain the vital activities of life, such as breathing and heartbeat, in a person at complete rest.

Benedict's test a simple biochemical reaction to detect the presence of reducing sugars.

beta blockers a group of drugs that prevent impulses affecting certain cells by blocking the receptor sites on the cell surface membrane. They are used to treat angina and *hypertension*.

biodiversity the range and variety of living organisms within a particular region.

biomass the total mass of living material, normally measured in a specific area over a given period of time.

biotic an ecological factor that makes up part of the living environment of an organism. Examples include food availability, competition and predation. See also *abiotic*.

Biuret test a simple biochemical reaction to detect the presence of protein.

B lymphocyte type of white blood cell that is produced and matures within the bone marrow. B lymphocytes produce *antibodies* as part of their role in *immunity*. See also *T lymphocyte*.

Bohr effect the reduced affinity of *haemoglobin* for oxygen in the presence of carbon dioxide.

bovine spongiform encephalitis (BSE) a degenerative disease of the brain in cattle caused by infectious particles called prions. Also known as 'mad cow disease'.

bronchitis a respiratory disease in which the bronchi of the lungs become inflamed and congested with mucus. *Acute* bronchitis is the result of a bacterial or viral infection while *chronic* bronchitis is frequently associated with smoking.

buffer solution with the ability to absorb hydrogen *ions* and which therefore does not significantly change its pH when moderate amounts of acid or alkali are added.

cambium dividing layer of cells in higher plants, parallel to the surface of stems and roots, which produces new cells leading to an increase in their diameter.

cancer a disease resulting from mutations that leads to uncontrolled cell division and the eventual formation of a group of abnormal cells called a tumour, from which cells may break away (*metastasis*) and form secondary tumours elsewhere in the body.

capsid protective protein coat around a virus particle.

carcinogen a chemical, a form of radiation or other agent that causes *cancer*.

cardiac cycle a continuous series of events which make up a single heartbeat.

cardiac muscle type of muscle found only in the heart. It has fewer striations than *striated muscle* and can beat continuously throughout life without stimulation by nerve impulses. See also *smooth muscle*.

cardiac output the total volume of blood which the heart can pump each minute. It is calculated as the volume of blood pumped at each beat (*stroke volume*) multiplied by the number of heart beats per minute (heart rate).

cartilage flexible supporting tissue found at the end of bones and between vertebrae, where it cushions the shocks and jolts that occur during movement.

Casparian strip a distinctive band of suberin around the endodermal cells of a plant root which prevents water passing into xylem via the cell walls. The water is forced through the living part (protoplast) of the endodermal cells.

cation positively charged *ion* which is attracted to the cathode during electrolysis. See also *anion*.

centrifugation process of separating out particles of different sizes and densities by spinning them at high speed in a centrifuge.

chloride shift the movement of negatively charged chloride *ions* from the plasma into red blood cells to replace the loss of negatively charged hydrogencarbonate ions during the transport of carbon dioxide. In this way the overall electrochemical neutrality of the red blood cells is maintained.

cholesterol lipid occurring in large quantities in the brain, spinal cord and liver. It is an intermediate in the formation of vitamin D and steroid hormones. Excess in the blood can lead to *arteriosclerosis*.

chromatid one of the two strands of a *chromosome* that are joined together by a single centromere.

chromatin the material which makes up *chromosomes*. It consists of DNA and proteins, especially *histones*.

chromatography technique by which substances in a mixture are separated according to their different solubilities in a solvent.

chromosome a thread-like structure made of protein and DNA by which hereditary information is physically passed from one generation to the next.

chronic illness an on-going ailment that arises gradually and may recur over a number of years.

chronic obstructive pulmonary disease name given to the lung condition where *emphysema* and *bronchitis* occur together.

clone a group of genetically identical organisms formed from a single parent as the result of asexual reproduction or by artificial means.

collagen fibrous protein that is the main constituent of connective tissues such as *tendons, cartilage* and bone.

collenchyma plant tissue which has cell walls thickened by cellulose especially in the corners. It provides mechanical support, especially in young stems and leaves.

colloid substances made up of fine particles or large molecules which are evenly dispersed within a second substance. Examples include emulsions (liquid dispersed in a liquid) aerosols (solid / liquid dispersed in a gas) and foams (gas dispersed in a liquid / solid).

community all the living organisms present in an *ecosystem* at a given time.

complementary DNA DNA which is made from messenger RNA using the enzyme *reverse transcriptase* in a process which is the reverse of normal transcription.

condensation reaction chemical process in which two molecules combine to form a more complex one with the elimination of a simple substance, usually water. Many biological *polymers*, such as polysaccharides and polypeptides, are formed by *condensation reactions*. See also *hydrolysis*.

consumer any organism that obtains energy by 'eating' another. Herbivores feed on plants and are known as primary consumers and carnivores feeding on herbivores are known as secondary consumers. See also *producer*.

coronary heart disease any condition, e.g. *atherosclerosis* and *thrombosis,* affecting the coronary arteries that supply *cardiac muscle.*

counter-current system a mechanism by which the efficiency of exchange between two substances is increased by having them flowing in opposite directions.

covalent bond type of chemical bond in which two atoms share a pair of *electrons*, one from each atom.

cystic fibrosis inherited disease in which the body produces abnormally thick mucus that obstructs breathing passages and prevents secretion of pancreatic enzymes. It is a recessive condition that leads to production of a non-functioning membrane protein needed to transport chloride *ions.*

cytokines chemicals secreted by T helper cells that stimulate an immune response in other white blood cells.

cytoskeleton network of microfilaments and microtubules that gives structural shape to *eukaryotic* cells.

decomposer a microorganism that breaks down the organic matter of dead organisms and waste products to form water, carbon dioxide and inorganic *ions.*

denaturation permanent changes due to the unravelling of the three-dimensional structure of a protein as a result of factors such as changes in temperature or pH.

denitrifying bacteria bacteria that convert nitrates to nitrogen gas as part of the nitrogen cycle.

diabetes metabolic disorder in which there is abnormal thirst and the production of large amounts of urine. Diabetes mellitus is caused by a reduction or absence of insulin production by the pancreas or insensitivity of insulin receptors on cells, leading to changes in the blood glucose level.

diastole the stage in the *cardiac cycle* when the heart muscle relaxes and ventricles fill with blood. See also *systole.*

dicotyledonous plants any member of the class of flowering plants called Dicotyledonae. Their features include: having two seed leaves (cotyledons), broad leaves, flower parts in rings of four or five, and vascular tissue arranged in a ring in stems. See also *monocotyledonous plants*.

dietary reference values sets of figures relating to the requirements for energy and nutrient intake of all healthy individuals in the UK.

diffusion the movement of molecules from a region where they are in high concentration to one where their concentration is lower.

diploid a term applied to cells in which the nucleus contains two sets of *chromosomes*. See also *haploid*.

disulphide bridges S–S bonds between two cysteine amino acids that are important in maintaining the structure of proteins.

diuretic drug used to increase the output of urine especially in the treatment of *hypertension* and *oedema.*

ecological niche role of a species within its *community*. It includes what the species is like, where it occurs, how it behaves, its interactions with other species and how it responds to the *abiotic* environment.

ecology study of the interrelationships of organisms with each other and their environment.

ecosystem all the living (*biotic*) and non-living (*abiotic*) components of a particular area.

electron negatively charged sub-atomic particle that orbits the positively charged nucleus of all atoms**.** See also *proton*.

electrophoresis technique for separating a mixture of charged particles in a fluid, e.g. a gel, by applying a voltage across the fluid. It is used in the analysis of mixtures of substances, especially proteins.

element one of just over 100 substances that cannot be split into simpler substances by chemical means.

emphysema a disease in which the walls of the alveoli break down, reducing the surface area for gaseous exchange, thereby causing breathlessness in the sufferer. See also *chronic obstructive pulmonary disease.*

endemic describes any disease that occurs regularly in a particular region or amongst a particular population. See also *epidemic* and *pandemic*.

endocytosis the inward transport of large molecules through the cell surface membrane. See also *exocytosis*.

endotherm an animal which uses physiological processes to maintain its body temperature at a more or less constant level. Birds and mammals are endotherms.

epidemic describes any disease that rapidly spreads through a population to affect a large number of individuals. See also *endemic* and *pandemic*.

epidemiology the study of the spread of disease and the factors that affect this spread.

epidermis the outermost layer of cells in a multicellular organism.

essential amino acid an amino acid that cannot be synthesised by the human body and which must therefore be included in the diet.

eukaryote an organism whose cells have a membrane-bound nucleus that contains *chromosomes*. The cells also possess a variety of other membranous organelles such as mitochondria and endoplasmic reticulum. See also *prokaryote*.

eutrophication term applied to an increase in plant nutrients, especially nitrates and phosphates, in fresh-water lakes and rivers, often as a consequence of 'run-off' of fertiliser from agricultural land. This often leads to a decrease in *biodiversity*.

exocytosis the outward bulk transport of materials through the cell surface membrane. See also *endocytosis*.

facilitated diffusion diffusion involving the presence of protein carrier molecules to permit the passive movement of substances across cell surface membranes.

gene length of DNA on a *chromosome* normally coding for a specific polypeptide. Recent evidence, however, suggests that a gene may code for more than one polypeptide.

gene technology general term that covers the processes by which *genes* are manipulated, altered or transferred from organism to organism. Also known as *genetic engineering*.

gene therapy a mechanism by which genetic diseases such as *cystic fibrosis* may be cured by masking the effect of the defective *gene* by inserting a functional gene.

genetically modified organism (GMO) organism that has had its DNA altered as a result of *gene technology*.

genetic engineering see *gene technology*.

genotype the genetic composition of an organism.

global warming the recent increase in average temperatures at the Earth's surface, thought to be the result of the increased production of greenhouse gases such as carbon dioxide and methane. These gases help to trap solar radiation at or near the Earth's surface.

glycoprotein substance made up of a carbohydrate molecule and a protein molecule. Part of the cell surface membrane and certain hormones are glycoproteins.

goblet cell mucus-producing cell found in the epithelium of the intestines and bronchi, so called because its shape resembles a wine glass or goblet.

gross primary production the total energy production in the form of *biomass* made by *producers* (plants) during photosynthesis. It is normally expressed as the biomass per unit area in unit time. See also *net primary production*.

Haber process industrial process in which nitrogen and hydrogen are passed over a heated catalyst, at a pressure of 1000 atmospheres, to produce ammonia.

habitat the place where an organism normally lives and which is characterised by physical conditions and the types of other organisms present.

haemoglobin globular protein in mammalian blood that readily combines with oxygen to transport it around the body. It comprises four polypeptide chains around an iron-containing haem group. See also *myoglobin*.

haploid term referring to cells that contain only a single copy of each *chromosome* e.g. the sex cells or gametes.

herbaceous term applied to non-woody plants.

histones proteins associated with the DNA in *chromosomes*. Their function is to condense *chromatin* and coil the chromosomes during cell division.

homologous chromosomes a pair of *chromosomes*, one maternal and one paternal, that have the same gene loci and therefore determine the same features. They are not necessarily identical, however, as individual *alleles* of the same *gene* may vary, e.g. one chromosome may carry the allele for blue eyes, the other the allele for brown eyes. Homologous chromosomes are capable of pairing during *meiosis* I.

human genome the totality of the DNA sequences on the *chromosomes* of a single human cell.

hydrogen bond chemical bond formed between the positive charge on a hydrogen atom and the negative charge on another atom of an adjacent molecule, e.g. between the hydrogen atom of one water molecule and the oxygen atom of an adjacent water molecule.

hydrolysis the breaking down of large molecules into smaller ones by the addition of water molecules. See also *condensation reaction*.

hydrophyte plant that is adapted to live in water or wet habitats. See also *xerophyte*.

hypertension persistent high blood pressure when a person is at rest.

immunisation an artificial means of producing *immunity*, either by injection of *antibodies* (passive immunity) or by inducing the body to produce its own antibodies (active immunity).

immunity the means by which the body protects itself from infection.

incidence the number of new cases of a disease in a population in a given time, e.g. in one month or one year.

introns portions of DNA within a *gene* that do not code for a polypeptide. The introns are removed from messenger RNA after *transcription*.

ion an atom or group of atoms that have lost or gained one or more *electrons*. Ions therefore have either a positive or negative charge. See also *anion* and *cation*.

ion channel a passage across a cell surface membrane made up of a protein that spans the membrane and opens and closes to allow *ions* to pass in and out of the cell.

isotonic solutions that possess the same concentration of solutes and therefore have the same *water potential*.

isotope variations of a chemical element that have the same number of *protons* and *electrons* but different numbers of neutrons. While their chemical properties are similar they differ in mass. One example is carbon which has a relative atomic mass of 12 and an isotope with a relative atomic mass of 14.

Krebs cycle series of biochemical reactions in most *eukaryotic* cells by which energy is obtained through the oxidation of acetyl coenzyme A produced from the breakdown of glucose.

leaching process in which chemicals are removed from soil by being dissolved in rainwater and washed away.

lignin a complex, non-carbohydrate polymer associated with cellulose in plant cell walls. Lignin makes the cell walls stronger, allowing them to resist tension and compression. It also makes them more waterproof.

low density lipoprotein (LDL) a compound containing both protein and lipid molecules that occurs in blood plasma and *lymph*. The proportion of protein is lower than in high density lipoproteins.

lumen the hollow cavity inside a tubular structure such as the gut or a *xylem vessel*.

lymph a slightly milky fluid found in lymph vessels and made up of *tissue fluid*, fats and *lymphocytes*.

lymphocytes type of white blood cell responsible for the immune response. They become activated in the presence of *antigens*. There are two types: *B lymphocytes* and *T lymphocytes*.

lysis the breakdown of a cell or compound.

meiosis the type of nuclear division in which the number of *chromosomes* is halved.

memory cells type of *lymphocyte* that circulates in the blood and *tissue fluid* long after the original *antigen* that caused them to develop has gone. They are reactivated when the same antigen returns, triggering an immediate secondary immune response.

menstrual cycle cycle in female humans, apes and monkeys of reproductive age during which the body is prepared for pregnancy. It differs from the oestrous cycle in other mammals in that, where fertilisation does not occur, the lining of uterus is shed along with a little blood in a process called menstruation.

mesophyll tissue found between the two layers of *epidermis* in a plant leaf comprising an upper layer of palisade cells and a lower layer of spongy cells.

messenger RNA form of ribonucleic acid that carries the genetic code from the DNA of the nucleus to the ribosomes in the cytoplasm, where it acts as a template on which polypeptides are assembled.

metastasis process in which cancer cells break free from the original (primary) tumour and are carried by blood or *lymph* system to other parts of the body, where they may start another (secondary) tumour.

microflora the population of various types of microorganism in a particular place, e.g. in the gut.

middle lamella layer made up of pectins and other substances found between the walls of adjacent plant cells.

mitosis the type of nuclear division in which the daughter cells have the same number of *chromosomes* as the parent cell.

monocotyledonous plant any member of the class of flowering plants called Monocotyledonae. Their features include having a single seed leaf (cotyledon) and leaves that are parallel veined. See also *dicotyledonous plants*.

mutualism an association between species in which both species benefit. Mutualism does not necessarily involve a close physical association.

myocardial infarction otherwise known as a heart attack, results from the interruption of the blood supply to the heart muscle causing damage to an area of the heart with consequent disruption to its function.

myoglobin red-coloured pigment found in muscle and used to store oxygen. See also *haemoglobin*.

myosin the thick filamentous protein found in *striated muscle*.

natural immunity resistance to disease that is either inherited or acquired as part of normal life processes, e.g. as a result of having an infection.

net primary production the rate at which material produced during photosynthesis is built up in a plant. Also known as the net assimilation rate, it is the *gross primary production* less the 20% or so used by the plant in processes such as respiration.

niche all of the ranges of environmental conditions and resources required for an organism to survive, reproduce and maintain a viable population.

nitrifying bacteria microorganisms that convert ammonium compounds to nitrites and nitrates.

nitrogen-fixing bacteria group of microorganisms that incorporate atmospheric nitrogen into nitrogen-containing compounds using the enzyme nitrogenase. They may be either free-living or act in conjunction with leguminous plants.

nucleotides complex chemicals made up of an organic base, a sugar and a phosphate. They are the basic units of which the nucleic acids DNA and RNA are made.

oedema a condition in which an excess of watery fluid collects in the tissues or cavities of the body.

oral rehydration therapy (ORT) means of treating dehydration, involving giving, by mouth, a balanced solution of salts and glucose that stimulates the gut to re-absorb water.

osmosis the passage of water from a region where it is highly concentrated to a region where its concentration is lower, through a partially permeable membrane.

osteoarthritis degeneration of the *cartilage* of the joints causing pain and stiffness of these joints.

osteoporosis condition in which bones lose some of their bony substance and so become brittle and fragile. Causes include a lack of calcium or vitamin D in the diet, old age, and a reduction in oestrogen levels in women, notably post-menopausal ones.

oxidation chemical reaction involving the loss of *electrons*.

oxygen debt the amount of oxygen needed to oxidise the lactate that accumulates during *anaerobic* respiration.

palisade mesophyll cells long, narrow cells, packed with chloroplasts, that are found in the upper region of a leaf and which carry out photosynthesis. See also *spongy mesophyll*.

pandemic describes any disease that spreads over vast areas of the world, e.g. AIDS. See also *endemic* and *epidemic*.

parasite an organism that lives on or in a host organism. The parasite gains a nutritional advantage and the host is harmed in some way.

parenchyma plant tissue made up of unspecialised cells with thin walls which are used to provide support through turgidity.

passive immunity resistance to disease that is acquired from the introduction of *antibodies* from another individual, rather than an individual's own immune system, e.g. across the placenta or in the mother's milk. It is usually short-lived.

pasteurisation heat treatment of food in order to kill certain microorganisms and named after its discover, Louis Pasteur.

pathogen any microorganism that causes disease.

phagocytosis mechanism by which cells transport large particles across the cell surface membrane.

phospholipid lipid molecule in which one of the three fatty acid molecules is replaced by a phosphate molecule. Phospholipids are important in the structure and functioning of the cell surface membrane.

phytoplankton small, often microscopic, photosynthesising organisms that live suspended in large bodies of water, such as oceans and lakes. See also *zooplankton*.

pinocytosis form of *endocytosis* by which cells take up liquids from their environment. Also known as 'cell drinking'.

plaque deposits of fatty material, fibrous tissue and dead cells that form large irregular patches on the lining of the inner walls of arteries causing a condition called *atherosclerosis*.

plasmid a small circular piece of DNA found in bacterial cells and often used as a *vector* in *gene technology*.

plasmodesmata fine strands of cytoplasm that extend through pores in adjacent cell walls and connect the cytoplasm of one cell with another.

platelets cell fragments found in blood which play an important role in blood clotting.

polymer large molecule made up of repeating sub-units.

polymerase chain reaction process of making many copies of a specific sequence of DNA or part of a *gene*. It is used extensively in *gene technology* and genetic fingerprinting.

polymerases group of enzymes that catalyse the formation of long-chain molecules (*polymers*) from similar basic units (monomers).

polyunsaturated fatty acid (PUFA) fatty acid that possesses carbon chains with many double bonds.

population a group of individuals of a particular species that occur in the same place at the same time and are capable of interbreeding.

prevalence the number of people in a *population* who suffer from a particular disease at a particular time.

producer an organism that synthesises organic molecules from simple inorganic ones such as carbon dioxide and water (= autotrophic organism). Most producers are photosynthetic and form the first *trophic level* of a food chain. See also *consumer*.

prokaryote an organism belonging to the kingdom Prokaryotae that is characterised by having cells less than 5μm in diameter which lack a nucleus and membrane-bound organelles. Examples include bacteria and blue-green algae. See also *eukaryote*.

protoctist an organism belonging to the kingdom Protoctista which is made up of single celled *eukaryotes* such as certain algae and *protozoa*.

proton positively charged sub-atomic particle found in the nucleus of the atom. See also *electron*.

protoplast the living portion of a plant cell, i.e. the nucleus and cytoplasm along with the organelles it contains.

protozoa a sub-group of the kingdom *Protoctista* made up of single-celled organisms such as *Amoeba*.

Purkyne tissue a region of specialist heart muscle that conducts a wave of excitation that causes contraction of the ventricles.

restriction endonucleases a group of enzymes that are able to cut DNA into shorter lengths at specific points. Found naturally in certain bacteria, they are important in *gene technology*.

reverse transcriptase enzyme capable of producing a DNA molecule from the corresponding *messenger RNA*. Produced naturally by many viruses, reverse transcriptase is used in *gene technology*.

rheumatoid arthritis autoimmune disease in which there is chronic inflammation of the joints leading to disability and disfigurement.

rickets a deficiency disease in children, caused by a lack of vitamin D, in which the bones are weakened due to a lack of calcium, leading to bowing of the legs.

root nodule swelling on the roots of certain plants, such as legumes (bean family), that contain *nitrogen-fixing bacteria* that live *mutualistically* with the host plant.

saprobiont also known as a saprophyte, a saprobiont is an organism that obtains its food from the dead or decaying remains of other organisms. Many bacteria and fungi are saprobionts.

sclerenchyma plant tissue whose cells have become rigid due to the presence of cell walls thickened with *lignin*. The cells are dead and function to provide support to the plant.

semi-conservative replication the means by which DNA makes exact copies of itself by unwinding the double helix so that each chain acts as a template for the next. The new copies therefore possess one original and one new strand of DNA.

serum clear liquid that is left after blood has clotted and the clot has been removed. It is therefore blood plasma without the clotting factors.

sieve plate the perforated end wall of the phloem component called the sieve tube element.

sieve tube phloem tissue made up of a series of sieve tube elements joined end to end to form a long column that carries organic material in the plant.

sinoatrial node an area of cardiac muscle in the right atrium that controls and coordinates the contraction of the heart. Also known as the pacemaker.

smooth muscle also known as involuntary or unstriated muscle, smooth muscle is not under conscious control. See also *striated muscle*.

spindle apparatus thread-like structure, made up of microtubules, that draws the chromosomes apart during *meiosis and mitosis*.

spongy mesophyll cells irregularly shaped cells in the lower half of a leaf. They have many air-spaces and are important in exchanging gases between the atmosphere and the rest of the leaf. See also *palisade mesophyl cells*.

stem cells undifferentiated cells from which specialised cells arise during development.

stoma (plural stomata) pore, mostly in the lower epidermis of a leaf, through which gases diffuse in and out of the leaf.

striated muscle the muscle that makes up the bulk of the body and which works under conscious control. Also known as voluntary muscle. See also *smooth muscle*.

stroke volume the volume of blood pumped at each ventricular contraction of the heart.

suberin complex waxy waterproofing material, made up of fatty acids, that is found in some plant cell walls.

supernatant liquid the liquid portion of a mixture left at the top of the tube when suspended particles have been separated out at the bottom during *centrifugation*.

sustainable yield the situation where the removal of produce from a renewable resource is equal to, or less than, the rate of growth of that resource. A sustainable yield can be taken indefinitely.

symbiosis the close relationship between two or more organisms of different species.

symplast pathway route through the cytoplasm and *plasmodesmata* of plant cells by which water and dissolved substances are transported. See also *apoplast pathway*.

systole the stage in the *cardiac cycle* in which the heart muscle contracts. It occurs in two stages: atrial systole when the atria contract and ventricular systole when the ventricles contact. See also *diastole*.

tendons tough, flexible connective tissue that joins muscle to bone.

thrombosis formation of a blood clot within a blood vessel that may lead to a blockage.

thylakoid series of flattened membranous sacs in a chloroplast that contain chlorophyll and the associated molecules needed for the *light dependent reaction* of photosynthesis.

tidal volume the volume of air breathed in and out during a single breath when at rest.

tissue a group of similar cells organised into a structural unit that serves a particular function.

tissue fluid fluid that surrounds the cells of the body. Its composition is similar to that of blood plasma except that it lacks some of the larger proteins, in particular those that cause the blood to clot. It supplies nutrients to the cells and removes waste products.

T lymphocyte type of white blood cell that is produced in the bone marrow but matures in the thymus gland. T lymphocytes coordinate the immune response and kill infected cells. See also *B lymphocyte*.

transcription the formation of *messenger RNA* molecules from the DNA that makes up a particular *gene*. It is the first part of protein synthesis.

transgenic animals animals that contain *genes* artificially transferred to them from another species.

translation process whereby the code on a section of messenger RNA is converted to a particular sequence of amino acids that will go on to make a polypeptide and ultimately a protein.

translocation the transport of substances from one part of a plant to another.

transpiration evaporation of water from a plant.

triglyceride an individual fat molecule made up of a glycerol molecule and three fatty acids.

trophic level the position of an organism in a food chain. See also *producer* and *consumer*.

vaccination the introduction of a vaccine containing appropriate disease *antigens* or *anitbodies* into the body, by injection or mouth, in order to induce *artificial immunity*.

vascular tissue any tissue that forms a network of vessels through which fluids are transported in organisms. Blood vessels are the usual vascular tissue of vertebrates, while in plants it is phloem and xylem.

vasoconstriction narrowing of the internal diameter of blood vessels. See also *vasodilation*.

vasodilation widening of the internal diameter of blood vessels. See also *vasoconstriction*.

vector a carrier. The term may refer to something such as a *plasmid*, which carries DNA into a cell, or to an organism that carries a parasite to its primary host.

virulent being strongly poisonous. The term is applied to a disease that spreads rapidly through a population.

vital capacity the total volume of air that can be breathed in and out of the lungs during one deep breath.

water potential the measure of the extent to which a solution gives out water. The greater the number of water molecules present, the higher (less negative) the water potential. Pure water has a water potential of zero.

xerophthalmia deficiency disease caused by a lack of vitamin A in the diet. Its symptoms are drying and scarring of the cornea of the eye, which may lead to blindness.

xerophyte a plant adapted to living in dry conditions. See also *hydrophyte*.

xylem vessels dead, hollow, elongated tubes with lignified side walls and no end walls, that transport water in most plants.

zooplankton small animals, often microscopic, that live suspended in large bodies of water such as oceans and lakes. See also *phytoplankton*.

Answers to summary questions

Summary Test 1.2

1 phospholipids
2 chromatin
3 mitochondria
4 glycogen
5 starch

6 cell wall
7 cellulose
8 cell sap
9 tonoplast

Summary Test 1.5

1 nuclear envelope
2 40–100nm
3 nucleolus
4 ribosomal
5 thylakoids / lamellae
6 chlorophyll
7
8 } Krebs cycle; electron transport
9 matrix
10 stalked / elementary
11 cristae

Summary Test 1.7

1 protein
2
3 } glycoproteins; cholesterol
4 basal body
5 microtubules
6 9

7 2
8 2
9 nuclear division
10 spindle
11 chromosomes

Summary Test 1.8

1 nucleus
2 bacteria
3 circular
4 plasmid
5 eukaryotic
6 70S
7 thylakoids

Summary Test 2.3

1 amylose
2 glycogen
3 amylopectin
4 α-glucose
5 amylose, amylopectin, glycogen
6 amylopectin
7 amylosc, amylopectin, glycogen
8 amylopectin, glycogen
9 α-glucose, β-glucose

Summary Test 2.4

1 β-glucose
2 1,4 glycosidic
3 hydrogen bonds
4 microfibrils
5 structural / support

6 bursting
7 osmosis
8 turgid / semi-rigid
9 photosynthesis

Summary Test 2.5

1 triglycerides
2 glycerol
3 polyunsaturated
4 2
5 hydrophobic

6 energy source / store
7 insulation
8 protection / padding
9 waterproofing

Summary Test 2.6

1 -COOH
2 amino
3 amphoteric
4 condensation

5 peptide
6 hydrogen
7 α-helices
8 β-pleated sheets

Summary Test 2.7

1 amino acids
2 polypeptide
3 hydrogen bonds
4 -CO
5 peptide
6 sulphur
7 disulphide bridges

8 ionic bonds
9 hydrophobic interactions
10 disulphide bridges
11 ionic bonds
12 hydrophobic interactions
13 prosthetic

Summary Test 2.8

1 globular
2 fibrous
3 glycine – proline – alanine
4 tendons
5
6 } strength; flexibility
7 4
8 haem
9 Fe^{2+} / ferrous ion

Summary Test 2.9

1 dipolar
2 electrons
3 hydrogen bonds
4 viscosity
5 surface tension
6 hydrolysis
7 photosynthesis
8 lubricant

9
10 } calcium; phosphate; magnesium
11
12
13 } sodium; potassium; chloride
14
15 nitrate
16 calcium

Summary Test 3.1

1 catalysts
2 globular
3 active site
4 activation energy
5 substrate
6 products
7 lock and key
8 induced fit

Summary Test 3.2

1 time course
2 disappearance
3 product
4 maltose
5 starch
6 increase
7 active site
8 denatured
9 40°C
10 2
11 salivary amylase

Summary Test 3.3

1 initial rate of reaction
2 increases
3 active sites
4 halved
5 be constant

Summary Test 3.4

1 non-reversible / permanent
2 active site directed
3 reduced
4 allosteric site
5 unchanged / unaffected / the same

Summary Test 4.1

1 bilayer
2 hydrophilic
3 hydrophobic
4 intrinsic / integral
5 extrinsic / peripheral
6
7 } cholesterol; glycolipids; glycoproteins
8

Summary Test 4.2

1 higher / greater
2 lower / less
3 kinetic
4 passive
5 less / reduced / lower
6 less / reduced / lower
7 more / greater / higher
8 carrier proteins

Summary Test 4.3

1 water
2 partially permeable
3 pressure
4 zero
5 lower / more negative
6 B
7 hypertonic
8 hypotonic
9 isotonic

Summary Test 4.4

1 higher/less negative
2 (solute) concentration
3 osmosis
4 protoplast
5 (cellulose) cell wall
6 pressure
7 ψ_p
8 higher
9 plasmolysis

Summary Test 4.5

1 against
2 ATP
3 mitochondria
4 respiratory
5 cytosis
6 endocytosis
7 exocytosis
8 phagocytosis
9 pinocytosis

Summary Test 4.6

1 0.6
2 0.3 / halved
3 increase
4 2
5 double
6 osmosis
7 less negative / higher
8 active transport

Summary Test 4.7

1 } large in size; maintain a high body temperature /
2 } high metabolic rate
3 trachea
4 bronchi
5 100–300μm
6 squamous epithelium
7 0.1–0.5μm
8 600 million (300 million in each lung)
9 70m^2

Summary Test 5.1

1 } nitrogen; phosphorus
2
3 pentose
4 5
5 } ribose; deoxyribose
6
7 pyrimidines
8 } cytosine; uracil
9
10 purines
11 5
12 } guanine; adenine
13
14 thymine
15 ribosomal RNA
16 transfer RNA
17 messenger RNA

Summary Test 5.2

1 pentose
2 deoxyribose
3 phosphate
4 adenine
5 cytosine
6 antiparallel
7 double helix

Summary Test 5.3

1 polynucleotide
2 helicase
3 replication forks
4 topoisomerase
5 DNA polymerase
6 complementary
7 DNA ligase
8 proofreading endonuclease
9 semi-conservative

Summary Test 5.5

1 gene
2 3
3 codon
4 3.2 billion (3.2×10^9)
5 degenerate
6 non-overlapping
7 nonsense / stop
8 Tyr (tyrosine)
9
10 } ATA; ATG (complementary to UAU and UAC on mRNA)
11 AUG

Summary Test 5.6

1 messenger RNA
2 cistron
3 helicase
4 transcribing template
5 RNA polymerase
6 uracil
7 guanine
8 adenine
9 introns
10 nucleus

Summary Test 5.7

1 translation
2 ATP
3 amino-acyl tRNA
4 ribosome
5 codon

Summary Test 6.2

1 interphase
2 prophase
3 centrioles
4 poles
5 spindle apparatus
6 nuclear envelope
7 nucleolus
8 metaphase
9 equator
10 anaphase
11 centromere
12 chromatids / daughter chromosomes
13 telophase

Summary Test 6.3

1 meiosis / reduction division
2 diploid
3 homologous
4 haploid
5 gametes / sperm and ova
6 mutation
7
8 } growth; repair; asexual reproduction
9

Summary Test 6.4

1 mutagen / carcinogen
2 tumor
3 malignant
4 metastasis
5
6 } skin; breast
7 cervical
8
9 } X-rays; gamma rays; ultra-violet rays (UVA and UVB) (any 2)
10 asbestos

Summary Test 7.1

1 ecology
2 biosphere
3 biotic
4 abiotic
5 community
6 population
7 habitat

Summary Test 7.2

1 trophic level
2 photosynthesis
3 producers
4 gross primary production
5 respiration
6 net primary production
7 herbivores
8 carnivores
9 food chain
10 decomposers

Summary Test 7.3

1 3
2 20–50
3 net primary production
4 herbivores / primary consumers
5 trophic level
6 biomass
7 energy

Summary Test 7.4

1 78
2 nitrogen fixation
3 mutualistic
4 *Rhizobium*
5 nodules
6 peas / beans
7 nitrates
8 amino acids / proteins
9 decomposers
10 ammonia
11 nitrites
12 *Nitrosomonas*
13 *Nitrobacter*
14 denitrifying

Summary Test 8.3

1 plasma
2 red (blood) cells / erythrocytes
3 8μm
4 120 days
5 haemoglobin
6 oxygen
7 white (blood) cells / leucocytes
8 phagocytes
9 monocyte / neutrophil
10 antibodies
11 lymphocytes

Summary Test 8.5

1 oxygen
2 carbon dioxide
3 $100–300\mu m$
4 $0.1–0.5\mu m$
5 diffusion
6 capillaries
7 $7–10\mu m$
8 red blood cells
9 Type II pneumocytes
10 surfactant
11 sticky
12 phagocytes

Summary Test 8.6

1 respiratory pigment
2 68 000
3 4
4 iron (Fe^{2+})
5 polypeptides
6 } alpha (α); beta (β) (polypeptide)
7 }
8 4
9 tension / partial pressure
10 oxygen dissociation curve

Summary Test 8.7

1 21%
2 one-third (or less)
3 acclimatise
4 hypoxia
5 erythopoietin
6 red blood cell production
7 higher / greater
8 liver
9 marrow of bones (e.g. ribs)

Summary Test 8.8

1 respiration
2 carbamino-haemoglobin;
3 hydrogen
4 10
5 5
6 plasma
7 85
8 hydrogencarbonate ions
9 carbonic acid
10 carbonic anhydrase
11 haemoglobinic acid
12 buffer
13 chloride shift
14 Bohr effect

Summary Test 9.1

1 cardiac
2 pericardium
3 atria
4 ventricles
5 left atrioventricular / bicuspid / mitral
6 right atrioventricular / tricuspid
7 tendinous cords (chordae tendinae)
8 pulmonary veins
9 left atrium
10 aorta
11 right atrium
12 pulmonary artery
13 coronary arteries
14 myocardial infarction

Summary Test 9.2

1 70
2 diastole
3 atrial systole
4 atrioventricular valves
5 ventricles
6 semi-lunar
7 pulmonary artery
8 aorta

Summary Test 9.3

1 myogenic
2 sinoatrial node
3 right atrium
4 atria
5 atrioventricular node
6 Purkyne tissue

Summary Test 10.1

1 main vein
2 petiole
3 xylem
4 phloem
5 cambium
6 sclerenchyma
7 lateral / bending
8 vertical / pulling

Summary Test 10.2

1 } sugars; amino acids
2 }
3 storage
4 sieve plates
5 phloem-protein
6 }
7 } nucleus; Golgi apparatus; ribosome
8 }
9 companion cells
10 plasmodesmata

Summary Test 10.3

1 phloem
2 sources
3 sinks
4 pressure flow
5 sieve tube
6 } symplast; apoplast
7 }
8 photosynthesising
9 lower/more negative
10 xylem
11 higher/less negative
12 osmosis
13 hydrostatic

Summary Test 10.4

1 tracheids
2 lignin
3 }
4 } annular; reticulate; spiral
5 }
6 lumen
7 pits
8 (xylem) parenchyma
9 (sclerenchyma) fibres

Summary Test 10.5

1 root hairs
2 lower (more negative)
3 vacuolar
4 apoplast
5 symplast
6 plasmodesmata
7 endodermis
8 Casparian strip
9 suberin
10 protoplast
11 mineral ions
12 lower (more negative)
13 osmosis
14 root pressure

Summary Test 10.6

1	transpiration	5	osmosis
2	stomata	6	cohesion
3	guard cells	7	capillarity
4	lower/more negative	8	root pressure

Summary Test 10.7

1	evaporation	7	decrease
2	stomata	8	decrease
3	cuticle	9	decrease
4	lenticels	10	increase
5	potometer	11	decrease
6	watertight	12	increase

Summary Test 10.8

1
2 } sand dunes; salt marshes
3 cuticle
4 lower/more negative
5
} hairy leaves; stomata in pits or grooves; leaves that
6 } roll up
7

Summary Test 11.1

1
2 } physical; mental; social
3
4 multifactorial
5 acute
6 chronic
7 arthritis
8 psychoses
9 agoraphobia
10 Creutzfeldt–Jacob Disease (CJD)
11 AIDS
12 kwashiorkor

Summary Test 11.2

1	parasite	8	cystic fibrosis
2	pathogens	9	recessive
3	infectious / communicable	10	self inflicted
4	degenerative	11	anorexia nervosa
5	Alzheimer's disease	12	deficiency
6	rheumatoid arthritis	13	rickets
7	inherited	14	kwashiorkor

Summary Test 11.3

1	epidemiology	5	tuberculosis
2	prevalence	6	influenza / AIDS
3	mortality	7	pandemic
4	endemic	8	100 000

Summary Test 11.4

1
2 } life expectancy; educational attainment; wealth (GNP)
3

4
5 } mortality rate; prevalence of disease; infant mortality rate; life expectancy
6
7
8 infectious (communicable) disease
9
10 } cholera; diarrhoea; dysentery; typhoid (any 2)
11 influenza / tuberculosis
12 accidents
13 cardiovascular disease
14 cancers

Summary Test 11.5

1 3.2 billion
2
} haemophilia; sickle cell anaemia; thalassaemia; cystic fibrosis (any 2)
3
4 gene therapy
5
} cancer; heart disease; Alzheimer's disease (any 2)
6
7
} genetic testing; designing more effective drugs
8

Summary Test 12.1

1 energy
2
} growth; repair
3
4 macronutrients
5
6 } carbohydrates; lipids (fats); protein
7
8 micronutrients
9 mineral (inorganic) ions
10 metabolic processes
11 metabolism / respiration
12 urine
13 roughage / dietary fibre
14 cell walls
15 cellulose / hemicellulose / pectins / gums

Summary Test 12.2

1 lower reference nutrient intake (LRNI)
2 reference nutrient intake (RNI)
3 estimated average requirement (EAR)
4 energy
5 safe intakes
6 alcohol
7 50 (47 including alcohol)
8 intrinsic
9 lactose
10 39 (37 including alcohol)
11 11

Summary Test 12.3

1 } calcium; phosphorus
2 }
3 bones
4 iron

5 } A; C; D
6 }
7 }
8 zinc

Summary Test 12.4

1 energy
2 } bread; pasta; potatoes; cereal (any 2)
3 }
4 } glucose; fructose
5 }
6 sucrose
7 fats / lipids
8 fatty acids
9 } growth; repair
10 }
11 8
12 essential
13 A
14 D
15 skin

Summary Test 12.5

1 under-nutrition
2 } stunting; wasting
3 }
4 protein
5 infections
6 glycogen
7 liver
8 skeletal muscle
9 protein
10 oedema
11 salt-fluid
12 energy
13 protein
14 anorexia nervosa

Summary Test 12.6

1 }
2 } milk; eggs; liver; fish liver oils; mango; papaya (any 3)
3 }
4 β-carotene
5 carrots
6 xerophthalmia
7 rods
8 epithelial
9 developing
10 } eggs; oily fish; meat; margarine (any 2)
11 }
12 calcium
13 bone(s)
14 rickets
15 osteomalacia

Summary Test 12.7

1 mass (in kilograms)
2 height (in metres)
3 30
4 20

5 blood pressure / hypertension
6 cholesterol
7 type II / mature onset
8 } osteoarthritis; rheumatoid arthritis
9 }
10 salt
11 saturated

Summary Test 13.1

1 horseshoe
2 goblet
3 } bacteria; pollen;
4 } dust (any 2)
5 cilia
6 bronchioles
7 smooth muscle
8 elastic fibres

Summary Test 13.2

1 0.5
2 tidal volume
3 0.35
4 dead space
5 vital capacity
6 1.5
7 residual
8 minute volume
9 tidal volume
10 one minute
11 spirometer
12 kymograph

Summary Test 13.3

1 heartbeats
2 wrist
3 neck / temple
4 140–150
5 70–75
6 } excitement; exercise; during digestion (any 2)
7 }
8 stroke volume
9 lower / less

Summary Test 13.4

1 16–18.5kPa
2 10.5–12.0kPa
3 rises / increases
4 systemic
5 aorta
6 arterioles
7 veins
8 fall / decrease
9 rise / increase
10 hypertension
11 kidney failure

Summary Test 13.6

1 $VO_2(max)$
2 cm^3 (or dm^3) $min^{-1} kg^{-1}$
3 65%
4 85%
5 treadmill
6 Douglas
7 heart (pulse) rate
8 12
9 2
10 12

Summary Test 13.7

1 cardiac
2 mitochondria

3 decreased / reduced / 52 beats min^{-1}
4 increased / 90cm^3
5 intercostal
6 alveoli
7 vital capacity
8 myoglobin
9 ⎫
10 ⎬ glycogen; triglycerides
11 cholesterol
12 atherosclerosis
13 ⎫
14 ⎬ coronary heart disease; strokes

Summary Test 14.1

1 27 (26–28)
2 20–24 years
3 nicotine
4 carboxyhaemoglobin
5 any figure in the range 5–15
6 breathlessness
7 ⎫
8 ⎬ bronchi; bronchioles
9 cilia
10 mucus
11 emphysema

Summary Test 14.2

1 chronic obstructive pulmonary disease
2 breathlessness
3 elastase
4 elastin
5 epithelial
6 goblet
7 cilia
8 80
9 carcinogens / benzopyrene (as an example)
10 tumour
11 metastasis
12 ⎫
13 ⎬ persistent cough; blood in sputum; shortness of breath
14 ⎭

Summary Test 14.3

1 3
2 20
3 25
4 10–15
5 emphysema
6 pneumonia
7 causal
8 tar
9 carcinogen
10 tumour

Summary Test 14.4

1 cholesterol
2 plaques
3 atheromas
4 lumen
5 thrombus
6 oxygen
7 myocardial infarction
8 cerebrovascular accident
9 aneurysm
10 haemorrhage
11 carbon monoxide
12 haemoglobin
13 adrenaline
14 blood pressure

Summary Test 14.6

1 49
2 one-third (33%)
3 ⎫
4 ⎬ saturated fat; salt
5 ⎫
⎬ antioxidants; non-starch polysaccharides (dietary
6 ⎭ fibre)
7 angioplasty
8 coronary bypass
9 pig
10 rejection
11 immune

Summary Test 15.1

1 pathogens
2 virulence
3 *Vibrio cholerae*
4 water
5 food
6 pandemics
7 120 000
8 Africa
9 sanitation
10 vaccination
11 antibiotics
12 oral rehydration therapy

Summary Test 15.2

1 protoctist (protoctistan)
2 *Anopheles*
3 vector
4 300
5 (sub-Saharan) Africa
6 larva / pupa
7 ⎫
8 ⎬ insecticide; oil
9 ⎫
10 ⎬ fish; bacteria (*Bacillus thuringiensis*)
11 nets
12 insect repellent / clothes
13 ⎫
14 ⎬ chloroquine; proguanil

Summary Test 15.4

1 *Mycobacterium tuberculosis*
2 *Mycobacterium bovis*
3 air
4 30
5 5–10
6 over-crowded
7 HIV positive
8 2
9 Bacille-Calmette-Guerin / BCG
10 6–9
11 (multi-drug) resistant / MRD

Summary Test 15.5

1 microorganisms
2 biostatic (bacteriostatic)

3
4 } nucleic acids; proteins

5 biocidal (bacteriocidal)
6 cell walls
7 broad spectrum
8 narrow spectrum
9 penicillinase

Summary Test 16.1

1 epithelial
2 mucus
3 blood clotting
4 hydrochloric acid
5 phagocytosis
6
7 } monocytes; neutrophils
8 marrow
9 long
10 opsonins
11 lysosomes

Summary Test 16.2

1 foreign / non-self
2 protein
3 immunoglobulins
4 B lymphocytes
5 heavy
6 light
7 disulphide bridges
8 binding sites
9 toxins
10 lysis

Summary Test 16.4

1 viruses
2 cell-mediated

3 T helper
4 cytokines (lymphokines)
5 macrophage
6 phagocytosis
7 B lymphocytes
8 plasma
9 T cytotoxic (T killer)

Summary Test 16.5

1 natural
2 passive
3 vaccination
4
5 } living attenuated; dead; genetically engineered
6
7 natural passive
8 artificial active
9 artificial passive

Summary Test 16.7

1 allergens
2 B lymphocytes / B / plasma
3 immunoglobulin E / IgE
4 mast
5 sensitised
6 histamine
7 cytokines
8 penicillin
9 hay-fever
10 goblet
11 fluid
12 bronchioles
13 narrow / constrict
14 bronchodilators

Acknowledgements

The authors and publishers are grateful to Oxford, Cambridge and RSA Examinations (OCR) for kind permission to reproduce the examinatiion questions.

Photograph acknowledgements

The author and publishers are grateful to the following:

Cover image: DNA double helix by Alfred Pasieka/Science Photo Library.

Biophoto Associates pp 10, 18, 19 (bottom), 91 (top); **Corbis** p 183; **Corel 451** (NT) p 126; **Corel 498** (NT) p 216; **Corel 759** (NT) p 164; **Digital Vision 1** (NT) p 234 (top); **Digital Vision 15** (NT) p 220 (bottom); **Digital Vision 17** (NT) pp 170 (top & bottom); **Getty** p 8; **Griffin** p 8; **John Birdsall** p 262; **John Walmsley** pp 194 (bottom), 260; **Natural Visions** p 155; **Oxford Scientific Films** p 23 (middle); **Photodisc 6** (NT) p 111; **Photodisc 18** (NT) pp 162, 172, 196, 206, 207, 220 (top); **Photodisc 24** (NT) p 256; **Photodisc 67** (NT) pp 194 (top), 198, 201, 202; **Photodisc 72** (NT) p 251; **Roche Sight and Life** p 182 (top), 182 (bottom – task force SIGHT AND LIFE); **Sally and Richard Greenhill** p180 (bottom – Sally Greenhill). **Science Photo Library** 64, 91 (bottom), 115 (bottom), 119, 126; 6, 151 Dr Jeremy Burgess; 9 (bottom), 146 (bottom), 152, 246 (bottom) Andrew Syred; 9 (top), 10, 193, 203 Biophoto Associates; 11 (bottom) Marilyn Schaller; 11 (top) Patricia Shulz, Peter Arnold inc; 14, 232 Biology Media; 16 Secchi, lecaque, Roussel, UCLAF, CNRI; 17 Professor P. Motta, T. Naguro; 19 (top), 222 (top) Dr Gopal Murti ; 21 Juergen Berger, Max-Planck Institute; 23 (top), 158, 191 (top) Alfred Pasieka; 23 (bottom) Sidney Moulds; 48 Bernhard Edmaier; 68 D. Phillips; 94 Dept of Clinical Cytogenetics, Addenbrookes Hospital; 102, 146 (top) Claude Nuridsany & Marie Perennou; 110 John Mead; 115 (top), 220 (top) CRNI; 118 National Cancer Institute; 122, 208 (bottom) Manfred Kage; 125 Francis Leroy, Biocosmos; 135 Professor P. Motta / G. Macchiarelli / University "La Sapienza", Rome; 142 (bottom) J.C. Revy; 160 (middle) Catherine Pouedras; 160 (bottom) Tim Beddows; 161 St Mary's Hospital Medical School; 166 James King-Holmes; 184 De Planne, Jerrican; 191 (bottom) Professor P. Motta , Correr & Nottola/ University "La Sapienza", Rome; 208 (middle bottom) Astrid and Hans Frieder Michler; 208 (middle top), 208 (top) James Stevenson; 213 (left) GJLP; 213 (right) Professor P.M. Motta, G. Macchiarelli, S.A. Nottola; 217 Deep Light Productions; 222 (bottom) Sinclair Stammers; 226 Eye of Science; 228 Charlotte Raymond; 242 Simon Fraser; 246 (top) Dr P. Marazzi; 246 (middle) Eddy Gray. **Still Pictures** pp 160 (top), 180 (top), 180 (middle). Picture research by Elisabeth Savery.

Index

Bold page references refer to illustrations, figures or tables.